Continuous Cover Forestry

Continuous Cover Forestry

Theories, Concepts, and Implementation

Arne Pommerening
Swedish University of Agricultural Sciences
Umea, Sweden

Registered Offices
John Wiley & Sons, Inc., 111 River Street, Hoboken, NJ 07030, USA
John Wiley & Sons Ltd, The Atrium, Southern Gate, Chichester, West Sussex, PO19 8SQ, UK

For details of our global editorial offices, customer services, and more information about Wiley products visit us at www.wiley.com.

Wiley also publishes its books in a variety of electronic formats and by print-on-demand. Some content that appears in standard print versions of this book may not be available in other formats.

Library of Congress Cataloging-in-Publication Data

Names: Pommerening, Arne, author.
Title: Continuous cover forestry : theories, concepts, and implementation / Arne Pommerening.
Description: First edition | Hoboken, NJ : Wiley, 2024 | Includes index.
Identifiers: LCCN 2023023698 (print) | LCCN 2023023699 (ebook) | ISBN 9781119895305 (cloth) | ISBN 9781119895312 (adobe pdf) | ISBN 9781119895329 (epub)
Subjects: LCSH: Continuous cover forestry. | Forest management.
Classification: LCC SD387.C67 P66 2023 (print) | LCC SD387.C67 (ebook) | DDC 634.9/2–dc23/eng/20230711
LC record available at https://lccn.loc.gov/2023023698
LC ebook record available at https://lccn.loc.gov/2023023699

Cover Design: Wiley
Cover Images: © feipco/Adobe Stock Photos; Cover image courtesy of Joss Everett

Set in 9.5/12.5pt STIXTwoText by Straive, Chennai, India
Printed and bound by CPI Group (UK) Ltd, Croydon, CR0 4YY

C9781119895305_120923

To my parents who provided me with a privileged start in life and to the whole Pommerening-Küter forestry family whose lives inspired me to choose what is a truly wonderful profession.

Contents

Foreword

When I first learned about this textbook project I immediately thought it was really high time that someone wrote a consistent, systematic English language textbook on continuous cover forestry (CCF), since we have been lacking this for quite a while. Therefore I am keen to endorse this fine accomplishment.

One of the most remarkable properties of forest ecosystems is the ability to autonomously prevail over other plant formations so that most areas in the world that are suitable for plant growth would naturally be occupied by forests without human aid. This is owed to the trees' ability to escape competition at ground level by growing tall, to their resilience and longevity. Forest ecosystems apparently have self-organisation capabilities allowing them to form stable states of equilibrium without our assistance.

Instead of maximising the production of certain ecosystem goods through genetically engineered species selection and other technical inputs in plantation forests, the processes in semi-natural forest ecosystems are regulated by making use of mechanisms of self-organisation. This is the fundamental change of paradigm that CCF or nature-based forest management, as it is also known, is based on. This concept is a low-cost alternative form of forestry ensuring ecological sustainability whilst fulfilling the expectations of modern societies. CCF began to develop more than a century ago in different parts of the world, and as a consequence much experience and silvicultural knowledge is now available. One of the most common concerns in this context is how highly artificial forests can be transformed without trade-offs to an equilibrium state of sorts and how to maintain such a steady state.

For successful implementation, CCF requires more pre-requisites than plantation forest management, namely a sufficient understanding of the natural principles of self-organisation, a good knowledge of the ecological requirements of native and introduced tree species, and particularly the ability to judge environmental conditions including soil properties correctly, since the successful development of forest ecosystems much depends on them. In addition, a good knowledge of how the trees involved respond to silvicultural interventions is required, e.g. the ability to successfully initiate processes of natural regeneration and make use of other ecosystem benefits without additional costs.

With ongoing climate change, energy crisis and increasing spare time, our societies are taking much greater interest in forests and forestry than in previous decades and are rightly concerned about long-term environmental sustainability. CCF is the best form of forest management to meet all these diverse expectations and simultaneously delivers a maximum of very different ecosystem goods and services across large landscapes.

Crucial to the success of CCF is the clever application of a solid silvicultural skill set including methods of biological rationalisation rather than an exclusive reliance on mechano-technical inputs and advances. Education and experience of silvicultural specialists play a central role in the successful implementation of nature-based forest management, but also the organisational freedom granted to forest managers for working directly with the forest stands that are put into their care. I am certain that this textbook will greatly contribute to building the necessary skills and knowledge for a successful uptake of CCF and its many variants.

Jean-Philippe Schütz
Zürich, Switzerland
February 2023

Preface

Творческая задача лесовода - суметь
законы жизни леса превратить в прин-
ципы хозяйственной деятельности.
Г. Ф. Морозов

It is the creative task of a silviculturist to figure
out how to turn the laws of forest life into the
principles of forest management.

G. F. Morozov

'A spectre is haunting Europe – the spectre of continuous cover forestry. All the traditional powers of forestry have entered into a holy alliance to exorcise this spectre.' Borrowing and only slightly modifying these words of The Communist Manifesto published by K. Marx and F. Engels in 1848 surprisingly well describes the situation at the end of the nineteenth and beginning of the twentieth century, when continuous cover forestry (CCF) was first proposed in publications. Since then forestry has come a long way and CCF has even become mainstream in quite many countries, although the quote still applies in others.

In recent years, CCF has received much attention for its potential to help mitigate climate change, for its usefulness in the context of forest conservation and for the increasing resistance of society to industrialised plantation forest management. Worldwide forestry students feel enthusiastic about courses and field trips related to CCF. Ironically, this is contrasted by a steady decrease of research funds in the core fields of silviculture and forest management. Hardly any national research council is prepared to fund the development of methods in forest management and the associated scientific experiments whilst forest industries often do not have sufficient means to invest in research. This situation leads to a continued decline of silvicultural research staff at universities and as a result, teaching of silviculture including CCF is in steady decline as well. Therefore, this book unwillingly acts as a lantern bearer of sorts but hopefully also as an eye opener at a time that is difficult for academic silviculture when the knowledge in this field is bizarrely more needed than ever.

Since the beginning silviculture and forest management were dominated by commercial forestry. Other objectives such as nature conservation and biodiversity were considered by-products or welcome side effects, if they were considered at all. This has fundamentally changed in the twenty-first century when other objectives of forestry have become at least as prominent as commercial forestry if not more important. This book is the first attempt to present a central part of silviculture, continuous cover forestry, in a neutral and balanced way so that commercial forestry is just one purpose of forest management among others but has no supremacy.

The contents of this book are based on more than 20 years of teaching in CCF in the United Kingdom, Switzerland, and Sweden. I had the good fortune to be educated in CCF as part of my own degree programmes at Göttingen University (Germany). When I took up my first faculty position in the United Kingdom, CCF was being introduced to the country and I was asked to contribute to CCF teaching, training of practitioners and setting up CCF research experiments. This provided me with invaluable experience of a kind I could never have achieved in my own country. During this time, I organised quite a few CCF field trips for British forestry students and practitioners. These academic activities I continued in Switzerland and there I had the opportunity to deepen my knowledge in the cradle of CCF. At the time of writing, CCF is being considered for introduction in Sweden, my current academic home, so the subject keeps following me around.

Chapter 1 sets the scene, gives vital definitions and context. Methods and options for getting started with CCF are discussed in Chapter 2. This is followed by an explanation of the principles of individual-based forest management in Chapter 3, which forms an important part of CCF. Forest structure is the key to understanding ecological processes and delivers CCF. These key relationships are explained in Chapter 4. Following on from Chapter 4, thinnings and silvicultural systems as the main methods of manipulating forest structure are introduced in Chapter 5. Demographic equilibrium and guidance models for identifying steady-state conditions of CCF woodlands are reviewed in Chapter 6. These modelling approaches were originally developed in the CCF research community. In Chapter 7, I discuss how general CCF management needs to be adapted to deliver different ecosystem goods and services. Finally, Chapter 8 is dedicated to the important topic of training for CCF management. Where appropriate, R source code is provided and a brief introduction to R can be found in Appendix C of Pommerening and Grabarnik (2019).

I particularly thank the members of my former Tyfiant Coed research team, Sharron Bosley, Dr. Owen Davies, Dr. Jens Haufe, Gareth Johnson, Dr. Steve Murphy, Mark Rogowski and Björn Welle for their support and cooperation when we made a humble contribution to the introduction of CCF in Wales. It has been fun to co-organise the annual one-week CCF field trips to Denmark and Germany together with my friend and colleague Professor Douglas Godbold (BOKU University, Vienna). These field trips have been a tremendous source of inspiration and together with my students I learnt a lot from the Danish and German CCF practitioners who kindly showed us around. I am also much indebted to my silvicultural mentors, Professors Hanns H. Höfle, Burghard von Lüpke (both Göttingen University, Germany) and Jean-Philippe Schütz (ETH Zürich, Switzerland). Dr. Thomas Campagnaro (Padova University, Italy) helped me with understanding the Susmel model (Chapter 6). Professor Hubert Sterba (BOKU University, Austria) engaged in many helpful discussions and kindly made photos and the Hirschlacke data available. I much enjoyed the discussions on biological rationalisation and silvicultural training I had with Dr. Peter Ammann (Fachstelle Waldbau, Switzerland). Dr. Lucie Vítková (Czech University of Life Sciences, Czech Republic) kindly shared her experience in graduated density thinning and thus contributed to Section 2.4.2. Professor Jorge Cancino (University of Concepción, Chile) joined a helpful discussion on the BDq approach (Section 6.3.1.1). Dr. Xiaohong Zhang (Chinese Academy of Forestry, Beijing) kindly made data collected at the Tazigou Experimental Forest Farm (Jilin Province, China) available. For some illustrations in this book, data were gratefully received from Andreas Zingg (Swiss Federal Institute for

Forest, Snow and Landscape Research WSL, Birmensdorf, Switzerland). Professor Áine Ní Dhubháin (University Colleague Dublin, Ireland) undertook the enormous task of proofreading and commenting on the text of the whole manuscript, for which I am most grateful. Professor Dan Binkley (Northern Arizona University, US) has kindly encouraged and advised me in the proposal phase of this book project. Zeliang Han and Dr. Gongqiao Zhang (Beijing, China) tremendously enriched the book project by preparing illustrations, often at short notice, for which I am very grateful. Any remaining errors and shortcomings in this work are mine alone.

Looking back and contemplating the long history of forestry one cannot help but think that there is some similarity between the evolution of civilisation in general and forestry in particular: Humankind has always strived to improve the constitutions of states up to the point when (social) justice and democracy finally were sufficiently developed so that even unprivileged members of society have a good chance to thrive and to live a full life. This evolution has taken a long time, taken many wrong turns and is by no means complete nor are current achievements secured forever. In a similar way, CCF is undoubtedly among the great achievements of humankind. With CCF, forestry has come a long way starting with agricultural practices that marked its beginnings and has now eventually arrived at a point where a type of land use is pursued that attempts to be fair to the forest ecosystems and landscapes we humans have inherited and will pass on.

Arne Pommerening
Umea, Sweden
February 2023

1

Introduction

Abstract

The terms 'forest management' and 'silviculture' refer to deliberate, professional human interaction with forest ecosystems, i.e. the direct hands-on human interference with tree vegetation for particular purposes – which, in one way or another, is almost as old as mankind. Over the last 200 years, a wide range of different techniques have been developed, refined, and formalised in forest practice and science with the objective to modify forest structure and thus to steer forest development into certain desired directions.

This chapter outlines the basic concepts and techniques of forest management and silviculture in a changing world. The text provides non-specialists with an easy access to this field of forest ecosystem management by explaining basic terms and definitions. The concept of continuous cover forestry (CCF) is introduced, and how the history of different societies has contributed to its shaping is described. Common misconceptions are explored, and important sustainability concepts are explained. Finally, the place of CCF in a changing world is outlined, and suggestions are made as to how to introduce CCF to a new region or country.

1.1 When Is a Forest a Forest?

Despite their very different plant compositions forests form very characteristic plant communities which set them aside from other vegetation forms (such as grasslands, shrublands, bog vegetation) not only in terms of visual appearance but also in terms of ecology. Trees, as large woody plants, are the most significant element of wooded landscapes. Forest trees are, unlike fruit trees, for example, only rarely genetically modified organisms, whether native to their sites or introduced from another, ecologically similar region. In order to form a forest, trees have to be so close to each other in a sufficiently large land area that they form a common crown canopy which shades the forest floor to a large extent. Only under these circumstances can life conditions be sustained which are very different from any other plant formation. After afforestation or replanting, it takes time for the elements which form the typical characteristics of a forest to appear. Such typical elements are the size of the forest area, canopy closure, forest soils and woodland microclimate and a specific vegetation of vascular plants on the forest floor which only occurs in forests. Natural disturbances and human interventions can destroy some of these elements.

But as long as the tree vegetation is sustained and can again develop a more or less closed formation, the corresponding land area remains a forest (Burschel and Huss, 1997).

'In the United Kingdom, a woodland or forest is defined as land with a minimum area of 0.1 ha under stands of trees with, or with the potential to achieve, tree crown cover of more than 20%. Areas of open space integral to the woodlands are also included. Orchards and urban woodland between 0.1 and 2 ha are excluded. Intervening land-classes such as roads, rivers or pipelines are disregarded if less than 50 m in extent'.

National Inventory of Woodland & Trees – Wales

There is no world-wide unique definition of forests. In developed countries, the lower limit in older forest stands is a crown canopy closure of more than 20%, while in many developing countries, only 10% is required. This affects the estimation of deforested areas around the world (Bartsch et al., 2020).

The IUFRO (International Union of Forest Research Organisations) terminology of forest management (Nieuwenhuis, 2000) provided an ecology and a management motivated definition:

Forest (ecology perspective)
'Generally an ecosystem characterised by a more or less dense and extensive tree cover. More particularly, a plant community predominantly of trees and other woody vegetation, growing more or less closely together'.

Forest (silviculture/forest management perspective)
'An area managed for the production of timber and other forest produce, or maintained under woody vegetation for such indirect benefits as protection of catchment areas or recreation'.

Helms (1998) in his dictionary of forestry put forward the following definition of a forest:

'An ecosystem characterised by a more or less dense and extensive tree cover, often consisting of stands varying in characteristics such as species composition, structure, age class, and associated processes, and commonly including meadows, streams, fish, and wildlife [...]'.

Finally, Thomasius (1990) characterised forests as ecosystems, in which life forms dominate, which

- reach such a height,
- are so close together,
- occupy a sufficiently large area,

so that the spatial occupation by living and dead biomass leads to

- a specific microclimate,
- characteristic soil conditions and
- interactions between the different organisms that influence their onto-, auxo- and morphogenesis.

Consequently, three conditions need to be fulfilled to deserve the status of a forest:

1. The dominant life form must be trees which reach a total height of at least 7 m in their adult stage (under extreme environmental conditions 3 m is often also acceptable).
2. The trees must form stands which shade at least 20% of the soil surface.
3. Stands of trees have to occupy an area A with a radius r which with full crown closure corresponds at least to their height.

These features create a habitat for plants and animals which require the specific microclimatic and/or soil conditions of a forest or which are dependent on other life forms that occur in forest ecosystems (Thomasius, 1990). A forest is a highly interactive community of organisms with the tree life form dominating among the energy-capturing green plants. These plants generate the cascade of energy that supports food webs composed of thousands of species (Franklin et al., 2018).

From the standpoint of forest management, the term 'forest' has a special meaning and denotes a collection of stands administered as an integrated unit, usually under one ownership. This division of forests into stands is especially important in regulating timber harvests as well as managing wildlife populations and large watersheds. One objective of stand management for timber products is usually the achievement of sustained yield. However, the forest (estate or district), and not the stand, usually is the unit from which this sustained yield is sought. Management studies of prospective growth and yield determine the volume of timber to be removed from the whole forest in a given period (Smith et al., 1997).

Thomas and Packham (2007) and Helms (1998) noted that the terms 'forest' and 'woodland' are commonly used almost interchangeably. In agreement with public perception, they suggested that a woodland is a small area of trees with an open canopy (often defined as having 40% canopy closure or less) such that plenty of light reaches the ground, encouraging other vegetation to grow beneath the trees. Since the trees are well spaced, they tend to be short-trunked with large canopies. The term 'forest', Thomas and Packham (2007) asserted, by contrast, is usually reserved for a relatively large area of trees forming for the most part a closed, dense canopy. Since the terms 'woodland' and 'forest' overlap, we have used them as synonyms in this book. A forest is made up of a series of more or less homogeneous stands, and the main focus of silviculture is on these forest stands with trees as their main components.

A forest stand is a contiguous community of trees sufficiently uniform in species composition, density, structure and/or age-class distribution, and growing on a site of sufficiently uniform site conditions and site quality, of a sufficient size to be a distinguishable planning and management unit (Thomasius, 1990; Helms, 1998; Nieuwenhuis, 2000; Franklin et al., 2018; Smith et al., 1997).

In continuous cover forestry (CCF), forest managers commonly deal with stands that can be quite diverse in terms of tree species, sizes and dispersal. Consequently, in this book, we used the term 'stand' in its broadest interpretation often denoting a distinctive population of trees (Franklin et al., 2018). In many countries, forest stands usually coincide with sub-compartments, as these administrative units were chosen to reflect the forest stand definition. Tree canopies or crowns of a forest stand can usually be found in one or more canopy levels. The uppermost canopy level is referred to as *overstorey*, the lowest layer of vegetation is the *understorey* and tree canopies in between over- and understorey form the *mid-storey* (Helms, 1998). For a possible quantitative definition of these forest canopy levels see Table 4.3. The vertical tree canopy structure of forest stands plays an important role in CCF, see, for example, Figure 2.15.

1.2   The Nature of Forestry and Forest Management

Nyland (2002) stated that forestry involves the science, business and practice of purposefully organising, managing and using forests and their resources to benefit people. Many authors adopted the anthropocentric view that it is the prime objective of forestry to satisfy the forest-related demands of society in a sustainable way with minimum input of scarce resources, e.g. energy and money (Köstler, 1956; Bartsch et al., 2020). These demands include the provision of various goods and services, namely of raw materials, environmental conservation and the conservation of aesthetical and spiritual properties (see Section 7.3.3). For a long time, the production of timber has been the only or at least the dominant objective of forestry. This has changed significantly in the last few decades in many parts of the world, although timber production will undoubtedly continue to be important. Even many academic educational programmes lag a little behind in this emancipation of forestry.

To achieve the general objectives of forestry, there are a number of academic and practical fields that help deliver them. The most central one is forest management or silviculture. Köstler (1956) wrote that silviculture is the kernel of forestry and forest science, for it includes direct action in the forest, and in it, all objectives and all technical considerations ultimately converge. Silviculture today is still the main academic discipline providing methods and techniques for goal-oriented management of forest vegetation for a wide range of ecosystem goods and services, i.e. the science of deliberate human–tree interaction. This is the main reason why this field is very attractive to students and largely contributes to their decision to enrol on a forest science rather than an ecology or nature conservation degree programme. Similar to the use of the terms 'ecosystem management' and 'conservation management', forest management is often used as a synonym of silviculture and sometimes even of forestry as a whole (Helms, 1998). In this book, both terms are used as synonyms. Silviculture has a long history and is the oldest and possibly most traditional subject of forest science and education. As such it also carries much traditional, inherited 'baggage' that sometimes is more a hindrance rather than a help (see Section 8.3) and hard to leave behind.

The purely natural forest is governed by no purpose unless it be the unceasing struggle of all the component plant and animal species to perpetuate themselves. Human intervention (be it, for example, for timber production or conservation) is characterised by preference for certain tree species, stand structures, or processes of stand development that have desirable characteristics for the purpose in mind. In silviculture, natural processes are deliberately manipulated to create forests that a majority of humans of a given society perceives as more useful than natural ones and to do so in as short a time as possible (Smith et al., 1997; Nyland, 2002). Given the lifetime of humans compared to that of trees, the acceleration effect achieved by silviculture is not unimportant. The primary objective of silviculture is therefore not necessarily the reconstruction of natural forests but, rather, the establishment and management of forests that can satisfy societary needs. This still requires, however, the retention of a forest's functioning as a viable ecosystem.

Silviculture, and particularly CCF, have gradually developed to use natural processes to a large extent in order to meet societary needs. The ability to utilise the forces of nature has improved during the history of mankind through an increased knowledge of natural laws (Thomasius, 1990). The gradual emancipation from agriculture has made room for employing natural processes in forest management. As a result, it is now entirely possible to conduct forestry indefinitely without the degradation of soils that is almost inevitable in most agriculture and in other 'higher' uses of land.

The Roman writer Pliny used the term *silvicultura* for the first time (Mayer, 1984). The German equivalent *Waldbau* was coined around 1764 in a time when large-scale forest restoration took place following long-term devastation (Hasel and Schwartz, 2006). A literal translation would be 'forest building', 'forest construction' or 'forest design'. This rather strange term was used in the Germany of the eighteenth century to label a new discipline, which was initially thought to be the forestry equivalent to *Feldbau*, i.e. agronomy (Cotta, 1816). The strong legacy of agriculture in forestry is still evident in the agricultural term 'crop' that is often used by foresters to refer to a forest stand or to a plantation grown for commercial purposes, whilst using the expression 'crop trees' to distinguish desired trees from less desired trees (Puettmann et al., 2009, see Chapter 3). Early concepts of ensuring sustainability such as the 'normal forest' (see Section 1.8) owe much to agriculture, and this legacy lives still on in plantation forestry (Heger, 1955).

Only gradually it became evident that silviculture had to follow very different lines than agronomy, since long production periods of 100 years (and more) and the limited possibilities of influencing production processes set forestry apart from agriculture (see Table 1.1). In agronomy, with its short production periods and with plenty of opportunities for technical manipulation of production, biological processes are part of an industrial framework. On the other hand, agricultural production is influenced much more by short-term weather fluctuations and extremes than forestry. The same applies to changes in the timber market: A forest company can choose to cut less in one year or even over the course of several years. When the market is ready to take larger quantities of certain timber assortments again, these timber reserves can then be reduced. Such flexibility is difficult to achieve in agriculture.

Successful long-term forest management has to be based on biological–ecological requirements, because there is little scope for technical manipulation which is also increasingly being restricted by law in many countries. Therefore, silviculturists by definition are bound to understand and to employ natural processes to meet economic objectives. Economic objectives have to be based on ecological and environmental site limitations rather than

Table 1.1   Characteristics for agronomy (agriculture) and silviculture (forestry).

Criteria	Agronomy (agriculture)	Silviculture (forestry)
Target plants	Short-lived grass, herb and perennial species	Long-lived trees
Production period	1 year	40–350 (120) years
Objectives	Meeting current well known demand	Speculative anticipation of future unknown demand
Amortisation of investments	Short with average interest rates	Long-term with low interest rates
Production risk	Low (insurance)	High (owner has to take the risk)
Scope for increasing yield	Larger	Smaller
Fertilisation	Standard. Short-term effect, increased yield rapidly available	Uncommon or prevented by law. Short-term effect with little influence on long-term production
Breeding results	Applicable after few years of testing	Applicable only after 1–2 forest generations (130–300 years)
Production	Cultivation, harvesting and processing mostly within one year	Harvest – management result of past forester generations, processing at present time, replanting for future generations, 4–6 generations of foresters in total
Historical aspects	Only present-day conditions matter	Past management has a long-lasting influence (forest development history)
Initial biological conditions	Artificial plant community with short life expectancy (4–6 months), which is kept alive artificially through soil preparation, fertilisation and chemicals. Maximum production expected	Near-natural to semi-natural biocenosis with long life expectancy (80–120 years). Only stable structures are able to provide long-term production, stands far removed from natural conditions can be at a very high risk

Source: Adapted from Mayer (1984).

keeping highly artificial, industrial and risky crops alive by tending symptoms through chemical forest protection, fertilisation and weeding (Mayer, 1984).

The German term, more than the English word, reflects the main task of silviculture – the active building and development of forests following a plan, which is largely determined by societary needs or preferences. Even forest conservation is anthropocentric, as it satisfies a particular societary need for protecting and sustaining nature. Morosov (1959) pointed out that silviculture bizarrely enough owes its existence to large-scale forest degradation and a painful lack of timber resources. Without this cataclysmic forest degradation and timber shortage, which came to the Europeans as a shock and turned into a long-term trauma, there would not have been a need to consider silviculture and to address sustainability. Dengler (1944) noted that silviculture is concerned with the building and design of forests by arranging their individual components, the forest stands, which significantly influence production, health and utilisation of the forest. In Central Europe, Schädelin, Leibundgut, Abetz and Pollanschütz made the next logical step and extended this idea by breaking these individual components even further down to the level of individual trees (Pommerening et al., 2021a, see Chapter 3).

Silviculture as we know it today is a process of forest engineering aimed at creating structures or developmental sequences that eventually serve the intended purposes, whilst being in harmony with the environment and withstanding the loads imposed by environmental influences. Because forests grow and considerably change with time, their design is more sophisticated and difficult to envision than that of static buildings. This complexity and the fact that considerable costs were involved have made foresters uneasy about their investments in the past and, as an expression of this, natural disturbances have often been labelled as 'calamities' and 'risks' (Puettmann et al., 2009).

Furthermore, forest stands alter their own environment sufficiently that the forester is partly creating a new ecosystem and partly adapting to the one that already exists (Smith et al., 1997). Based on Helms (1998) and Nyland (2002), we can summarise these aspects in the following definition:

> The term 'silviculture' or 'forest management' denotes the main activity of forestry to establish new forests and the management and regeneration of existing ones as healthy communities of trees and other vegetation. Silvicultural activities should aim to maintain and improve site quality and growth, resilience, quality and diversity of forest vegetation to meet the targeted diverse needs and values of landowners and other members of society on a sustainable basis.

Silviculture is the oldest conscious application of the science of ecology and is a field that was recognised even before the term 'ecology' was coined. It is concerned with the technology of growing tree vegetation. Silviculture is also a major part of the biological technology that carries ecosystem management into action (Smith et al., 1997) and silvicultural activities reflect the forester's efforts to imitate natural succession and disturbance. As such silviculture is fundamental to sustainable forestry (Nyland, 2002).

Silviculture and forest management heavily draw on a wide range of basic sciences as does practical silviculture carried out by forest managers in the field. These basic sciences include soil science, climatology, geology, dendrology, hydrology, ecophysiology and many more. When talking about silviculture, it can be practical to distinguish different hierarchical levels. These are

- Tree,
- Stand,
- Forest,
- Landscape.

For planning purposes, public and private forest services have other or additional hierarchical levels such as for example sub-compartment, compartment, local area and district.

1.3 Silvicultural Regimes and Types of Forest Management

Silvicultural regimes are technically different methods which give whole forests their specific character (Köstler, 1956). In the Anglo-American literature, both silvicultural regimes and forest regeneration methods are usually lumped together by the term 'silvicultural systems' (see, for example, Matthews, 1991; Nyland, 2002). In this tradition, silvicultural systems are perceived as 'plans for management' (Nyland, 2002) or 'planned programmes of silvicultural treatment' (Smith et al., 1997) including and fully integrating the three main domains of silviculture, i.e. establishment, thinning and harvesting, and extending throughout the life of a forest stand or its present generation. This concept differs from the European view where most silvicultural systems are purely seen as regeneration methods, whilst coppice, coppice with standards and high forest are fundamental silvicultural regimes operating at a higher conceptional level than silvicultural systems, see, for example, Burschel and Huss (1997) and Bartsch et al. (2020). Particularly in the central parts of Europe, it was felt that silvicultural regimes are structural and management archetypes that give all stands subjected to the same regime a similar, general appearance and long-term character. This difference in long-term fundamental appearance sets them apart from the short-term silvicultural systems that they were separated from as a matter of principle (Köstler, 1956; Figure 1.1).

One of the oldest silvicultural regimes is *coppice* or *low forest* (in some of the European terminologies like, for example, in German and Spanish), because they were perceived as being limited in stand height due to vegetative reproduction and short rotation (Smith et al., 1997), i.e. a short production period, cf. Figure 1.4. It is believed that this silvicultural regime was first discovered in the Neolithic as a natural consequence of clearing forest vegetation for settlements (Matthews, 1991). When felled near ground level, some of the trees re-sprouted from dormant, adventitious buds in the cambial layer of the remaining tree stumps (also

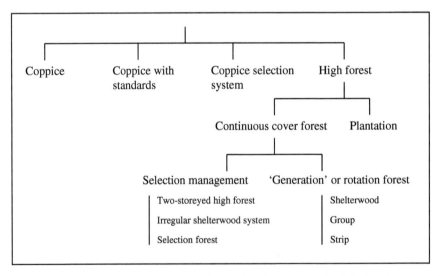

Figure 1.1 The relationship between silvicultural regimes and basic silvicultural systems. Source: Adapted from Pommerening and Grabarnik (2019).

termed *stools*) or at the roots (root *suckers*) and these *shoots* gave rise to a secondary forest that eventually developed into a coppice forest. Layering of trees is another, but rare, technique of coppicing. It naturally occurs in peat bogs where *Sphagnum* spp. moss tends to overgrow the lower branches of trees in open stands so that roots start to form on them with time. Artificial layering is one of the techniques of inducing branches to form roots while they are still firmly attached to the trees. Later, they can be severed and left *in situ* or planted elsewhere (Smith et al., 1997). These processes of vegetative regeneration are collectively referred to as *coppicing* (Nyland, 2002).

Coppicing was well developed in Roman times and often coupled with the introduction of the species *Castanea sativa* Mill.. The method was perfected towards the High Middle Ages (Hasel and Schwartz, 2006). Coppice forests (sometimes also referred to as *copses*) typically have a short rotation length, i.e. all trees are cut every 15 to 25 years because the sprouting ability of trees decreases with age and because traditionally the focus was on small-sized timber assortments (Rittershofer, 1999; Bartsch et al., 2020). New trees of seedling origin occasionally replace worn-out stools. In England, hazel (*Corylus avellana* L.) grown in coppice forests for thatching spars is cut every five to seven years which suggests that many other shrub species may be suitable as well. After coppicing, which essentially is a clearfelling operation, the forest then regenerates from stool shoots and root suckers (Figure 1.2). Coppice woodlands are rarely thinned, but on occasion there may be some merit in removing abundant shoots, thereby improving the growth of the remaining stems (Hart, 1995). It is likely that first concepts of rotation and timber sustainability were developed for coppice forests before they were applied to other silvicultural regimes (Hasel and Schwartz, 2006). Due to their fairly short lifetime, the tree diameters involved remain small. In contrast to trees in high forests, each tree of a low forest typically tends to have multiple stems. These stems form vegetation clusters and share similarities with the basitonic growth pattern of shrubs.

In coppice forests, broadleaved species are predominantly used not least for their prolific resprouting ability. In Europe, tree genera included in coppice forests are *Acer*, *Alnus*, *Betula*, *Castanea*, *Carpinus*, *Corylus*, *Fraxinus*, *Populus*, *Prunus*, *Quercus*, *Robinia*, *Salus*, *Sorbus*, *Tilia* and *Ulmus*. Some species of *Eucalyptus* apparently show phenomenal coppicing potential (Matthews, 1991; Nyland, 2002). The resprouting ability typically decreases with repeated coppicing. Regionally different types of coppice forest exist that favour

Figure 1.2 Resprouting of *Catanea sativa* Mill. trees from coppiced stools in the Swiss canton Ticino. Source: © Arne Pommerening (Author).

different species, are managed in different ways or have specialised on delivering particular timber products. The main products of coppice forests included firewood, fencing material, posts, staves for barrels, tanning agents for leather production, roofing materials, charcoal, bark and other small-sized items. A special form of coppice management is *pollarding*, in which the tops of trees are removed, thus inducing them to sprout at points above the reach of browsing animals (Smith et al., 1997), i.e. coppicing can be carried out at different stem heights. In conservation, the pollarding of *Fraxinus excelsior* L., *Carpinus betulus* L. and *Salix* spp. plays an important role in Europe. Whilst being a simple and ancient regime, coppice forests are currently re-visited as a source for animal fodder and fuelwood contributing to renewable energy. Short-rotation coppice (SRC) forests relying on fast-growing species such as *Salus* and *Populus* on agricultural land have in recent years been much considered throughout Europe for the production of sustainable biofuels and pulp (Bartsch et al., 2020).

By ways of gradual evolution and diversification coppice with standards or middle forest (in some of the Central European terminologies) were developed when the need for larger-sized products arose along with the traditional coppice products. As such this regime is on a continuous scale somewhere between coppice and high forest, but in terms of tree morphology, it is closer to coppice forests. Coppice with standards dates to at least the thirteenth century, and such woodlands were often managed by rural communities. It is likely that the origin of this silvicultural regime lies in France (Bartsch et al., 2020). Essentially, coppice with standards (cf. Figure 1.3) is a combination of trees grown from coppice shoots and from seedlings. Whilst the former would remain comparatively short, the standards (sometimes also referred to as reserves) would become large trees. This combination gives rise to two *permanent* canopy layers. Some main-canopy trees may also originate from former coppice trees that were not coppiced any more. The lower canopy (referred to as *underwood*) serves the same purpose as the trees in coppice forests, while the standards (termed *overwood*) are harvested for producing construction timber and furniture. The standards have shorter stems and a larger proportion of branchwood than trees grown in a comparable close-canopied high forest (Matthews, 1991). Still, Wilhelm and Rieger (2018) pointed out that overwood *Fagus sylvatica* trees of former coppice with standards woodlands often produce high-quality timber.

Since the standards are not coppiced, it is possible to include conifers in the main canopy, e.g. *Pinus*, *Larix* and occasionally also *Picea*. The main canopy in coppice with standards woodlands is comparatively open with distances of 15–20 m between the standards on average to ensure sufficient light conditions to maintain the lower canopy layer (Bartsch et al., 2020; Hart, 1995). Traditionally, management interventions first take place in the lower canopy before considering the main canopy (see Figure 1.3). Coppice with standards follows longer rotations of between 20 and 40 years. There have also been agroforestry variants of coppice, and coppice with standard woodlands where either agricultural crops were temporarily planted between the stools (Mayer, 1984), or cattle or pigs were temporarily driven into the forest to feed on acorns and/or young coppice shoots (Matthews, 1991). Coppice forests with standards have received much attention in recent years because they usually have a considerable biodiversity and conservation value and simultaneously provide fuelwood (see Section 7.3.6) and other timber and non-timber products. Regionally,

Figure 1.3 Traditional coppice with standards with *Catanea sativa* Mɪʟʟ. trees in the Swiss canton Ticino where the underwood has just been coppiced whilst the overwood remained untouched. In Ticino, management of coppice with standards is currently re-introduced in areas where it was used in the past for many centuries. Source: © Arne Pommerening (Author).

like in many parts of France, coppice with standards forests are important characteristics of rural landscapes (Hasel and Schwartz, 2006). Coppice with standards and related coppice selection forests (Matthews, 1991) as traditionally practised in the Swiss canton Ticino (Schweizerischer Forstverein, 1925) can clearly be a management option within CCF, whilst simple coppicing shares many similarities with rotation forestry and clear-felling. Uniform shelterwood systems are often applied for regenerating coppice with standards woodlands (cf. Figure 1.3). Schütz (2001b) stated that many selection systems (see Section 5.4) in Central Europe had their origin in coppice with standards woodlands.

High forest regimes can be characterised as forests regenerated from seeds or planted seedlings (sometimes traditionally referred to as virgin trees). This contrasts with coppice or low forest systems that originate vegetatively through natural sprouts from the stools of felled trees. The name of this regime comes from the fact that trees grown from seeds usually develop larger total heights than those that have regenerated vegetatively. Also, the production period or lifetime of a forest stand (= rotation) is usually longer. In many parts of Europe, coppice and coppice-with-standards forests were converted to high forests towards the end of the nineteenth century, a process which is currently being reversed in some regions.

Many high forests are traditionally dominated by conifers. High forests came into fashion when increasingly a need for construction and sawn timber arose. At the same time, the demand for fuel wood declined due to the discovery and the increasing availability of fossil fuels. High forests also offer a greater choice of species, since it is not necessary

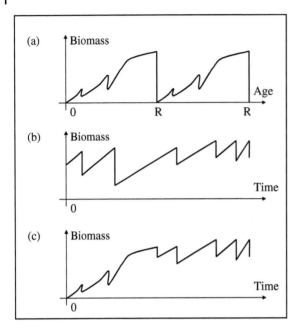

Figure 1.4 Biomass development in different management approaches. Plantation (a), CCF (b) and transformation management (c). R is rotation age. Source: Adapted from Pommerening and Grabarnik (2019).

to rely exclusively on species that are able to regenerate vegetatively (Rittershofer, 1999). The structure of high forests most closely resembles that of natural forests, since trees are usually allowed to have longer lifespans and larger total heights than in the other two regimes. High forest trees typically have single stems that show more or less acrotonic growth in contrast to shrubs and coppiced trees (Pommerening and Grabarnik, 2019). High forest regimes can be further subdivided into plantation forest management and CCF or near-natural forest management, based on the development and continuity of tree biomass over time (Pukkala and Gadow, 2012). Associated with the first type of management is a rather abrupt or sudden transition from one forest generation or rotation to another one, usually by clearfelling all trees of a stand (see Figure 1.4).

In commercial forestry, *rotation* is the period during which a generation of a single forest stand is allowed to grow or in other words the period between establishment and final cutting. It is also known as either economic or natural maturity (Smith et al., 1997). The length of a rotation may be based on many criteria including species, mean size, age, increment culmination, growth rates, wind hazard and biological condition among others. Typical properties of plantation forest management include

- A very short period of stand establishment with instant artificial regeneration,
- The origin of seedlings is often not local and in many cases includes 'genetically improved' material,
- Low genetic diversity,
- Hardly any age difference between trees of the same stand,
- Clearfelling is the predominant harvesting method,
- Highly industrial use of timber products (e.g. pulp and paper, fibreboards).

(a)   (b)

Figure 1.5 Examples of broadleaved plantations involving native tree species. (a) A *Betula* spp. plantation in Galicia (Spain) Source: © Arne Pommerening (Author). (b) A *Quercus robur* L. plantation in the Neuhaus Forest District (Lower Saxony, Germany). Source: Courtesy of Paul Cody.

Plantation forest management is the most widespread type of forest management on the planet, but compared with natural forests – that are largely unaffected by human management at least for some time – it is also the most extreme one in terms of species composition and structure. Both tree species and size diversity are low in plantations, as commonly only one species is planted, and the size structure is deliberately homogenised through forest management (see Figure 1.5).

The share of the global timber production coming from plantations is high and increasing (Pommerening and Grabarnik, 2019). However, this forest type remains a small portion of the earth's total forest cover. Plantations often involve non-native species, particularly non-native conifers, for example in Britain, but they do not necessarily need to (cf. Figure 1.5): In the United States, which has a sizeable portion of the world's planted forests, practically all plantations are of native species. In contrast, CCF is usually characterised by selective thinnings and natural regeneration, resulting in diverse horizontal and vertical tree structures and frequently, multi-species forests (see also Section 1.5.4). Since final harvesting in CCF is also selective, the difference between thinning and harvesting operations is often unimportant, particularly in CCF stands with complex structures (cf. Figure 2.15). Stand age is typically undefined in such forests and the growing stock usually oscillates about a specified level (Pukkala and Gadow, 2012). Interestingly, CCF shares similarities with the horticultural concept of *permaculture* (Whitefield, 2013). Both approaches have in common that the soil is never completely exposed at large scale and is always covered by some level of vegetation. In every country, as part of defining CCF certain thresholds for allowable gap size have entered legislation. For example, in Britain, the maximum size of an area permitted to be cleared from tree vegetation is 0.25 ha (Hart, 1995). In this context, it is remarkable that already Anderson (1953) referred to the *permanent forest* when writing about what is known as CCF and near-natural forestry today (Pommerening and Grabarnik, 2019).

Depending on how abruptly the transition from one forest generation to the next is carried out, there are grey zones and overlaps between the *age-class* or *generation*/rotation forest system and the selection forest system at the far end of the range, where no distinct tree generations and rotations exist. In the former, processes of natural regeneration are also used, but forest development still progresses in distinctive generations or rotations usually allocated to two distinctive storeys. The basic variants of shelterwood, group and strip systems are silvicultural systems that more or less propagate age-class or generation forests (see Section 5.3). As a result, the tree biomass development over time is fairly similar to the top graph in Figure 1.4 with the difference that the transition from one generation or rotation to another is smoother and less abrupt than in plantation management. Still there can be considerable age and size differences in such forests. This is contrasted by plantation forestry that can be considered an extreme form of rotation forest management with no temporal overlap at all between two successive forest generations. Also, regeneration in plantation forestry is usually (but not exclusively) established by replanting. The over-storey removals in silvicultural systems often reflect geometric shapes, i.e. in group systems overstorey trees in circular areas are removed whilst in strip systems overstorey trees in rectangles are cut (Pommerening and Grabarnik, 2019). More complex or combined silvicultural systems can include patterns of natural disturbances (cf. Section 5.3).

Selection or '*plenter*' forests (cf. Section 5.4) are a very specific type of CCF management with a wide range of age and size classes, and tree canopies present throughout the vertical growing space (Schütz, 2001b). There is evidence that any attempt to remove the age-class or generation structure of forests, for example, by maintaining a high forest with two per-manent storeys (see Section 5.5), ultimately converges towards the structure of a selection system (Sterba and Zingg, 2001). Irregular shelterwood systems usually also come very close to the structure of selection forests. Conifer species appear to play a decisive role in achiev-ing a diverse vertical stand structure, which is more difficult in broadleaved forests without any conifers (Schütz, 2001b). The selection system is the only silvicultural system where sustainability of timber resources applies at stand and not at estate or district level. Also, the selection system is the only silvicultural system that comes close to the Anglo-American view of 'planned programmes of silvicultural treatment' (Smith et al., 1997) because the overall structure of the resulting forest more or less stays the same at all times. It should also be mentioned that the *coppice selection system* (Matthews, 1991, cf. Section 5.4) as a rare exception is an interesting combination of coppicing and selection system. In contrast to conventional coppicing, not all trees are cut at the same time like in a clearfelling operation, but rather individual trees and small groups of trees are coppiced at any one time. This type of short-rotation (12–15 years) selection system was, for example, traditionally practised in the Swiss canton Ticino to produce larger stems and to avoid soil erosion (Schweizerischer Forstverein, 1925).

Often silviculturists are concerned with the *transformation* of one silvicultural regime or system to another (Matthews, 1991), see Section 2.4. As mentioned before, a widespread forestry task of the nineteenth and early twentieth century in many European countries was that of transforming coppice forests to high forests, whereas for environmental and conser-vation reasons, this process is now being reversed in some countries. The transformation of plantation forests to continuous cover woodlands is currently a frequent activity of silviculturists in countries where CCF has recently been introduced. Some authors give transformation forest management the same attention as plantation and CCF management (Pukkala and Gadow, 2012). This group of activities concerns the active gradual change of woodland structure and or species composition. The biomass curve of transformation management consequently has elements of both, plantation management at the beginning and continuous cover management later on (cf. Figure 1.4).

As shown in Section 1.5, there is altogether a wide range of silvicultural possibilities within CCF including many combinations of silvicultural systems and thinning treatments (Pommerening and Murphy, 2004).

1.4 Silvicultural Analysis and Planning

Silvicultural analysis and planning describe the process of appraising the development of a forest stand to date and prescribing forest management for the next 5–10 years. Long-term plans for each woodland community occurring in the area under study usually exist, and they provide the general framework, see Section 7.2. Operational details are added by the forest manager through silvicultural analysis and planning. An important pre-requisite of any silvicultural plan is a site visit for sampling and for a full appreciation of a stand's condition and potential. The basic method of forest management planning at stand level includes four distinctive steps:

1. **Analysis of current state**
   - Description of site (elevation, exposition, area, climate, soil), vegetation and fauna (if possible and relevant).
   - Expected ecosystem goods and services?
   - Logistics: Extractions racks, access to forest roads.
   - Current state of the forest, stand development stage, stand description, inventory and quantitative analysis.
   - Frame trees (cf. Chapter 3) already selected, time since last frame-tree release?
2. **Anticipation of forest dynamics**
   - 'Anamnesis': What has influenced forest dynamics in the past? Any clues from historical records or current tree morphology? Stand origin, thinning history, silvicultural objectives, past forest development. When was the last intervention?
   - How will the ecosystem change? In which direction and what is the likely pace of change?
3. **Assessment of forest development pathways**
   - Reference scenario: What would happen without human interventions (see Section 2.6)?
   - What vegetation types and forest structures are possible in the future? What should be the target diameter range (cf. Section 3.4), the anticipated lifetime of the current forest generation, if any?

4. **Selecting the silviculturally appropriate techniques to meet the objectives**
   - What are important sustainability criteria for this site?
   - What forest structure should this site develop in the long run?
   - What tree species mixture is intended?
   - What should be the minimum distance between frame trees?
   - How much input is necessary (compared to zero intervention as a reference, see Section 2.6) and justifiable? Do frame trees need to be released? Intervention types, intervention cycles, intervention intensity.
   - When will be the next intervention following this one?
   - Optimising the solution.

A key element of silvicultural planning, and in fact of many aspects of research, is the description of the current state of a given forest unit. This is always the beginning and the basis for subsequent planning and research. After careful observation in the field and a detailed analysis of environmental factors, the analyst starts with a qualitative description of the current state. This provides the reader of the silvicultural plan/the research work with an easy access to the woodland in question (Pommerening and Grabarnik, 2019). There are many possible schemes and templates for stand descriptions, see, for example, Appendix A in Pommerening and Grabarnik (2019).

At the same time, up-to-date sampling data are a great help to analyse the current state of the respective planning unit (Figure 1.6), see also Section 4.4. In the same step, the processes that have led to the current state have to be taken into consideration. Any information on the history of stand development may prove useful in order to get a feeling for the velocity of the dynamics involved. Considering past disturbances and management also ensures that management prescriptions are avoided which may destabilise a forest stand. Particularly, the questions of how the given forest ecosystem is likely to develop with and without forest management, in what direction and how fast, are important in this context.

Information on stand density, height-diameter ratios (Eq. 2.2), crown ratios (Eq. 2.3), and crown damage offer insights on both past management and disturbances. For example, if h/d ratios are high and c/h ratios are low, it is quite likely that there have been high tree

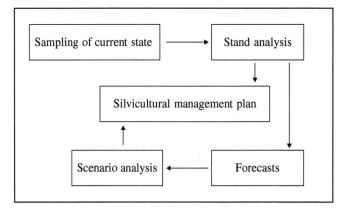

Figure 1.6 The general process of silvicultural planning. Source: Adapted from Gadow (2005).

densities and few interventions in the past. In CCF, the 'do-nothing' option is a crucial consideration, as part of the definition of CCF is to carry out interventions only if they are absolutely necessary; otherwise, one should leave woodland development to natural courses. This strategy is often termed *biological automation* or *biological rationalisation*, see Section 2.6. Based on the analysis, projections of potential forest development paths are undertaken which can be model aided (Coates et al., 2003; Wikström et al., 2011). Usually a number of different forest management possibilities or scenarios are explored in such scenario analyses and the best option under the given legal and environmental constraints and management objectives is identified. This process can be formalised by methods of optimisation and operations research that are usually presented in forest planning textbooks (see Bettinger et al., 2017). The silvicultural scenarios include key processes and methods of forest management such as species choice, silvicultural regime, silvicultural system and forest management types (Pommerening and Grabarnik, 2019).

Both the results of the sampling analysis and scenario analysis then feed into the silvicultural management plan. The core part of this plan is a set of silvicultural prescriptions usually for the next 5–15 years. A *silvicultural prescription* is a planned series of treatments designed to change current stand structure to one that meets management objectives (Helms, 1998) and can be part of the planning of a larger forest district or estate. In CCF, the development of silvicultural prescriptions begins with knowledge of the forest ecosystem that is to be managed and its constituent components (Franklin et al., 2018).

After reaching a conclusion, the forest management plan needs to be translated into a clear, transparent and reproducible work description for practical implementation. The practical implementation should ideally be followed by a control or appraisal element. This follow-up is very often neglected in practice, especially in times of excessive work pressure and scarce resources. However, the best scientifically based silvicultural planning is useless if delegated to staff and operators who are not able to implement it because of insufficient training and need advice (see Chapter 9). Follow-ups can help reveal such issues. Control should not be thought of as having negative connotations and can already be taking place while the actual work is under way. In plantings and thinnings, for example, major mistakes can be identified and mitigated during on-site visits. On the other hand, some results will only become evident several years after the actual work has finished, such as the survival rate of plants on a replanting or underplanting site. Long-term monitoring is therefore important, particularly when a type of silviculture new to an area is introduced, and it is crucial to record even seemingly minor details, which help understand *cause and effect relationships* (Pommerening and Grabarnik, 2019). Also, in practical or applied forest management, the benefit of monitoring or control can be enhanced by detailed record keeping that allows further statistical analysis (see Section 1.10). Statistical analysis can provide a better and deeper understanding of advantages and disadvantages of individual components used. In the long run, this can lead to a forest enterprise specific optimisation of silvicultural planning through adaptive management (Thomasius, 1990). Other aspects of silvicultural control include recording the density and height of natural regeneration, the impact of felling and extraction on remaining trees, the impact of browsing and bark stripping by animals, the growth performance, the impact of natural disturbances, and the natural mortality of and the interaction among trees.

The methodology employed for monitoring the implementation of silvicultural plans (see, for example, Gadow and Stüber, 1994) can also be used for quantifying the impact of disturbances and vice versa. Recently, some of this methodology has also been adopted for use in forest management training which has received much attention in forest practice in France, Switzerland, Britain and Ireland. For this purpose, all trees inside a rectangular or circular sample plot are marked with tree numbers and are callipered. These training sites are frequently referred to as *marteloscopes* (Bruciamacchie et al., 2005; Poore, 2011; Susse et al., 2011) (from French *martelage* – marking). Participants of such training seminars are then provided with a list of all trees and are asked to mark trees for thinnings. Methods and strategies that are discussed in Section 8.3 in detail are used after the exercise to assess how each participant has carried out the silvicultural objectives compared to others in their group by quantifying how the marking will modify the current state of the forest after implementation (Messier et al., 2013). Based on these, individual feedback is provided for every participant.

Finally, it should be emphasised that this traditional procedure of preparing silvicultural plans, implementing, checking up on the implementation, and revising the plan after 10–15 years is a good example of what is often described as *adaptive management*, cf. Section 7.2. For more details on silvicultural planning for CCF see Chapter 7. More general information on forest planning and optimising management can be found in specialist literature such as Bettinger et al. (2017) and Franklin et al. (2018).

1.5 Continuous Cover Forestry – Definitions, Terms and Semi-synonyms

CCF or near-natural forestry is not a new phenomenon, but over the last two decades, there has been a renewed world-wide debate regarding its position in forest management (Brang et al., 2014; Cairns, 2001; Gadow et al., 2002; Guldin, 2002; O'Hara, 2002; de Turck-heim, 1999; Lähde et al., 1999). CCF has been a standard in Central Europe for more than 50 years. Its comparatively long history, however, stretches over more than a 100 years. In some countries and world areas, CCF is still relatively new, e.g. in the United Kingdom, Ireland, Scandinavia, North America, Australia and China. Currently, CCF is re-visited in many countries around the world for its potential to mitigate climate change, to increase or at least maintain biodiversity in forest ecosystems, to provide valuable tools for forest conservation, and to enhance the appeal of woodlands used for recreation.

Recently, the EU forest strategy for 2030 (European Commission, 2021) stated clearcutting should be 'used only in duly justified cases, for example, when proven necessary for environmental or ecosystem health reasons' and the strategy promotes 'the creation or maintenance at stand and landscape level of genetically and functionally diverse, mixed species forests, especially with more broadleaves and deciduous trees and with species with different biotic and abiotic sensitivities and recovery mechanisms following disturbances'. These political statements clearly support CCF, and a recent policy paper published by the European Forest Institute provided more explicit definitions and implementation guidelines for this strategy (Larsen et al., 2022).

Early concepts of CCF can be regarded as a naïve understanding and application of silviculture practised on an ecological or environmental basis, see the historical review in Section 1.7. The different terms and the wide variety of, sometimes conflicting, definitions can potentially cause considerable confusion among politicians, practitioners, students and scientists and need to be clarified when used. Pommerening and Murphy (2004) grouped the terms and semi-synonyms under six general headings which often highlight only a particular aspect or focus of CCF rather than broadly defining the concept as a whole (Table 1.2). These six categories are *continuity of forest cover, ecosystem/natural management, structural diversity, retention, thinning/harvesting methods* and *programmatic semi-synonyms*. The last group of terms include those that are particularly difficult to relate to CCF without an explanation and appear to suggest a new agenda or programme. The semi-synonyms in this category are somewhat philosophical and emphatic. The groups of ecosystem/natural management and retention have received particular attention in the literature. Not all semi-synonyms interchangeably denote exactly the same type of forestry; however, most of them share much overlap and many similarities.

Specific definitions exist for some of the semi-synonyms. For example, Fedrowitz et al. (2014) suggested that *retention forestry* aims to reduce structural and functional contrasts between managed forests and natural forests, mainly by increasing the abundance in harvested stands of key structures important for biodiversity, such as old and dead trees (Figure 1.7). This is contrasted by the term *managed retention* implying that a stand in question is exempt from clearfelling for reasons of biodiversity, visual impact or danger of landslides (Forest Enterprise, 2000). While the former definition implies retention of individual trees, the latter focuses on the retention of whole forest stands. Some distinctions between CCF semi-synonyms have been made somewhat artificially based on misunderstandings or on the authors' understandable desire to coin their own terms and concepts.

Naturalness plays a prominent role in the semi-synonyms of CCF and in general definitions of the concept. Naturalness is a popular term; however, it is very difficult if not impossible to define (Peterken, 1996; Schirmer, 1998). A very common approach is to assess the differences between a forest's current state and the potential natural vegetation (PNV) on the same site in order to define the degree of naturalness. The general idea of near-natural forestry or CCF is to promote managed forests with structures, management practices and/or species compositions that are more akin to the potentially natural stages of development and to the potentially natural processes of tree vegetation on any particular site than those that are commonly observed in rigid plantation management (Pommerening and Murphy, 2004).

It is interesting to note that authorities in countries where CCF has only recently been introduced, often coined a term different from CCF or one of the other more commonly known semi-synonyms. The Forestry Commission in the United Kingdom, for example, adopted the terms *LISS* (low-impact silvicultural systems), *ATC* (alternatives to clearfelling) and *managed retention*, whereas the Forest Authority (Skogsstyrelsen) in Sweden uses the term *non-clearcut forestry* (hyggesfritt skogsbruk). Coining their own semi-synonyms allows these organisations to retain the liberty to deviate from seemingly rigid or too liberal CCF criteria, e.g. in terms of the allowable size of felling coupes or to set

Table 1.2   Synonyms and semi-synonyms used in connection with near-natural forestry or CCF.

Synonym or semi-synonym	Source
Continuity of forest cover	
Alternatives to clearfelling, alternative silvicultural systems to clear cutting, alternative silvicultural practices	Penistan (1952), Hart (1995), Beese and Bryant (1999), Lencinas et al. (2011)
Continuous forest	Troup (1928) and Hart (1995)
Continuous cover silviculture	Yorke (1998)
Dauerwald	Möller (1922), Troup (1928), and Helliwell (1997)
Low-impact silviculture, ~ management approaches	UKWAS Steering Group (2000), Mason et al. (1999)
Permanent forest	Anderson (1953), Häusler and Scherer-Lorenzen (2001)
Ecosystem/natural management	
Close-to-nature forestry/forest management	Mlinšek (1996) and Mason et al. (1999)
Close-to-nature silviculture	Schütz (2001a), Kenk and Guehne (2001) and O'Hara (2016)
Ecoforestry	Drengston and Taylor (1997)
Ecological forestry, ~ forest management	Mason et al. (1999), Seymour and Hunter (1999), Franklin et al. (2018)
Ecosystem (friendly, oriented) management	Thomasius (1992), Grumbine (1994) and Salwasser (1994)
Ecological silviculture/forestry, ecologically oriented silviculture	Benecke (1996), Lähde et al. (1999), Palik et al. (2021)
Forest management based on natural processes	Pro Silva Europe (1999, 2012)
Nature-based forestry/forest management	Diaci (2006) and Larsen and Nielsen (2007)
Natural disturbance-based management	Bose et al. (2014)
Near-natural forestry/forest management	Benecke (1996) and Gadow et al. (2002); Larsen (2005)
Nature-orientated silviculture	Lähde et al. (1999) and Koch and Skovsgaard (1999)
Naturalistic silvicultural systems	Mitchell and Beese (2002)
Restoration forestry	Pilarski (1994)
Sustainable forestry	Maser (1994)

Table 1.2 (Continued)

Synonym or semi-synonym	Source
Thinning/harvesting methods	
Green-tree retention (GTR) harvest	North et al. (1996) and Craig and Macdonald (2009)
Retention harvesting	Craig and Macdonald (2009) and Baker et al. (2013)
Selective cutting/selective timber management	Curtis (1998)
Structural diversity	
Diversity-orientated silviculture	Benecke (1996) and Lähde et al. (1999)
Irregular structure forestry/silviculture	Johnston (1978), Lord Bradford (1981) and Pryor (1990)
Irregular forestry/forests, management of ~	Susse et al. (2011)
Uneven-aged/multi-aged/multi-cohort management/silviculture/forestry	Anderson (1953) and Oliver and Larson (1996)
Retention	
Managed retention	Forest Enterprise (2000)
Overstorey retention	Dovčiak et al. (2006) and Halpern et al. (2012)
Retention forestry, retention harvesting	Craig and Macdonald (2009), Gustafsson et al. (2012), Baker et al. (2013) and Fedrowitz et al. (2014)
Structural retention	Dovčiak et al. (2006)
Tree retention, GTR	Franklin (1989), North et al. (1996), Vanha-Majamaa and Jalonen (2001), Rosenvald and Lohmus (2008), Johnson et al. (2014), Halpern et al. (2005) and Gustafsson et al. (2010)
Variable retention	Mitchell and Beese (2002)
Programmatic semi-synonyms	
Back to nature	Gamborg and Larsen (2003)
Common sense forestry	Morsbach (2002)
(Complex) adaptive systems approach	Bončina (2011a) and Messier et al. (2013)

(Continued)

Table 1.2 (Continued)

Synonym or semi-synonym	Source
Programmatic semi-synonyms	
Excellent forestry	Robinson (1994)
Free-style silviculture	Bončina (2011a) and O'Hara (2014)
Holistic forestry	O'Keefe (1990), Peterken (1996), Koch and Skovsgaard (1999) and Mason et al. (1999)
Lübeck model	Sturm (1993)
New forestry	Franklin (1989) and O'Keefe (1990)
New perspectives	Kessler et al. (1992)
Positive impact forestry	McEvoy (2004)
Systemic silviculture	Ciancio and Nocentini (2011) and Nocentini et al. (2021)

Source: After Pommerening and Murphy (2004) and O'Hara (2014).

their own CCF agenda. Often, these new terms come with a set of more or less strict rules, which are sometimes linked to national legislation or certification.

Of the many other CCF definitions eight examples may be quoted which reflect aspects already identified by the group headings:

1. According to the IUFRO Multilingual Forest Terminology Database the term continuous cover forestry describes a highly structured forest ecosystem managed to maintain continuous tree cover over the total forest area (Nieuwenhuis, 2000).
2. The British silviculturist Professor R. S. Troup was very interested in the Dauerwald idea of the 1920s, and visited German CCF stands. He defined this form of silviculture in the following way (Troup, 1928): 'The [German] term Dauerwald, one of the semi-synonyms (see Table 1.2), may be translated briefly as "continuous forest," that is, forest treated in such a manner that the soil is never exposed, the forest cover being continuously maintained over every part of the area'. Möller (1922) applied the term 'Dauerwald' in general not to any one particular method of treatment, but to any system not involving clearcut and the exposure of soil (Troup, 1928).
3. Mason et al. (1999) stated 'continuous cover is defined as the use of silvicultural systems whereby the forest canopy is maintained at one or more levels without clear felling'.
4. Gadow (2001) came to the conclusion that 'CCF systems are characterised by selective harvesting; the stand age is undefined and forest development does not follow a cyclic harvest-and-regeneration pattern'.
5. Hart (1995) put forward a very detailed definition stating that continuous forest cover [sic] is 'a general term covering several silvicultural systems which conserve the local forest canopy/environment during the regeneration phase. Coupe size is normally below 0.25 ha (50 × 50 m) in group systems; and in shelterwood – where used – is retained for longer than 10 years. The general aim of all systems within the concept is the encouragement of diversity of structure and uneven age/size on an intimate scale'.

Figure 1.7 Retention forestry/green tree retention is often only but a small step from clearfelling towards CCF with very few residual trees (standards) that can hardly maintain woodland climate and forest soil conditions. Example from Järvselja Forest near Tartu in Estonia involving *Betula* spp. and *Pinus sylvestris* L. with closed forest stands in the background. Source: © Arne Pommerening (Author).

6. Franklin et al. (2018) stated that *ecological forestry*, one of the semi-synonyms of CCF, see Table 1.2, utilises ecological models from natural forest systems as a basis for managing forests. The concept incorporates principles of natural forest development including the role of natural disturbances in the initiation, development and maintenance of forests and forest landscape mosaics. Most importantly, ecological forestry recognises that forests are diverse ecosystems and seeks to maintain their fundamental integrity.

7. Palik et al. (2021) defined *ecological silviculture*, another semi-synonym of CCF, see Table 1.2, as an approach to managing ecosystems, including trees and associated organisms and ecological functions, based on the emulation of natural models of development and that explicitly incorporates principles of continuity, complexity/diversity, timing and context.

Most definitions focus only on parts of what the whole concept can include, e.g. on selective interventions and undefined stand age. The definition put forward by Palik et al. (2021) makes an attempt to describe the whole concept of CCF as the authors see it. Many definitions stress the idea of the continuity of woodland conditions over time, hence the name 'continuous cover forestry', a term which has given rise to many misunderstandings. Often, it has been misunderstood as suggesting an intention to create and maintain dense forests with closed canopies, see Section 1.6. The term, however, does not imply any degree of canopy closure and is fully compatible with the creation of gaps for promoting natural regeneration, the conservation of wildlife, or for the establishment of viewpoints or vistas should the necessity arise. CCF does not imply a lack of management but emphasises the need to avoid clearfelling over large areas and within this broad concept, a range of silvicultural systems and thinning methods are possible. However, CCF is more than the mere avoidance of clearcutting (Lähde et al., 1999; Kenk and Guehne, 2001; Nabuurs, 2001; Vanha-Majamaa and Jalonen, 2001), and any particular CCF concept requires a clear description of goals and principles to be transparent (Grumbine, 1994; Brunner and Clark, 1997). Some of the definitions above highlight other important components such as the selective removal of trees, allowable gap size, suitable silvicultural systems and vertical structure. In particular, CCF is being seen as compatible with a holistic approach to forestry with multi-purpose management objectives (Häusler and Scherer-Lorenzen, 2001; Niedersächsische Landesregierung, 1991; Landesanstalt für Wald und Forstwirtschaft Thüringen, 2000; Forestry Commission, 2001; Palik et al., 2021).

Most of the semi-synonyms in Table 1.2 are suitable for describing a general concept of silviculture based on ecological and natural processes. In this book, I have adopted *continuous cover forestry* (CCF) for its appeal and popularity; however, other terms would certainly serve as good descriptors, too. With the choice of name, I do not follow one particular variant of silviculture on an ecological basis but rather all of them collectively. After all, what matters most is the general concept itself and not so much the label assigned to it. Whilst CCF and the practices labelled by other semi-synonyms can and should differ in different parts of the world, there are related principles that form a common denominator. Differences reflect the variation in forest types, in intensity and scale of natural disturbances and in the management history to name but a few (Larsen et al., 2022).

The concept of CCF is usually broken down into themes or components. Palik et al. (2021) referred to them as *tenets*. Many of these tenets or components reflect the general goals of ecosystem management (Grumbine, 1994). In the following, the general tenets of Figure 1.8, which were first published by Pommerening and Murphy (2004), are introduced and outlined in more detail (see also Appendix A for a summary). By arranging them in groups with common headers, the context is indicated.

1.5.1 Continuity of Woodland Conditions

This is the oldest and most important part of the definition of CCF. The idea of the tenet and the associated wording is to encapsulate a number of important ecological conditions. Troup (1928) stated that CCF may be said to include all those silvicultural systems which involve continuous and uninterrupted maintenance of the forest. Franklin et al. (2018) stressed the continuity in forest structure, function and biota between pre- and post-intervention ecosystems. As natural disturbances leave significant biological legacies from the pre-disturbance forest that provide for continuity, CCF emulates that process through the selective retention of tree vegetation and other biological legacies in all silvicultural treatments. This includes the retention of seed dispersal mechanisms, nutrient and water cycles as well as the exchange of genetic material typical of forest ecosystems (Larsen et al., 2022). From an ecological perspective, some tree species require the continuity of woodland conditions with only moderate changes of their habitats, imposed by forest

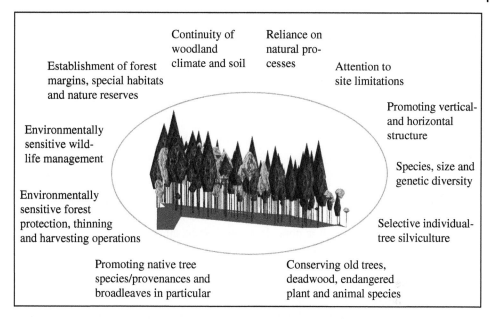

Continuity of woodland climate and soil

Reliance on natural processes

Establishment of forest margins, special habitats and nature reserves

Attention to site limitations

Environmentally sensitive wildlife management

Promoting vertical- and horizontal structure

Species, size and genetic diversity

Environmentally sensitive forest protection, thinning and harvesting operations

Selective individual-tree silviculture

Promoting native tree species/provenances and broadleaves in particular

Conserving old trees, deadwood, endangered plant and animal species

Figure 1.8 The main components or tenets of the contemporary international continuous cover forestry debate. Source: Adapted from Pommerening and Murphy (2004).

management or natural disturbances, to ensure their survival. The same continuity is also an important feature of protection forests securing and stabilising watersheds, mountain slopes, coastlands and woodlands managed for amenity and recreation (Rittershofer, 1999). Historically continuity of woodland conditions has always been an important factor in the timber production of traditional selection forests on the small-sized farm holdings of Slovenia, Switzerland, France, Germany and Austria (Burschel and Huss, 1997).

Almost half of the total organic carbon (C) in terrestrial ecosystems is stored in forest soils. Harvesting, particularly clear-cut harvesting, generally results in a decrease in soil C stocks, with highest C losses occurring in the forest floor and the upper mineral soil (Mayer et al., 2020). The continuity of woodland conditions is therefore crucial for carbon sequestration both in trees and soils and for the conservation of soil processes (Ontl et al., 2019). There is, however, some debate on the size of gaps (i.e. small, localised clearfells) allowed especially because current regulations do not distinguish between tree species and site types involved (see Figure 1.7).

Since the late 1980s and the beginning of the 1990s, new forest legislation in Germany requires special permission by the forestry authority of the respective federal state, if clear-fell coupes exceed a certain size. In Germany, the regulations governing the maximum clearfelling size vary from state to state. For example in the state of Northrhine-Westfalia, clear cutting is limited to a size of three hectares, whereas in Baden-Württemberg, permission is required for clearfelling areas larger than one hectare (Häusler and Scherer-Lorenzen, 2001). Similar legislation is implemented in Austria and Switzerland.

Gresh and Courter (2021) have argued that the North American pursuit of ecological forestry has embraced a natural disturbance paradigm. Europe, in contrast, according to Gresh and Courter (2021), is pursuing ecological forestry by requiring low-intensity harvest protocols which continuously protect the forest canopy. Perhaps prompted by the term 'continuous cover', it is clearly one of many misunderstandings (see Section 1.6) that European ecological forestry or CCF allegedly does not incorporate or attempts to mimic natural disturbances. Continuity of woodland conditions is stressed in European CCF policies because clearfellings have in the past been the dominant feature of European forestry.

1.5.2 Reliance on Natural Processes, Promoting Vertical and Horizontal Structure

In CCF, forest managers aim to create a varied horizontal and vertical structure of individual trees and groups of trees in a stand. By allowing a varied amount of horizontal and vertical structural elements, it is possible to save establishment and tending costs apart from obvious diversity and habitat benefits. Experience has shown that forest structures can be managed in such a way that natural processes such as natural regeneration, natural pruning, the development of good stem form and self-thinning (natural stem number reduction) in the early growth stages are stimulated, see Section 2.6. In general, shading provided by tree shelter results in thinner branches, higher wood density and reduced stem taper (Burschel and Huss, 1997; Messier et al., 2013; Rittershofer, 1999; Schütz, 2001a). This form of steered self-regulation is often referred to as biological automation or biological rationalisation and is related to the conservation principle of passive restoration (Newton, 2007, see Section 7.3.3). Smith et al. (1997) referred to this principle as 'imitating nature through silviculture'. Reliance on natural processes is facilitated by introducing more structural diversity. In terms of stem form and timber quality, a similar effect can be achieved by managing dense plantations with small initial spacing, but there is general consensus that the benefits cannot compensate for the additional costs incurred. Therefore, managing vertical and horizontal structure in conjunction with natural regeneration is a clear alternative to controlling the level of competition through management (Klang and Ekö, 1999). The three tenets of continuity of woodland conditions, reliance on natural processes, and promoting vertical and horizontal structure imply a reliance on natural regeneration. The use of natural regeneration is an advantage where the native parent trees are site-adapted, have a high genetic diversity, and have other desirable qualities (Larsen et al., 2022). Natural regeneration established over long periods from many parent trees maintains high genetic variation (Finkeldey and Ziehe, 2004). The individual-tree and collective resilience of woodlands can be enhanced, and another positive side effect of this strategy is a greater biodiversity. There is a close link between forest structure and diversity, thus diversifying the size and species structure of forests increases the availability of habitats and ecological niches (Pommerening and Grabarnik, 2019). There is also evidence for the fact that forests with high structural diversity are more resistant to wind and pests (Dvorak and Bachmann, 2001; Hanewinkel et al., 2014; Seidl et al., 2018) and enhance adaptive capacity. Well-structured forests like those managed as selection systems are appealing to visitors and have therefore an important recreation and amenity value (Jephcott, 2002; Nyland, 2003) which makes them ideal for community and urban forests. This element of CCF is what Palik et al. (2021) described by the tenet 'seek complexity and diversity'.

1.5.3   Attention to Site Limitations

As with all good forestry practice, tree species/provenance choice should be made dependent on site. This ensures that species/provenances are used that are well adapted to the particular environmental conditions and therefore can resist biotic and abiotic damage and have high growth rates (Pro Silva Europe, 1999; Burschel and Huss, 1997). With ongoing climate change, this tenet is even more important than ever before. Site conditions should guide species selection in any approach to forest management, but this is particularly important for CCF, given its demanding requirements in terms of stand resilience, diversity, use of natural processes and the idea to deliver multiple ecosystem goods and services (Davies et al., 2008). Traditional CCF techniques have largely been developed for and applied to native species. However, there is usually no reason why CCF should not be practised successfully with non-native species, as long as they are adapted to local site conditions. A tree species is site adapted, if it is able to grow vigorously and to reproduce over a number of generations under the specific climatic, pedological and topographical conditions of the site in question (Bartsch et al., 2020). This implies that

- the species must be capable of reaching its natural life expectancy,
- it should achieve its natural growth and size range,
- it must produce viable natural regeneration,
- it should not lead to any form of site deterioration (e.g. through litter accumulation),
- it must not show invasive behaviour.

Depending on management objectives, the choice of species should take into account both ecological (maintenance of site productivity, biological diversity) and management (stability, yield, resilience) criteria. All involve great uncertainties, as they are partly based on future conditions that may be difficult to predict. In particular, the consequences of climate change are likely to have an impact on the suitability of a species. Although the suitability of some species can already be gauged on the basis of the most likely scenarios for climate change, in general, it will be best to favour species with a greater amplitude of environmental requirements, and to make use of a wider range of species to minimise any risks (Davies et al., 2008). Currently experiments are underway where species provenances that evolved in warmer climates are transplanted into areas which currently still have cooler climates to anticipate future climate change (Brang et al., 2014; Frischbier et al., 2019). Future results from the long-term monitoring of these experiments will reveal whether this is an appropriate strategy for mitigating climate effects. The principle of attention to site limitations is also in the spirit of the tenet 'maintain options for the future' proposed by Palik et al. (2021).

1.5.4   Species, Size and Genetic Diversity

There is potentially a range of benefits to be gained by encouraging mixed coniferous/broadleaved woodlands. These benefits include the reduction of biotic, abiotic and economic risk, for example, diseases and insect calamities cannot spread as easily as in monospecific stands (Duchiron, 2000; Nyland, 2003). One of the aims of CCF is therefore the diversification of monospecies coniferous plantations, which are outside their natural range. Recent research in monospecific and mixed stands of *Picea abies* (L.) H. KARST. and *Fagus sylcatica* L. showed that in such mixed woodlands total volume production does not decrease as stand density approaches the maximum but remains constant (Griess et al., 2012; Pretzsch and Biber, 2016). This contrasts with monospecific stands where total volume production usually decreases after reaching a maximum. These studies also showed that the resilience of predominantly single-species *P. abies* stands can be improved, if unstable spruce trees are replaced by resilient *F. sylvatica* trees. According to the study an important element of resilience to disturbances in these mixed stands is the presence of sub-dominant and co-dominant trees which, therefore, should be retained. These findings revealed that mixed-species stands are much better able than monospecific stands to compensate for impacts on the stand density, such as windthrow or heavy thinnings, through an accelerated growth of the residual stand (Pretzsch and Biber, 2016). The work of Pretzsch and Biber even suggests extending the occurrence of mixed stands for the sake of increasing forest resilience. Therefore, two conclusions can be drawn from this research: (i) Ecosystem productivity increases with increasing biodiversity and (ii) biodiversity enhances ecosystem resilience (Messier et al., 2013). Conclusion (2) is supported by the *insurance hypothesis*, an ecological theory suggesting that species that might be functionally redundant in a given ecosystem increase in numbers to compensate for the reduction in performance of the dominant species thus providing 'insurance' for community productivity (Yachi and Loreau, 1999; Matias et al., 2013).

Mixed species forests also provide a wider range of size classes and timber products allowing flexible and rapid response to market conditions in commercial forestry. They contribute to greater biodiversity and therefore provide more habitats. Studies have shown that mixed woodlands are believed to be more appealing to visitors (Jephcott, 2002; Petucco et al., 2018; Arnberger et al., 2019). Mixtures may also help reduce risk from global or regional climate change (Lindner and Cramer, 2002). Swedish and Danish investigations on the nutrient status of *Picea abies* in monospecies and in mixed-species stands revealed that spruce needles from mixed stands had higher concentrations and ratios to N of K, P and Zn than needles from pure spruce stands. Among the mixed stands, the K status appeared to be positively correlated with the percentage of deciduous tree basal area. Soil samples from mixed stands had a higher Mg concentration and base saturation than soil samples from monospecies stands. The authors came to the conclusion that the positive effects on *P. abies* nutrient status in the mixtures may promote total stand productivity in the long run, increase the resistance to adverse effects of air pollution, and limit the need to counteract ecosystem nutrient imbalance with direct treatments like fertilisation or liming (Thelin et al., 2002).

1.5.5 Selective Individual-Tree Silviculture

In managing stands for CCF, trees are individually selected for thinning and harvesting in an attempt to move a forest stand into a desired direction. This is often accomplished as a compromise between silvicultural, economic and conservation needs (Rittershofer, 1999), see Section 7.1. Originally, individual-based forest management using frame trees (cf. Chapter 3) developed independently of CCF (Schädelin, 1926, 1934), but already Möller (1922) saw a connection between both. In Central Europe, most CCF silvicultural systems now aim at a combination of frame-tree selection, crown thinnings and target diameter harvesting (cf. Section 3.4) (Abetz and Klädtke, 2002, see Chapter 3). Thinning and harvesting decisions are made on a tree-by-tree basis and depend on individual tree size (Messier et al., 2013). In commercial scenarios, they include considerations of size premiums, penalties for very large stem diameters, size-dependent growth rates and production risks (e.g. wind damage, pathogens). The principle implies that rotation ages or periods do not apply. Forest stands become slightly less important as management units and individual-tree management can be considered a bottom-up approach that potentially offers greater flexibility. Interventions typically focus on a subset of trees whilst other parts of the forest largely remain undisturbed (Pommerening et al., 2021a). The individualisation of management also promotes individual-tree resilience through a gradual change of morphology in released trees allowing them to better withstand the forces of wind and snow. This is particularly important in tree species with shallow root systems (e.g. *Picea*, *Betula*). Selective fellings may increase timber quality and are a pre-requisite for many other elements of CCF. As the economic benefits of tourism become better understood, it is important to realise that selective individual-tree silviculture much contributes to winning potential tourists when compared to the effect of clearfelling large patches of trees (Jephcott, 2002; Nyland, 2003; Arnberger et al., 2019).

1.5.6 Conserving Old Trees, Deadwood, Rare and Endangered Plant and Animal Species

Many CCF guidelines suggest retaining a certain amount of downed and standing deadwood in each forest stand for biodiversity reasons. It is also recommended that a certain percentage of old scenic trees and similar natural features are kept for their amenity value. Forest management should also aim at promoting endangered plant and animal species along with other management objectives (Niedersächsische Landesregierung, 1991). Remnant broadleaved trees in conifer plantations should be secured by releasing them from conifer competition. This principle also includes the protection of special biotopes within forests such as wetlands, rocky outcrops and dunes (Otto, 1994). The biodiversity aspect gains some importance because continuous high forests with native species such as European beech (*Fagus sylvatica*) can have a negative effect on tree species diversity because rare and less competitive species tend to be extinguished as opposed to in coppice and coppice-with-standards forests (Kausch-Blecken von Schmeling, 1992). Typical victims of this effect are the native European *Sorbus* species, i.e. whitebeam (*Sorbus aria* (L.) CRANTZ), true service tree (*Sorbus domestica* L.) and wild service tree (*Sorbus torminalis* (L.) CRANTZ) and yew (*Taxus baccata* L.). In some countries, policymakers are aware of this effect and therefore explicitly included the conservation of rare and endangered plant species and provenances in their CCF guidelines (Niedersächsische Landesregierung, 1991; Landesanstalt für Wald und Forstwirtschaft Thüringen, 2000; Nyland, 2003).

1.5.7 Promoting Native Tree Species/Provenances and Broadleaves

For centuries fast-growing conifer species were promoted in large areas of Europe beyond their natural range on sites naturally dominated by broadleaves to increase commercial timber production. These secondary coniferous stands mainly in areas below 1000 m asl turned out to be extremely sensitive to environmental stress factors and are highly susceptible to progressive loading by air pollution and climate change. Thus, the restoration of such forest ecosystems originally dominated by broadleaves is an important silvicultural task (Hasenauer and Sterba, 2000).

As there can be silvicultural problems with native tree species, for example, in British upland forestry and in Mediterranean countries, this issue is traditionally contentious, though it is a clear element of international definitions (Pro Silva Europe, 1999, 2012). The idea comes from the assumption that native tree species and provenances are better adapted to local site conditions and have co-evolved with other plant and animal species in a particular country or region whilst exotic species often show faster forest growth and are more productive. Numerous animal and plant species are directly connected to native trees in this co-evolutionary development. The introduction of exotic species potentially disrupts this symbiosis and results in a reduction of biodiversity (Pro Silva Europe, 1999, 2012). Species are considered native, if they have not been introduced by humans, either recently or in the distant past (Peterken, 2001). However, this definition presents some difficulties and the natural range of species cannot always be sharply delimited. There will always be considerable debate over the 'nativeness' of one species vis-à-vis another, and it is questionable whether species which have just been excluded from a certain region by the last ice age, but before that had a long co-evolution with other plant and animal species of the same area, should be termed non-native (Niedersächsische Landesregierung, 1991). The boundaries of areas, which can be described as the native range of a species, change over time even without human influence. This, for example, applies to Douglas fir (*Pseudotsuga menziesii* (Mirbel) Franco) that did not survive the ice ages in Europe, but may be better adapted to climate change than some of the native European conifers. Also, the 'site' at which a species is native can be defined at various scales: From a large region through a discrete wood to each patch of contrasting soil within a wood (Pommerening and Murphy, 2004). Mixing native and site-adapted non-native species in semi-natural woodlands may enhance their adaptive capacity, whilst limiting the potentially negative effects of non-native species and provenances (Peterken, 1996; Vitali et al., 2018). In this context, there is an ongoing debate about *genetically improved* or *genetically engineered* tree provenances. These are tree species provenances resulting from tree breeding, not genetically modified (transgenic) trees. Breeding mainly aims at growth and timber quality properties, and given the concept of CCF, it is questionable whether such provenances should be included in near-natural forest management. In addition, it is uncertain how genetically improved or genetically engineered tree provenances will perform with ongoing climate change. Another consideration here is the consequent removal of invasive non-native tree species. According to Pro Silva Europe (1999) and Pro Silva Europe (2012), exotic tree species should only be planted in situations where this is an economic or other necessity, and then only if the exotics can be mixed with the indigenous vegetation pattern within certain qualitative and quantitative limits. However, the need to move towards a greater use of native tree species is a clear requirement of forest restoration, which forms a part of CCF. The removal of non-native and especially invasive species among natural regeneration might be necessary and underplanting with native species plays an important role in this context (Thompson et al., 2003; Niedersächsische Landesregierung, 1991; Pommerening and Murphy, 2004).

1.5.8 Environmentally Sensitive Forest Protection, Thinning and Harvesting Operations, Environmentally Sensitive Wildlife Management

The idea of this tenet is to reduce human disturbances in the forest ecosystem to a minimum by carrying out only limited forest protection from diseases and promoting biological methods instead. In the same way thinning and harvesting should be conducted in a manner which disturbs the remaining trees and the ecosystem (especially the soils and the ground vegetation) as little as possible. This also includes the avoidance of harvesting and extraction damage. Part of this tenet is, for example, the emphasis on permanent extraction racks, at least 40 m apart, in some countries (see Figure 1.9) and the requirement for harvesters and forwarders not to leave these racks when carrying out their operations. There is a need for highly trained workers and specialised low-impact machinery.

Health and safety considerations may shift and require special attention in CCF, since the average size of trees to be felled increases while visibility within forest stands decreases with increasing structural complexity. There is also a need for a sufficiently dense forest road network (Larsen et al., 2022). Artificial liming of forest soils is often conducted to compensate for soil degradation. Whole-tree harvesting and stump removal are clearly not compatible with CCF. Branches, twigs, barks and tree tips of selectively harvested trees should also be left in the forest, as they contain most of a tree's stored nutrients (Otto, 1994).

Figure 1.9 Example of the marking of a permanent extraction rack on the stem surface of trees bordering the rack in a mixed broadleaved forest in the Reinhausen Forest District (Germany). Source: Courtesy of Paul Cody.

The density of populations of deer and other grazing and bark-stripping animals needs to be in balance with the carrying capacity of the site in order to make the use of natural regeneration feasible. This will also keep fencing costs low which could otherwise be quite high (Palik et al., 2021). In Britain, apart from deer, this also applies to the grey squirrel (*Sciurus carolinensis* GMELIN), the control of which is absolutely crucial to the success of CCF, particularly to forms of CCF involving broadleaved trees.

1.5.9 Establishment and Conservation of Forest Margins, of Other Special Habitats Inside Forests and Networks of Protected Forests

Adopting a holistic approach to forest management by managing the ecosystem rather than just crops, is a very prominent feature of CCF. This tenet also considers the landscape context. One way to contribute to this objective is the establishment of forest margins as transition zones between the open landscape and woodlands.

Here, a certain stage of natural succession is artificially and permanently preserved. Although there would be a strip of 25–30 m at the stand boundary with low or no timber production, this could significantly increase wind firmness and diversity (Burschel and Huss, 1997; Otto, 1994), see Section 7.3.3. Siebenbaum (1965) reported the successful use of such forest margins for the purpose of increasing forest resilience in the windy climate of the German federal state Schleswig-Holstein. In a similar way, CCF guidelines often

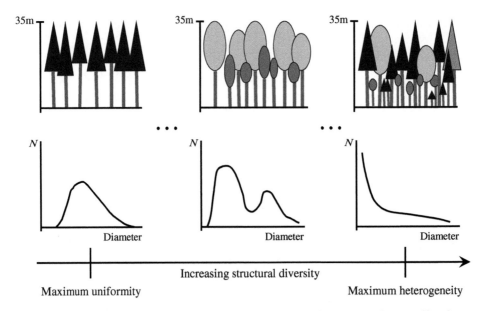

Figure 1.10 The continuum of continuous cover forestry stretching from maximum uniformity (e.g. in an even-aged coniferous woodland managed on a non-clearfelling basis) to maximum heterogeneity (e.g. in a single tree selection forest). Source: Modified from Pommerening and Murphy (2004).

demand that riparian forests and streamsides are restored by appropriate and sensitive conservation management (Forestry Commission, 2000; Nyland, 2003).

This can contribute to decreasing catastrophic flooding events, but most importantly secures water quality and important habitats (Greger, 1998). Other authors additionally suggest the establishment of a network of protected woodlands, including some non-intervention areas, as 'stepping stones' in large areas of commercial forests which would provide areas of retreat and almost undisturbed development for flora and fauna (Niedersächsische Landesregierung, 1991; Otto, 1994). This and the previous element support the fifth tenet proposed by Palik et al. (2021) to practise silviculture in the context of landscapes, i.e. to consider the connectivity between adjacent ecosystems and cumulative impacts. Franklin et al. (2018) proposed the tenet of 'restoring and sustaining the integrity of forest and associated ecosystems'. This is a complex tenet involving that ecosystems in all their complexity and variety are the source of ecosystem goods and services. In the CCF aspects presented here and in Figure 1.8, the *integrity* tenet proposed by Franklin et al. (2018) has been broken down into a number of themes.

1.5.10   In Conclusion

A short overview of these most common CCF principles can be found in Appendix A. Having reviewed these elements or tenets of CCF, it is obvious that many different silvicultural systems can be included under the broad umbrella of CCF, and Figure 1.10 gives an impression of how wide the spectrum of structures and their resulting diameter distributions is. It stretches from an even-aged coniferous woodland managed on a non-clearfelling basis (representing maximum uniformity) to a selection forest (representing maximum heterogeneity), cf. the illustrations of the two extremes in Figure 1.11. This stresses the need for

(a)   (b)

Figure 1.11   Illustrations of the two extremes of the continuum of continuous cover forestry shown in Figure 1.10. (a) A stand of *Picea sitchensis* (Bong.) Carr. at Clocaenog Forest (Cefn Du, plot 2, Wales, UK) where clearfelling was replaced by a shelterwood system providing abundant regeneration. (b) A stand of *Picea abies* (L.) H. Karst. and *Fagus sylvatica* L. near Bad Gandersheim (Lower Saxony, Germany) approaching the structure of a selection forest. Source: Courtesy of Stephen T. Murphy.

clear definitions of interventions and envisaged management scenarios when transformation of even-aged stands or afforestation/restocking is proposed. Ultimately, no matter how we define our personal understanding of CCF, the important feature of any CCF variant is the maintenance of continuous woodland cover in space and time whereby natural disturbances such as windthrows, fires and insect calamities can be incorporated.

1.6 Common Misconceptions Dispelled

The term 'continuous cover forestry' and also other semi-synonyms have attracted a number of misconceptions that are summarised in this section (Davies et al., 2008). Most responses presented here can be derived directly or indirectly from the texts in other parts of this book, but it can be useful to study these summaries when preparing oneself for discussions with students, stakeholders, policymakers, or practitioners.

CCF results in dense, unbroken forests without open spaces or vistas. This misunderstanding is partly related to the term 'continuous cover'. Novices often feel that the term suggests dense, unbroken forests. This is wrong; the term is rather a reflection of the continuity of woodland climate and soil preservation as opposed to the climate on bare land such as clearfelling sites. In most cases, CCF stand structures vary in space and time, and this includes differences in tree density and the opening of temporary gaps. Permanent open spaces can and should also be maintained for conservation and amenity. In the same way, natural disturbances are accepted and included. CCF does not imply permanently maintaining mature forest cover either.

CCF results in natural forest structures. CCF is sometimes considered to be synonymous with 'close-to-nature' or 'near-natural' forestry, see Section 1.5, but this statement is potentially misleading. While CCF management makes use of natural processes such as natural regeneration and natural pruning (cf. Section 2.6), and the resulting forest structure often looks more natural than that of plantations, CCF does not necessarily result in natural forest stand structures: Single tree selection, the classical and most extreme CCF method, gives rise to structurally very heterogeneous stands, often of mixed species and may therefore visually convey the impression of very natural forest stands, whereas some European native forests, e.g. involving *Fagus sylvatica*, rather tend towards homogeneity with time (Schütz, 2001a, 2002). Although CCF management is closer to natural forest ecosystems than plantation forest management, a careful examination of each individual case is required when considering this statement.

CCF is possible only with single tree selection. The single tree selection forest is often described as the archetype of CCF implying maximum heterogeneity of tree size and species that many other variants of CCF often converge towards (Pukkala and Gadow, 2012). This is an extreme view and the small area world-wide managed in accordance with this silvicultural system suggests that there are limitations. The single-tree selection system (see Section 5.4) was originally devised by farmers as a special, localised variant of farm forestry or agroforestry. CCF, however, has a very wide range of possible realisations (see Section 1.5) including more or less even-aged forests that can be established under the canopy of existing even-aged stands using shelterwood systems, and it is important to retain this flexibility.

CCF must rely entirely on natural forest processes. There is certainly a strong tendency towards using natural forest processes in CCF (see Section 2.6), and this may alleviate some costs of management, but it is by no means compulsory or exclusive. For example, if there is a desire or a need to introduce new species or provenances into a stand, there is no reason why regeneration should not be achieved by underplanting (cf. Section 2.4.1) rather than naturally. It is also possible to combine natural with artificial regeneration for diversifying a forest stand.

CCF is possible only with native tree species. CCF can be practised with native or non-native species, but all species used should be suitable for a given site and not develop invasive behaviour.

CCF is only possible with shade-tolerant species. In CCF, the main canopy is typically kept much more open than in a plantation forest allowing light-demanding and intermediate species to survive in the under- and mid-stories of light canopies. It is also possible to cut canopy gaps for the regeneration of light-demanding species genera such as *Quercus*, *Pinus* and *Larix*. There are many good examples of CCF in *Pinus* forests. Having said that, CCF involving intermediate and shade-tolerant species may be easier to achieve, particularly for managers new to CCF, and tree density can then be higher than in comparable woodlands that include more light-demanding species (cf. Appendix B).

CCF forests are more natural and therefore require less management. CCF does include an element of biological rationalisation (see Section 2.6), but the use of natural processes does not obviate the need for management input. In particular, continuous cover management is all but impossible without a programme of regular thinning interventions to develop individual-tree resilience among other reasons.

CCF management is more costly than plantation management. The relative costs of CCF have been widely debated without resulting in firm evidence or conclusions that can be generalised (Dedrick et al., 2007; Vítková et al., 2021). Some have suggested that costs will be lower because replanting costs are avoided; others suggested that they will be higher because of higher management inputs, increased needs for motor-manual felling and losses of economies of scale associated with clearfelling. While the costs of establishment are lower, costs of respacing (early thinnings) can potentially be higher when too much regeneration has established. Fellings may be less concentrated than with clearfelling whilst the switch from thinning from below to crown thinning improves the financial returns from early thinning. Selectively managed frame trees should be of greater quality and value.

CCF leads to better timber quality. Research has shown that CCF does not automatically enhance timber quality or results in more trees with good timber quality. If anything then timber quality is more varied. However, by stratifying trees into different functional groups (e.g. frame trees, competitors and matrix trees, see Chapter 3) early in stand development, it is possible to focus on a number of high-quality trees, and the prices they fetch on the timber market often more than compensate for the sacrifice of overall tree density compared to plantations of the same species on the same site (Wilhelm and Rieger, 2018; Pommerening et al., 2021a).

CCF is a silvicultural system. No, it is not. CCF is a fundamental forest management type, see Figure 1.1. Most silvicultural systems are regeneration methods that help a forest manager to move a forest stand from one generation to the next without resorting to clearfelling. As such silvicultural systems play a key role in the transformation of plantations to CCF. In CCF, a forest manager has a wide range of silvicultural systems to choose from. Therefore, CCF is not a silvicultural system, CCF rather involves many different silvicultural systems, but also thinning and even planting methods.

Silvicultural systems deliver CCF. Close to the truth, but this statement still misses the point. Most silvicultural systems (apart from the selection system and the two-storeyed high forest) are short-term regeneration methods and as such replace clearfelling and replanting of rotation forest management (RFM) by selective fellings and natural regeneration. Accordingly, they do not deliver CCF *per se*. Admittedly natural regeneration is a key element of CCF. However, silvicultural systems also provide shelter and other ecosystem services, and it is the long-term, envisaged structure that really delivers and maintains CCF. This long-term forest structure often differs from that adopted in the short-term silvicultural system. Forest structure ensures processes of biological automation (including, for example, regeneration, natural pruning) that are so important to CCF, see Section 2.6.

1.7 The Societies that Shape Us: Contrasting History of Forestry

The question of how much forest cover exists in a given area of the world and what kind of forest management prevails often is the result of complex societal processes that lie in the past and are therefore beyond our influence. These processes typically differ from region to region. Even on the comparatively small European continent, the historical processes that have shaped today's forest management in various countries are vastly different from one another. Societies as they have formed in different countries have not only an influence on the individuals they are composed of and vice versa, but also shape general attitudes towards vital topics such as the environment, climate change, human health, and democracy. They also provide a base opinion on forest management. Depending on this base opinion, certain types of forest management are more acceptable than others. For example, CCF much agrees with the way how Central European societies currently define their relationship with the environment so that it would be utterly impossible to re-introduce industrialised plantation management including clearfelling any time soon. The societies involved would not have it, public resistance would be considerable. In Britain and in Scandinavia, however, it is much more accepted that there are production forests that are clearfelled at the end of their economic lifetime whilst valuable, semi-natural forest ecosystems are set aside and preserved elsewhere in the country. Particularly plantation management of non-native conifers is largely accepted in some of these countries. A society's hold on individual opinions is often so strong that even scientific arguments can be influenced and biased by the opinion of the general public that researchers are naturally part of. It takes great strength of character to develop an alternative view and to communicate it. The strong influence of and indoctrination by society and upbringing can to some degree be overcome when individual researchers spend a few years working abroad in other societies or foreign researchers are invited to work in another country. Therefore, acceptance or rejection of forestry practices needs to be viewed in the context of society. The comparison of the forest histories in Britain and Central Europe attempted in this chapter may help appreciate the influence of society better and serve as a case study.

Contrary to many other European countries, Britain has a relatively short history of state-organised forest management and education. Forest authorities and services as representatives of state-organised forestry usually provide a national or regional agenda promoting the case of trees and woodlands and the comparatively long absence of a forestry agenda and a dedicated forest authority may have significantly contributed to the low woodland cover in Britain and Ireland at the beginning of the twentieth century. The first forest services on the European continent were formed as part of state-owned mines in the sixteenth century in order to provide the recently emerging industry with timber in a sustainable way. First national and regional forest services were founded in Central Europe in the eighteenth century (Hasel and Schwartz, 2006).

Towards the end of the eighteenth century, a shift from exploitation forestry and timber mining to organised sustainable forest management took place in many Central European countries. 'They went into the hills and mountains to fell trees like others ladle water from a stream' was apparently a famous expression in canton Neuchâtel in Switzerland to describe the reckless timber-mining attitude towards forests shortly before the introduction of sustainable forestry (République et Canton de Neuchâtel, 2001). As part of this introduction, the new field of silviculture rose from the poor remnants of devastated forests as a 'child of need', as Russian forestry professor G. F. Morosov wrote in his textbook (Morosov, 1959). The paradigm shift was initiated by changes in agriculture and the industrial revolution. The changes in agriculture led to a spatial separation of agriculture and forestry, which proved to be very beneficial for forestry. In the wake of the industrial revolution, the demand for timber in general, and for quality construction timber in particular, increased which gave rise to a large-scale transformation of coppice woodlands to high forests. The increased reliance on fossil fuels made the production of wood fuel in coppice woodlands redundant. By contrast, early professionalism in forestry in the Britain of the eighteenth century was built up on private estates, especially in Scotland. As in Germany, forestry in Scotland was often a family profession, in which son followed father. Incidentally, this was an effective and very successful way of passing forestry knowledge and experience down the generations while formal forestry education did not exist or was in its infancy.

Long before sustainable forestry was established in Central Europe and Scandinavia, French foresters pursued a thinning method that is now considered a crucial component of CCF management, *thinning from above* or *crown thinning*. This thinning method, 'eclaircie par le haut' in French, targets dominant trees when selecting trees for removal and typically leads to a markedly greater diversity of tree sizes than the traditional method of thinning from below or low thinning (see Section 5.2). According to Bauer (1968), the crown-thinning method was introduced by Tristan de Rostaing in 1560 and promoted by

Duhamel du Monceau in his instructions for quality oak timber production in 1755. Heger (1955) mentioned Varenne de Fenille as an important forestry representative supporting crown thinnings in France. This method has become a bit of a 'national' thinning type in France and was accepted in Central Europe (Germany, Austria, Switzerland) only in the twentieth century after a long domination of the low-thinning tradition. Denmark, however, has also a comparatively long crown-thinning tradition. In other European countries such as Britain and Ireland, the crown-thinning method is still largely a novel method.

While forestry education on the European continent was established in the eighteenth century (e.g. 1763 in Wernigerode by H.D. v. Zanthier; 1785 in Zillbach/Tharandt by H. Cotta) in the wake of industrialisation and an increased timber demand, the first School of Forestry in Britain was established at the University College of North Wales at Bangor in Wales only in 1904 (Linnard, 2000). Other forestry departments at Oxford, Aberdeen, Edinburgh and Newton Rigg followed later. This, however, was preceded by the introduction of forestry courses at the Royal Indian Engineering College at Coopers Hill near Egham in Surrey in 1885. These courses were only available to students who intended to enter the Indian Forest Service (James, 1990).

During the nineteenth century, the increasing sense of professionalism in forestry resulted in the formation of a number of societies devoted to its furtherance in Britain. With the approach of the twentieth century, continental forestry practice began to exert an increasing influence in Britain. Up to this time, English forestry had chiefly consisted of allowing trees to grow and then felling them, and forest management, as it is known today, was non-existent. As a new interest in forestry began to develop in Britain, so a demand arose for more information, improved techniques and a new approach to growing timber.

The obvious source of help lay in the continent, where the French and the Germans had applied themselves to the management of their forests over a long period. However in India, the government in that country had already been actively concerned with the creation of a forest service and, in 1855, a permanent policy for forest administration had been laid down. According to James (1990), the success of the Indian Forest Service was largely due to three German foresters, Dietrich Brandis, William Schlich and Berthold Ribbentrop. It is probable that many of the principles and practices which were followed in German forestry were introduced into the Indian Forest Service and when members of the Service returned to Britain, it is very likely that they were instrumental in dispersing these ideas.

It was Sir William Schlich who made the greatest impact on British forestry in the early years of the last century. When Coopers Hill closed in 1906, the forestry section was transferred to Oxford, where Schlich continued to lecture, while from time to time, he advised landowners on forestry matters and prepared working plans for several large estates. In this way, he was able to pass on much that was based on German forestry practice. Books provided another channel through which German forestry practice was disseminated in Britain and towards the end of the nineteenth century many German forestry textbooks were translated into English (James, 1990). Despite the growing climate of informed opinion favouring major state involvement in forestry, no definite measures were taken until the pressure of the First World War forced rapid action. In 1914, over

90% of Britain's supplies of wood were imported. Britain was not only heavily dependent on imported timber, but most of the supply lay outside the Empire. By 1913, Russia had been supplying Britain with nearly half its total imports of wood (Pringle, 1994). When increasingly cut off from its overseas timber supplies as a result of submarine warfare, the British government hastily established a Timber Supplies Department. Compulsory felling operations were conducted on many British estates.

Out of the trauma of the First World War the Forestry Commission, Britain's state forest service was formed in 1919, initially like a military operation and many former army officers were recruited as staff. Its aim was 'the regeneration of British forestry' (Pringle, 1994) with the primary objective of building up a strategic reserve of timber for the nation. New planting was to be with conifers since the overwhelming demand was for softwood timber. Such demand as there was for home-grown hardwoods could be met by the replanting and appropriate management of existing broadleaved woodland areas (Pringle, 1994). This gave rise to the establishment of large, single-species, even-aged plantations dominated by exotic species. The main tree species of these plantations are Sitka spruce (*Picea sitchensis* (Bong.) Carr.), Scots pine (*Pinus sylvestris* L.) and Japanese larch (*Larix kaempferi* (Lamb.) Carr.) – only Scots pine being native to parts of Britain. This initial forest policy has created a major conifer woodland legacy – sometimes informally referred to as the 'Great Spruce Project'.

The national forest policy, defined in 1919, put forward two objectives. The ultimate objective was to create reserves of standing timber sufficient to meet the essential requirements of the nation over a limited period of three years in time of war or other national emergency. The immediate objective was a ten-year scheme for state afforestation of new land, plus assistance to local authorities and private owners for afforestation and reforestation.

This development finds a parallel in the Central European forest management of the nineteenth and twentieth centuries, when large even-aged coniferous plantations were established to meet the increasing timber demand of growing populations. Coniferous plantations enabled forestry services to provide more timber volume in less time and also helped to put an end to illegal logging and timber thefts due to the uniformity of plantations which make it easier to detect missing trees. Towards the end of the nineteenth century, the disadvantages and dangers of creating large coniferous plantations became more and more evident on the European continent and the advance of soil and site sciences helped to develop a deeper understanding of the adverse effects of rigid plantation forestry. Concerned about the resilience of commercial forests German Professor Karl Gayer (1886) emphasised the advantages of uneven-aged, mixed forests in his seminal book 'Der gemischte Wald' (The Mixed Forest) and encouraged mixtures in groups by applying group shelterwood systems (Bauer, 1968). In the North of the country, his colleague Alfred Möller coined the famous term 'Dauerwald' (continuous or perpetual forest) in 1913. Apparently, at the beginning of the twentieth century, there was a considerable interest in a change of forestry methods in many parts of Europe and North America. For many years, leading continental silviculturists including Hartig and Cotta had vehemently expressed their disapproval of any kind of forest management not following the rigid idea of the normal forest with clearfelling as the only harvesting method. According to Schütz (2001b), one of the main reasons for this was that forest management following the normal forest idea could be controlled more efficiently. This was in fact an important point at a time when illegal logging was still very common in Europe and foresters were poorly paid.

Uno Wallmo, for example, a Swedish forester, promoted 'blädningskog' at this time, often translated as selection forest (or 'Plenterwald' in German); however, what he advocated was selective tree removals as opposed to clearfellings (Krutzsch, 1952; Lundmark et al., 2013). In Switzerland, preceded by forester Henry Biolley in canton Neuchâtel (cf. Figure 1.12) silviculture Professors Arnold Engler and Walter Schädelin acted as pioneers of selective tree management and CCF. In France, towards the end of the nineteenth century, an influential CCF pioneer was Adolphe Gurnaud working in the Vosges Forest (Alsace). This was also the time of Carl Schenck's 'Biltmore Lectures on Sylviculture' and other works in North America where he described forms of selective forest management. He was joined by other American authors such as Graves and Hawley (O'Hara, 2002). In Russia, G. F. Morosov finished the first draft of his silviculture textbook already in 1912. In his text, he pointed out that silviculture first started off completely empirically without the benefit of knowledge from other fields of natural sciences such as soil science or plant science that came into existence only much later. Morosov was one of the first to firmly base silviculture on biological principles. As a consequence he came to the conclusion that working against the nature of a woodland community on a given site is not what silviculture should attempt to do (Morosov, 1959).

In 1922, Möller published his famous book 'Der Dauerwaldgedanke: Sein Sinn und seine Bedeutung' (The Dauerwald idea: Its meaning and significance) which initiated a long running debate. He was among the first forest scientists to understand managed forests as forest ecosystems in a holistic way (Thomasius, 1996) and also was among the first in Germany to consider an individual-tree approach (see Chapter 3) along with the size-control principle (see Section 1.8). This was before ecology officially became an academic field of science. Möller discovered that most of his ideas and principles were

(a)　　　　　　　　　　　　　　　　　(b)

Figure 1.12　Memorial stone erected in honour of CCF pioneer Henry Biolley in Couvet Community Forest, Val-de-Travers, canton Neuchâtel, Switzerland (a) and a typical snapshot of Couvet forest including *Picea abies* (L.) H. Karst., *Abies alba* Mill. and *Fagus sylvatica* L (b). A marked educational trail dedicated to Biolley's work leads through the forest. Source: © Arne Pommerening (Author).

being practised in the *Pinus sylvestris* forests of the Bärenthoren Estate near Dessau in the modern German state of Sachsen-Anhalt. Clearfellings were replaced by continued thinnings following the principles of individual-tree management (Bauer, 1968). He used these forest estates as management demonstration sites for his Dauerwald concept, which gave rise to a considerable number of 'pilgrimages' by forest managers as well as members of the academia. Möller's ideas were discussed in detail by the British silviculturist Troup in the first volume of the journal *Forestry* in 1927 (Troup, 1927). He even paid a visit to the Bärenthoren Estate and included descriptions and conclusions in his book on silvicultural systems (Pommerening and Murphy, 2004; Troup, 1928). Later other large-scale management demonstration sites were organised in Germany such as Bärenfels (Saxony, south of Dresden), Hohenlübbichow (Pomerania, east of the Oder River) and the Göttingen community forest (Lower Saxony, south of Hanover) (Krutzsch, 1952). The continued professional interest in these management demonstration areas shows how important it is to offer examples of best practice in the field, see Section 1.10 and Chapter 8.

With only a few exceptions, such as the CCF initiatives of Lord Bradford and his forest manager Phil Hutt in Devonshire (Timmis, 1994; Kerr et al., 2017), see Section 2.3.1, and the group selection trials at Glentress in Scotland (Anderson, 1960; Wilson et al., 1999; Kerr et al., 2010), see Section 2.3.2, there was a decline in interest in CCF, following Möller's untimely death in 1922. In Germany, the critics of CCF – especially Afred Dengler and Eilhard Wiedemann – succeeded in discrediting the approaches of Möller and his colleagues. Also, during and after the Second World War, there was considerable resistance from foresters in Germany to CCF, as it had been adopted and compulsorily imposed by the national socialist government during the 1930s and 1940s (Huss, 1990). Only in East Germany, Krutzsch, Blankmeister and Heger continued and refined the CCF principles proposed by Möller and termed them 'stock maintenance forestry' ('vorratspflegliche Waldwirtschaft' in German), while in West Germany, it was mostly private woodland owners and managers organised in a private forestry association, the 'Arbeitsgemeinschaft naturgemäße Waldwirtschaft (ANW)', who kept Möller's Dauerwald ideas alive and opposed state forestry that largely returned to rotation forest management (Bauer, 1968). Interestingly, in those years, clearfelling, as a hot topic, even made it into the communist class-struggle theory:

'Rotation forest management including clearfelling is an expression of capitalistic production. In capitalism, forests constituted capital that had to produce maximum interest rates within the shortest time possible. In pursuit of this doctrine, any principles of sustainable silviculture were neglected' (Eisenreich and Nebe, 1967).

In Britain, the first large-scale coniferous plantations were established in various parts of the country in the 1920s. Through considerable land acquisition and planting in Wales, Scotland and England, new commercial forests were created. Sadly ravaged again, as a result of the Second World War, many woodlands had to be restored and land acquisitions for forestry peaked again in the 1950s (James, 1990).

The one-sided commercial character of the Forestry Commission's early work was lamented by Peterken (1996) and echoed the harsh nature of plantation forestry in many countries at that time:

> 'It has been a great pity that the forestry profession developed such a narrow view of its responsibilities after the 1950s. At the height of their imperialist phase, foresters were regarded as "tree farmers" – i.e. those foresters who advocated highly mechanised, short-rotation cropping of genetically engineered, uniform stock. Official forestry talked down the widespread public concern about the dark character and deadening effects of conifer plantations. [...] Increasingly, it failed to deliver the kinds of woodland that people prefer. Not surprisingly, "forester" almost became a term of abuse in some parts of the conservation world'.

Alternative forms of forest management developed largely beyond the influence of British state forestry, sponsored by conservation trusts and councils. Representatives of this alternative forestry are, for example, not only the Woodland Trust and Coed Cymru in Wales but also many conservation bodies (e.g. Countryside Council for Wales, Scottish Natural Heritage, English Nature). Under their influence, many woodland owners rejected plantation forestry and worked within natural constraints (Peterken, 1996).

Outstanding academic British silviculturists include Professors R. S. Troup (Oxford University) and M. L. Anderson (Edinburgh University). Troup compiled a very influential book on silvicultural systems in 1928 which was later revised by Jones (1952) and Matthews (1989). He was steeped in the literature of continental forestry, and his varied experience of silviculture in Western and Central Europe gave authority to his teaching and publications. Anderson pointed out that, in consequence of British foresters' concentration on creating and managing forest on the clearcutting even-aged system, 'there has been little appreciation of the importance of the stand and the site, considered together as a single producing agent and not separately, and of the need for maintaining the production of that combination in perpetuity, or even of increasing it'. He came to the conclusion that a specific modification of the selection system, which he referred to as *group selection system* (see Section 2.3.2), would be the most suitable system for the Scottish uplands. Anderson proposed that dense groups of *Betula*, *Larix* and *Alnus*, spaced quite widely, should be established to suppress the ground vegetation and to provide shelter for the later planting of *Picea abies*, *Abies alba*, *Acer pseudoplatanus* L. and other broadleaved species between the spaced groups in a progressive development towards an uneven-aged structure (Anderson, 1960). Anderson also translated many continental forestry textbooks into English.

The findings of research into forest decline from the 1960s onwards and the certification process starting at the end of the 1980s applied a corrective to academic and practical silviculture on the European continent and turned them into disciplines firmly rooted in forest ecology (Pommerening and Murphy, 2004). Since the 1980s British silviculturists have taken an increasing interest in conservation work and habitat management through research, grants and active management (Peterken, 1996).

Interest in CCF was revived in the 1980s when the discussions on acid rain, forest decline, restoration and certification stimulated debate on concepts of more environmental forestry (Hasenauer and Sterba, 2000). The philosophy, goals and forest management practices of certification schemes such as the *Forest Stewardship Council* (FSC) and the *Programme for the Endorsement of Forest Certification* (PEFC) broadly match those of CCF (Franklin et al., 2018). In 1989, *Pro Silva Europe* was founded as an association of foresters practising management which follows natural processes. Most European countries have joined Pro Silva and established national and regional sub-organisations such as the Continuous Cover Forestry Group in Britain (founded in 1991) and Pro Silva Ireland (founded in 2000). Pro Silva supports the uptake of CCF by exchanging information within regional working groups, by establishing demonstration forests, (inter)national meetings, field trips and through a cooperation with educational and scientific institutions. The association also publishes statements, resolutions and guidebooks.

In parts of Central Europe, the catastrophic storm events of 1984, 1990, 1999 and 2007 repeatedly raised the question of ecosystem resilience; CCF has been identified as a way to increase this resilience (Dvorak and Bachmann, 2001; Messier et al., 2013; Hanewinkel et al., 2014) and the gales thus contributed much to a further promotion of the principles of CCF (Knoke and Plusczyk, 2001). In a similar way, the large-scale flooding events (Otto, 1994) of recent years across Europe and the global climate change debate have stimulated the adoption of CCF (Hasenauer and Sterba, 2000). The revival of the CCF debate also owes much to the United Nations Commission on Economic Development (UNCED) summit at Rio in 1992 when the terms and scope of sustainable forest management were re-defined, and it was suggested that they become an integral part of modern forestry practice world-wide. Naturally, the term 'sustainability' is at the centre of the CCF idea. In the narrow sense of sustainable timber yield, sustainability is an old forestry maxim which goes back as far as the sixteenth century (Hasel and Schwartz, 2006). It has received a new emphasis since the summit at Rio. The new definition states that forest resources and forest lands should be sustainably managed to meet the social, economic, ecological, cultural and spiritual needs of present and future generations (United Nations, 2001). Although initially raised as a separate issue, the forest restoration debate is in many regards similar to the continuous cover idea, especially with issues involving native species and stabilisation. Transformation or conversion as methods of establishing CCF woodlands are also techniques used in forest restoration (Thompson et al., 2003).

Subsequently, the Rio–Helsinki process, and requirements of certification also began to stir up concerns about traditional forest management in Britain. This process was given added impetus in 1997 by the start of devolution which led to the establishment of regional parliaments in Scotland and Wales. Devolution has paved the way for new and distinctive agendas for the woodlands of the three nations in Great Britain and separate woodland strategies have been launched for England, Scotland and Wales (Forestry Commission, 1998, 2000, 2001). New CCF programmes were also compiled in Germany, Ireland, Luxembourg and other European countries at this time (Larsen et al., 2022).

The Welsh woodland strategy was the last in this series and contained the strongest commitment to CCF. The 'Woodlands for Wales' strategy 'aimed to convert at least half of the National Assembly woodlands to continuous cover and encourage conversion in similar private sector woodlands'. The document also stated that efforts would be undertaken to

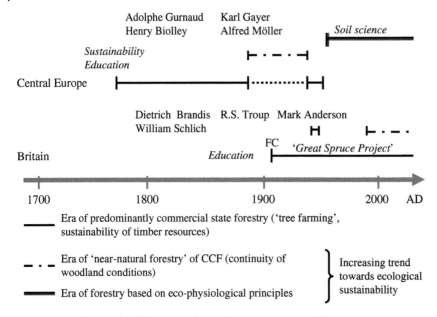

Figure 1.13 Contrasting British and Central European history of forestry.

gather information about continuous cover systems and how best to manage these systems. By 2002, the National Assembly for Wales had established three large-scale trial areas in state and private woodlands, to pioneer techniques for transformation to continuous cover systems and to collect information to guide future transformation in all woodland types (Forestry Commission, 2001), see Section 1.10.

Figure 1.13 summarises this section. British history of organised state forestry often echoes that of continental Europe taking place with a considerable time lag. The concern over sustainability of timber resources led in both places to the creation of a national agenda. In a sense, Britain repeated the course of evolution of European forestry and forest science but in a much shorter time period with less time available for a natural evolution of professional forest management methods. Rigid management of coniferous plantations was the predominant remit of European silviculture of the eighteenth and nineteenth centuries, a phase British silviculture is still very much concerned with. When, with the advent of soil sciences, it became clear that such management can lead to serious soil deterioration and other adverse environmental consequences, academic silviculture went through a serious process of re-thinking. As a result, it has now become a discipline based upon principles of forest ecology. While Britain and Ireland are emerging from a typical afforestation silviculture, a transition to a silviculture of applied ecology has yet to come.

CCF now constitutes the forestry normality in Denmark, Germany, Slovenia, Luxembourg, Switzerland and Austria and is an important part of forest legislation in these countries. CCF is also very popular in France and in Northern Italy. Anglo-American countries and most Scandinavian countries share a lasting RFM legacy, and the uptake of CCF is slow here (Puettmann et al., 2015; Hertog et al., 2022), although fuelled by climate-change concerns, there is an ongoing societal debate in these countries. Recently, China has taken great interest in CCF methods. It is an interesting question why CCF is more prominent in some countries and not in others. In our experience, the question of uptake is not much related to environmental or geographic conditions as one might think. Despite occasional professional prejudices, CCF is in principle possible wherever forest ecosystems can sustain themselves naturally. As alluded to at the beginning of this section, the uptake of CCF is more determined by societal contexts and traditions. In those countries where CCF is dominant now, a strong societal support for environmentally friendly processes has gradually evolved in general and any attempt to change forest policies in favour of RFM would certainly meet fierce public resistance.

1.8 Ensuring Sustainability: Area Control Versus Size Control

When timber shortage and increasing transportation costs became a reality all over Europe at the beginning of early industrialisation, companies and local governments were forced to face the problem of timber sustainability. Bettinger et al. (2017) emphasised that the term 'sustainability' often refers to the general ability to maintain a resource indefinitely into the future, with no decline in quality or quantity, regardless of outside influences. In the eighteenth and nineteenth centuries, a straightforward system was required that would replace forests fast where they were harvested for industrial purposes and ensured that timber sustainability could be monitored easily. The bottom line of any such system was to strike a balance between growth and harvesting, i.e. not to extract more timber than would regrow in any time period.

The question of how to ensure timber sustainability was not trivial and given the technical means at the time rather difficult to implement. Inspired by agriculture (see Section 1.2), the easiest way to achieve timber sustainability was by applying the concept of the *area-control method*. According to this concept, human interventions were spatially disaggregated. Regeneration (artificial [planting] or natural), thinning and harvesting, the main forms of human interventions in forests, were concentrated in designated areas of a larger forest, e.g. a forest estate (Figure 1.14a), whilst in CCF, when moving from left to right on the axis of Figure 1.10, these three types of interventions increasingly merge in space and time (Figure 1.14b, c).

The concept of spatial segregation of human interventions was later refined by the conceptual model of the normal forest (Gadow and Bredenkamp, 1992; Bettinger et al., 2017; Franklin et al., 2018; Smith et al., 1997). According to this model, the total forest area is basically divided into as many equally productive units as there are years or age classes in the planned rotation, and one unit is harvested by clearfelling each year or every five years depending on the width of the age classes.

The system ensured that more or less equal amounts of timber could be harvested every *x* number of years. Each unit contained monospecies stands where all trees were of the same age; hence, the term 'even-aged stands'. Several stands were often lumped together in spatial units that stretched over a range of stand ages, the so-called 'age classes' (with bins of 5–20 years). This was a rigid system, but it was felt that even-aged monospecies forest stands were easier to comprehend and manage than more complex forest stands that seemed 'irregular' and somewhat 'out of control' at that time. The main advantage was in

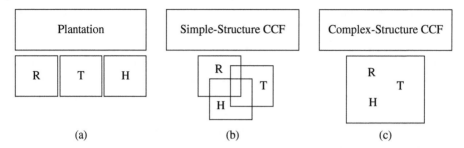

Figure 1.14 Spatio-temporal separation of regenerating (R), thinning (T) and harvesting (H) with increasing complexity of stand structure from (a) to (c). For the definition of simple (b) and complex (c) structure see Figure 2.15. The complex activities regenerating, thinning and harvesting are explored in great detail in Section 5.2. Source: Adapted from Assmann (1970).

the spatial allocation of forest operations to clearly defined areas (Knuchel, 1953). There was no room for doubt which areas had to be treated in what way and when. All trees in the designated areas were managed in more or less the same way. The system also ensured that timber thefts were easier to detect at a time when poaching and illegal logging, particularly in rural areas, was still quite common. This concept was finally formalised and implemented in Germany at the beginning of the eighteenth century and forest scientists such as Hartig, Cotta and Hundeshagen perfected the sustainability principle based on the area control method in the eighteenth and nineteenth centuries.

With time forest management experience grew and the staff involved realised that setting up such a rigid spatial arrangement required great sacrifices compared to the advantages. The method requires uniform growing conditions and thinning regimes throughout the entire forest (Gadow and Bredenkamp, 1992). The regularly formed stands in the long run did not deliver what had been expected of them (Knuchel, 1953). In addition, the area control method proved to be inflexible and could not easily handle unexpected events such as natural disturbances and even benefits such as natural tree regeneration.

The area-control method has been through a considerable process of evolution in many countries (Aplet et al., 1993; Toman and Ashton, 1996). Gradually, the rigid system of spatial arrangement was replaced by a system of individual-stand maturity where population characteristics such as standing volume, stand growth rates, basal area, stand height or economic growth rates were used as criteria for optimum harvesting time (Mantel, 1990). The concept, for example, focussed on how much total stem volume to cut each year and the areas to cut were then selected to satisfy this volume (Franklin et al., 2018). This approach can be referred to as *size-control method*. Also, combinations of area and size control methods were pursued (Franklin et al., 2018). Particularly, in Central Europe subsequently less and less importance was placed on the spatial arrangement of stands and balancing of age classes.

Beginning with the introduction of individual-based forest management by Schädelin, Leibundgut, Abetz and Pollanschütz (see Chapter 3), the next logical step was taken and the size control method was delegated further down to the level of individual trees (Pommerening et al., 2021a). The concept of individual-tree size control was first developed in areas with diverse topography, where area methods were difficult to implement because of variable environmental conditions at small spatial scales. Initially, individual-based forest management was an independent development, but it is now increasingly associated with CCF. Individual-based forest management is sometimes also referred to as *free-style silviculture*, but the term can also apply to a combination of elements from different silvicultural systems (Bončina, 2011a; O'Hara, 2014). Individual-based size control is now the dominant sustainability control method in Central Europe, and a typical application of this method is *target-diameter harvesting* (cf. Section 3.4), where for each site and species stem-diameter thresholds are given which determine when a tree can be considered for harvesting (Figure 1.15).

In conclusion, the advantage of area control is that it is very clear where in the forest (stand) silvicultural input is required at any given time. Also, the type of work necessary is fairly straightforward to explain and more or less comes with the area. Therefore, work in RFM and under the conditions of area control can mostly be delegated even to a lay person. However, this advantage comes at the expense of flexibility due to the rigidity of the area-control method. This flexibility is often needed to respond to rapidly changing, partly unforeseen stand conditions, as, for example, required with ongoing climate change. The flexibility of the size-control method is what makes adaptive forest management adaptive. Naturally, it is possible to combine both approaches temporarily and/or spatially.

Figure 1.15 Current diameters sprayed (in blue paint) onto the stem surface of *Larix decidua* Mill. and *Pseutotsuga menziesii* (Mirbel) Franco trees in the Rosengarten Forest District (Lower Saxony, Germany) for training apprentice forest managers to be able to visually recognise target diameters (cf. Section 3.4) without measuring trees. Source: Courtesy of Paul Cody.

In many countries where plantation silviculture is the dominant forest management type, the area-control method still has a considerable influence on the landscape and on the mindset of forest managers. This is particularly common in Anglo-American countries. Sometimes in these countries, even attempts have been made to apply the method of area control to CCF (Timmis, 1994; Kerr et al., 2010; O'Hara, 2014, p. 56), see Section 2.3. One can understand that the free arrangement of mature, mid-storey and regeneration trees can challenge and confuse CCF novices. Therefore, they can feel the urge to fall back on elements of the area-control method that they are familiar with.

Since area control and size control are fundamental concepts of forest management, they help understand why foresters and researchers think differently in various parts of the world. In the remainder of this book, we will often come back to these principles.

1.9   CCF in a Changing World

In Section 1.7, we traced the eventful history of CCF through the last two hundred centuries and concluded that the concept has seen a gradual evolution with many changes along the way. CCF deliberately broke off from agriculture-influenced types of forestry and pursued a path towards forest management based on ecological principles. The more we will learn about the ecophysiology of trees and other plants in forest ecosystems and about how forest ecosystems including forest soils and tree–soil interactions work in general, the better forest managers are potentially in a position to fine-tune and perfect CCF. Change, uncertainty and surprise will likely dominate the future. Climate change, globalisation, changing policies, social trends and many other changes will affect the future in ways that we can now dimly imagine (Franklin et al., 2018).

A traditional motivation for CCF was the need to diversify timber products. Because of decreasing profits, forest owners and state forestry policymakers intended to move away from maximising overall volume production to maximising the volume production of a smaller subset of trees in each stand whilst rationalising management input at the same time. This subset of trees was selected as the best-quality trees, and their quality was maintained and improved in every management intervention. All other trees in these stands were considered by-products that eventually were sold on the pulp and pallet market. As briefly pointed out in Section 1.6, CCF does not necessarily lead to an improvement of timber quality in a given stand, but by stratifying trees into different functional groups (e.g. frame trees, competitors and matrix trees, see Chapter 3) it is possible to focus on a number of high-quality trees and the prices they fetch on the timber market often more than compensate for the sacrifice of overall tree density compared to plantations of the same species on the same site (Wilhelm and Rieger, 2018; Pommerening et al., 2021a).

In the past 20 years, human-induced climate change has become a reality, and its dramatic consequences have been understood by the overwhelming majority of people on earth. Different measures and policies have been proposed to mitigate the threat to life on our planet. Increases in carbon dioxide (CO_2) in the atmosphere have been identified as a major contributor to climate change. Forests are global carbon sinks and among these measures much emphasis has therefore been placed on increasing woodland cover and on avoiding any kind of forest management that leads to a sudden, large-scale exposure of forest soils to the atmosphere so that carbon emissions are minimised. CCF seeks to create forest ecosystems and landscapes that will be resilient in the face of climate change (Franklin et al., 2018). A standard financial response to risk and uncertainty is diversifying investments. Similarly, ecologists call for heterogeneity at different spatial scales, creating redundancy and increasing resilience to face potential disturbances. Through the diversification of forest structure and tree species that typically is part of CCF, it is perceived that CCF woodlands can be more resilient to destabilising effects of ongoing climate change. This diversification also contributes to increased biodiversity or at least allows a maintenance of current diversity levels in forest ecosystems. A recently debated dilemma is that climate change may challenge the CCF tenet of favouring the regeneration of native and local parent trees. In the past, the convincing rationale for this has been that local provenances are potentially best adapted to a given site for their long co-evolution with the abiotic and biotic environment. As climate change now proceeds at a rate that is considered faster than natural rates of climate change in the past, scientists have suggested introducing provenances of native tree species that have developed in warmer and dryer climates (Köhl et al., 2010; Brang et al., 2014; Frischbier et al., 2019). CCF has also gained wide-scale recognition as an opportunity to respond to unplanned natural disturbances (Messier et al., 2013). With rising sea levels, coastal forests are also likely to play a more important role as sea defences than in the past.

Carbon forestry has now become an established term to denote forest management with the explicit objective of optimising carbon sequestration (Pukkala, 2018). On closer inspection, the strategies suggested for increasing carbon sequestration markedly overlap with the elements of CCF discussed in Section 1.5. These, for example, include increasing the resilience to natural disturbance and enhancing forest recovery following disturbance (Ontl et al., 2019) thus reducing the probability of sudden, large carbon losses. Part of carbon forestry is also increasing structural complexity and increasing thinning frequency and intensity among other strategies such as disfavouring species that are distinctly maladapted. Therefore, carbon forestry can clearly be considered a part of CCF. In carbon forestry, it may also be beneficial to concentrate sequestered carbon in a few large and long-living trees rather than maximising overall carbon sequestration by spreading carbon across many small and medium-sized trees.

At the same time, researchers in conservation science discovered that CCF includes many sophisticated techniques that come very handy when engaging in woodland restoration. In this context, forest restoration refers to the process of assisting the recovery of a forest ecosystem that has been degraded, damaged or destroyed (Mansourian, 2005, cited in Newton, 2007; Franklin et al., 2018). CCF can, for example, contribute to forest restoration by thinning stands that have become overgrown because of fire suppression. Removing small patches of trees can create compositional and spatial heterogeneity in uniform, single-species plantations on ancient woodland sites (Franklin et al., 2018). While restoration measures and silvicultural techniques have had no connection for a very long time, it is now increasingly acknowledged that many forest management methods, particularly the silvicultural systems (see Chapter 5.3), can support restoration and that completely new forest restoration methods often do not necessarily need to be devised entirely from scratch. In contrast to ecology, both conservation and silviculture have in common that they involve at least some degree of management, i.e. the informed, goal-oriented interference with natural processes based on scientific knowledge. Methods of transformation and conversion to CCF have been successfully applied in conservation projects.

CCF can also contribute to keeping groundwater levels low by intercepting rain and snow and taking up the water for transpiration and photosynthesis (Sarkkola et al., 2010). This is different to clearfelling sites, where the water level rises after all trees have been removed. In CCF, tree removals are comparatively moderate in each intervention, and the residual forest canopy keeps the groundwater table unchanged and therefore hardly any leaching of organic compounds, nutrients and sediments occurs. In addition, the residual trees take up any excess nutrients (Nieminen et al., 2018). This improves water quality and the state of aquatic ecosystems. Furthermore, the choice of tree species adapted to a given site reduces the risk of raw humus accumulation and acidification of groundwater. For these reasons, CCF is increasingly favoured in areas where maintaining good water quality is crucial.

There is also a trend towards increasing spare time and a need for outdoor recreation opportunities particularly near urban centres. In some countries, mountain biking has become a thriving business that both private and public forest owners increasingly include in their portfolio as part of business diversification through providing the necessary infrastructure. Recreation in forest landscapes is also beneficial to human health and interaction with trees is often included in therapies. CCF is increasingly considered as a way to make recreation forests more attractive to visitors by introducing more size structure and by increasing tree species diversity. As discussed in Section 1.5, both tree diversity aspects are elements of CCF. Visitors often enjoy seeing large, scenic trees (which they perceive as old trees) next to medium-sized and small trees. Forests with diverse structure also have the benefit that selective forest operations are difficult to notice and stumps of felled trees disappear in the understorey. This is very useful in community and recreation forests near urban centres where visitors often have a negative attitude towards any kind of forest management. Mixed forests that include broadleaved trees, particularly those with beautiful autumn colours, are much appreciated by visitors.

The recent energy crisis has emphasised once more that the quest for sustainable energy is crucial to the survival of mankind and to the ecosystems we inhabit. Here, energy wood production can clearly contribute to reducing our dependency on fossil fuels. Recent advances in wood-burning technology involving low emissions and high efficiency have suggested that wood fuel combustion is a realistic strategy for providing energy (Johann, 2021; Vanbeveren and Ceulemans, 2019). Wood fuel can be sourced from thinning by-products and from road-cuttings, but there is also the possibility of employing traditional wood-fuel production methods such as the coppice-with-standards and coppice-selection system described in Section 1.3.

Another trend of some importance is that of forest cemeteries. In the same way as people enjoy spending time walking through forests, they feel consolation in the thought of eventually being buried in a peaceful, natural environment such as a woodland. To make this possible, cemetery companies have set themselves up and lease forest land from the state forest service. This land is then managed by the cemetery companies. After cremation, people can be buried in biodegradable urns near trees that they themselves or relatives have chosen for them. These trees act as grave markers, and the names of the deceased along with their dates and some comforting words are inscribed on a metal sign that is attached to these trees. The cemetery company ensures that the grave marker trees continue to thrive in the woodland, and that they are not felled or damaged during the contract time of the grave. Such burials have become very popular, since they answer people's longing for nature, and they are comparatively low in maintenance. The latter is increasingly required, since nowadays people are more mobile than in previous decades and tend to flexibly change place of residence. This makes it harder to maintain traditional gravesites, and a forest burial may be a good solution. The relatives of the deceased come and visit the graves of their loved ones from time to time, and these visits should be a pleasant experience, if at all possible. Visits can be combined with leisurely strolls through the woods. Therefore, also here visitors appreciate a continuous, naturally looking forest with a mix of tree species and sizes. Again, this is something CCF can provide.

These different contexts in which CCF is of potential benefit explain why CCF is currently re-visited once again and has been identified as a universal solution for a number of quite different challenges. The question of how CCF management has to be adapted or fine-tuned to present solutions to these very different challenges, is explored in Section 7.3.

1.10   How to Introduce CCF to a New Region or a Country?

In countries with a comparatively long tradition of CCF such as Austria, France, Germany, Slovenia and Switzerland, to name but a few, this forest management type does not need to be introduced and usually is part of the standard silvicultural toolbox. Silviculture textbooks and university lectures in these countries implicitly refer to CCF as the prevailing forest management type without drawing much attention to the concept.

The situation is different in countries where forestry is or until recently used to be dominated by plantations and rotation forest management. Here forest managers, machine operators and researchers with experience in CCF are still comparatively few and continue to be so for some time. The transition towards CCF is so fundamental that even the organisation of large forest companies and state service needs to adapt. In countries where rotation forest management dominates, forestry has often been compartmentalised in an attempt to increase efficiency, i.e. individual staff do not any longer oversee the whole management and production process as the 'traditional forester of old' did but are highly specialised. Some staff, for example, manage Christmas tree sales, others only organise clearfellings, others manage thinning operations in large areas and yet again others only oversee ground preparation and replanting in vast pieces of land. This possibility of organising forestry work is much supported by the area method of control discussed in Section 1.8. However, this specialisation is far from ideal for carrying out CCF, since, for example, the main operations such as regeneration, thinning and harvesting take place in the same stands, and it is crucial to understand the interactions between different management activities, see Figure 1.14. For the same reason, it makes sense to maintain 'sessile' staff, i.e. to give up on frequent staff rotation so that forest managers are granted the opportunity to understand how the trees in their local area respond to site conditions and new interventions. Local promotion schemes can help maintain job satisfaction so that staff rotation is reduced.

Machine operators that are experts in clearfelling or systematic row removals in plantations will not find it easy to change to selective thinnings. Without appropriate training, it is hard for them to perceive differences among the trees. In this situation, training, as outlined in Chapter 8, is a very important part of introducing CCF to a new region or a new country. In general, CCF needs more knowledge, particularly about tree ecology and physiology, and forester skills than plantation management (O'Hara, 2014; Franklin et al., 2018). Models that predict plantation tree growth well, may give less precise results with CCF and traditional sampling designs may also need to be adapted.

One consequence of CCF, for example, is a trend to longer production cycles in managed forests. This implies that at least some trees in such forests are increasingly allowed to approach natural life expectancies. Naturally, trees managed under such conditions also reach larger sizes than those grown in standard plantation regimes even if the target diameter (cf. Section 3.4) is kept low (Bartsch et al., 2020). There is potentially a need to communicate this to the regional timber-processing industry. Yet another consequence of CCF for commercially managed forests is a move away from *maximising volume production* to *maximising timber quality* or other traits in individual trees. Both of these consequences may lead to a species change on many sites. Considering all these implications, it is wise to proceed in small steps and not to set too ambitious targets. For example, it may be a good idea to initially move away from plantation structure only slightly which can be achieved by replacing clearfelling with a simple silvicultural system such as the strip or uniform shelterwood system (cf. Section 5.3). Also, the transformation period (cf. Section 2.4) adopted should not be too short to allow for beginners' mistakes and other unfortunate circumstances. Once the first instances of success have strengthened self-confidence, more ambitious targets can be set.

Because the concept of near-natural forestry or CCF was comparatively new to the United Kingdom and to help improve knowledge and understanding of these new approaches to silviculture, the Forestry Commission established large-scale trial and demonstration areas in the three countries of Great Britain between 2000 and 2002 (see Table 1.3). This is a good way to gain national or regional experience in CCF comparatively quickly and can be recommended to other countries as well. The British CCF trial areas were anticipated to be in excess of 500 ha in forests managed by the Forestry Commission for commercial purposes. A notable exception is the Welsh trial area of Trallwm, which is much smaller in size and privately owned. Here, the owner saw an opportunity to diversify his business by adopting CCF. The Trallwm estate not only produces timber but also includes mountain-bike trails and holiday cottages inside the forest. Another important criterion for the selection of trial sites was easy access and a good representation of major species across a variety of site types. The areas should have transformation to CCF as management objective. In Wales, all trial sites have an almost exclusive focus on non-native *Picea sitchensis*, the main commercial species, and very much reflect initial plantation conditions rather than that they give examples of advanced irregular structures and visions for the future. The main and most intensively studied trial site in Wales is Clocaenog Forest. The trial sites were intended to serve as a long-term resource with appropriate recording and monitoring enabling the Forestry Commission to evaluate the operational, economic, social and

Table 1.3 The GB CCF trial areas.

	Trial site	Established
England		
Wykeham	North York Moors FD	
Whinlatter	North West England FD	2000
Dartmoor	Peninsula FD	
Wales		
Clocaenog	Coed y Gororau FD	
Trallwm	Midwales, privately owned	2002
Cwm Berwyn	Llanymyddfri FD	
South Scotland		
Glentress/Cardrona	Scottish Borders FD	
Lochard/Strathyre	Cowal & Trossachs FD	2002
Loch Eck	Cowal & Trossachs FD	
North Scotland		
Inshriach	Inverness FD	
Craigvinean	Tay FD	2002
Morangie	Dornoch FD	

Source: Adapted from Forestry Commission (2008). FD – Forest District.

environmental impacts of CCF and to compare it with standard plantation management (Forestry Commission, 2008). The trial areas were also envisaged to demonstrate the potential of CCF and the different approaches being used.

In the 2003 version of Forestry Commission (2008), an operational guidance booklet, a wise statement was made that is even more important in a time where silvicultural research attracts less and less research funding:

> 'We are unlikely to have meaningful comparative information [on CCF management] in the next ten years. So for some time to come, management decisions will have to be based on local experience and professional judgement supported by forest research information where available. An essential principle is a close relationship between researchers and managers and a willingness from both parties to make changes as knowledge grows'.

In this context, it is crucially important that forest managers engaging in CCF keep good records on why they decided to adopt CCF, management objectives and plans, past and future interventions and monitoring information. These records should be kept safe and be passed on to successors.

Generally speaking, it is a good idea to choose trial areas for diverse environmental conditions, which are typical of a particular region or country, e.g. lowland and upland sites. Including different ownerships and associated problems is also a good way of identifying useful trial sites. This can be combined with differing general objectives such as timber production, conservation, or carbon forestry.

If any mistake was made in the British selection of trial areas, then that often sites were selected that were very uniform and only in early stages of transformation to CCF. As discussed in Section 2.4, the more uniform and plantation-like a stand is, the longer the transformation takes. This implies that meaningful results and demonstration forests are not available for some time. Therefore, it may be useful to also include forest stands in the selection of trial areas, where transformation or conversion started earlier or where 'accidents' (natural disturbances; seemingly inappropriate, heavy management interventions) happened that have given rise to a diverse stand structure, see Section 2.2. Such sites are ideal for demonstration of future CCF situations, deliver early results, and promise success.

In the author's experience, it has also proved useful to offer field trips to regions and countries, where CCF has been taken up and implemented earlier so that the participants can develop a better vision of what is possible and what lies ahead. This usually gives forest owners and managers vital inspiration and help. Here, it can be recommended to choose neighbouring countries where CCF has been introduced not a very long time ago. The reason for this is that the CCF sites in the host country are then more comparable to the situation in the home country or region and the field trip participants can better understand the road ahead. Visiting areas with refined, long-standing CCF application can discourage novices and make them conclude that these concepts will never work at home. Such visits, however, should be coupled with a network of management demonstration sites in the home country or region, since otherwise the perception may develop that certain methods work fine in the country the participants visited, but it is uncertain whether they are suitable to their home range. The Association Futaie Irrégulière (AFI), for example, is an officially registered association set up by a group of forestry experts and forest owners in 1991 in France in order to study and to develop continuous cover forest management techniques. The association has established a network of permanent management demonstration sites to illustrate and study best-practice silvicultural techniques. AFI now has a network of roughly 90 research stands spread over 15 French regions, Belgium, Great Britain, Ireland and Luxembourg (Susse et al., 2011).

Trial areas and field trips should be accompanied by regional and national workshops and conferences. Any research project related to CCF that attracted national funding should be associated with a dissemination part where the latest findings are thoroughly and comprehensively communicated to forest owners and forest managers. Since silviculture is a subject area that has suffered from recent academic trends in many countries, the field received decreasing funds and lost staff. As a consequence, important forest-management skills and knowledge are often no longer taught at universities. This is particularly unfortunate for the uptake and continuation of CCF. Since silviculture is a core competence of forest managers, the Swiss federal government, for example, counteracted this development by funding two silvicultural competence centres, one for lowland forestry at Lyss and another one for upland forestry at Maienfeld. Both competence centres are supposed to maintain the high professional standard of CCF in Switzerland by advising practitioners, publishing silvicultural guidelines and most importantly by running forest management courses. Many of these courses include marteloscope exercises, see Section 8.3, and the competence centres established a national network of marteloscope plots. The Technical Development Department of Forest Research at Ae (Scotland, UK) is a similar institution and regularly carries out training courses and marteloscope exercises for forest practitioners.

Since thinnings are a crucial instrument of bringing CCF about by modifying forest structure, a government considering the introduction of CCF can substantially contribute to the uptake of this management type by contributing to the costs of thinnings through an appropriate grant scheme. The introduction of such a scheme has considerably helped make transformation to CCF possible in Britain and Ireland (Forestry Commission, 2008).

2

How Do I get Started with CCF?

Abstract

How do I select land suitable for CCF? What are the typical starting situations of CCF that one can find in any forest? What are the steps that need to be considered to take these initial situations forward and to develop them towards CCF or to maintain good CCF conditions? These crucial questions are often raised and are not always answered in a satisfactory way. The complexity of these questions can put forest managers off, if they remain unaddressed. In this chapter, we discuss how to identify land suitable for transformation and present simple indices that help make informed decisions. Then we turn to the three most common CCF situations of instant new CCF, transformation/conversion and maintenance along with a number of example applications. This section also includes underplanting techniques. Even what seem to be 'accidents' caused by natural disturbances or excessive management can often be prolific starting points for CCF. Biological rationalisation is a crucial part of CCF and includes a number of methods that minimise expensive management input and reduce its impact whilst ensuring satisfying outcomes. This chapter sets the scene for important planning considerations in CCF by defining fundamental techniques.

2.1 Introduction

A valid and frequently asked question is what the 'entry requirements' for continuous cover forestry (CCF) are. Can I start with CCF right away or does my forest need to meet certain pre-requisites? Is CCF at my location possible at all?

In this context, it is important to remember that theoretically CCF is possible wherever forest ecosystems naturally exist. Since CCF by definition is near-natural forestry based on ecological principles, there are no limitations to its uptake other than the limitations that exist for forest ecosystems in general, e.g. the availability of fresh water, sufficient soil oxygen and the temperature required for tree growth.

According to Gadow (2001), there are three basic CCF situations, i.e. *establishment* on bare land or '*instant new CCF*', *transformation* of even-aged plantations, and *maintenance* of existing CCF systems (see Figure 2.1). These are the most common starting configurations.

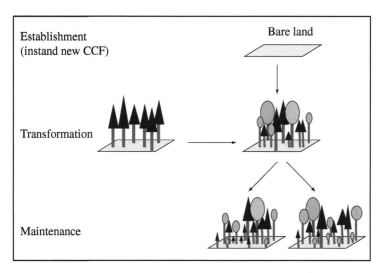

Figure 2.1 The three basic CCF situations. Source: Adapted from Gadow (2001).

The establishment of mixed-species CCF woodlands directly from scratch on bare land, irrespective of whether it has been recently under forest, for example, on a replanting site after clearfelling, or not, is one way of moving forward. This scenario is very common in countries with low woodland cover where afforestation is still high on the agenda. However, instant new CCF can also be a good solution for replanting clearfelling or windthrow sites. Direct establishment of mixed forests based on the *nurse-crop technique*, also referred to as the *artificial shelterwood method* (cf. Section 5.3), has been very successful especially where severe climatic conditions, such as strong winds and high precipitation, along the North Sea coast of Denmark and Germany made other methods impossible (Siebenbaum, 1965; Kramer, 1970; Pommerening and Murphy, 2004). This technique has often been applied on degraded and on former agricultural land. However, even today, the strategy of establishing CCF directly on bare land is still much underused.

In the traditional silvicultural literature, the transformation of existing monospecific even-aged conifer plantations to diverse uneven-aged continuous woodlands or the restoration of native woodlands on ancient woodland sites, are considered the most common starting points (Schütz, 2001a; O'Hara, 2014). Here, the existence of some forest, usually a commercial plantation, is a pre-requisite. In the British literature (Troup, 1928), the terms 'transformation' and 'conversion' are often treated as synonyms. They are regarded as very general terms implying the active change from one silvicultural regime or system to another, e.g. from coppice to high forest (see Ciancio et al., 2006), from high forest to coppice (see Section 1.3), from systems involving clearfelling to CCF, from coniferous plantations to semi-natural broadleaved woodlands and the other way round. More recently, suggestions have been made to distinguish between the two terms (Spiecker et al., 2004; Vítková and Ní Dhubháin, 2013):

> **Transformation**
>
> Active gradual change of woodland structure and/or species composition (in the German terminology referred to as *Überführung*) to eventually result in CCF management, e.g. the transformation of even-aged *Picea sitchensis* plantations to uneven-aged *P. sitchensis* woodlands (with possibly some broadleaved enrichment).
>
> **Conversion**
>
> Active abrupt change of species composition (thus coinciding with the German term *Umwandlung*) used in the context of ecosystem restoration, e.g. the conversion of coniferous plantations to restored broadleaved woodlands on ancient woodland sites.

Transformation currently is the most common pathway to CCF. It appeals to many forest owners and managers, since it does not involve an abrupt change that is potentially associated with great uncertainties. Schütz (2001a) noted that in forestry run for commercial purposes, there is often a requirement that transformation should be achieved without productivity losses. This requirement implies that the transformation process should not be hastened. Patience is a very important part of the skill set that forest managers in CCF need to possess. In some cases, however, fairly rapid conversion is necessary, e.g. where there is an urgent requirement for removing an abundant invasive tree species and replacing it by a native species. In forest conservation, conversion is a frequently used method.

Having succeeded in transformation or conversion, the uneven-aged forest needs to be maintained, i.e. managed in a way that sustainable and uninterrupted harvesting, regeneration and recruitment are possible. This is a management task that is not necessarily easier than transformation/conversion and requires a great deal of knowledge and experience.

2.2 Identifying Land Suitable for CCF

In Section 2.1, we already discussed that CCF is theoretically possible anywhere where forest ecosystems establish naturally. Where contexts and management objectives are more specific, e.g. CCF applied in commercial forestry, the question can be raised whether CCF is economically viable compared to RFM. Dependent on environmental and economic conditions, answers can be very different.

Since CCF covers a wide range of possible stand structures, silvicultural systems and thinning regimes (see Section 1.5), it is good practice to specify first what exact variants of CCF forest owners and/or managers plan to implement. Depending on the specifics of these variants, the identification of suitable sites may considerably vary. For example, a more or less single-storey forest regenerated through a uniform shelterwood system (see Figure 1.11a), a simple structure according to Mason and Kerr (2004, see Figure 2.15), can be accommodated and achieved on many more sites than a single-tree selection system (see Figure 1.11b). The reasons for this is not only the comparative ease of application of uniform shelterwood systems (cf. Section 5.3.1) and their windfirmness but also the fact that many simple-structured forests exist in countries where RFM dominate and then uniform shelterwood systems appear to be a natural choice for achieving regeneration. Therefore, it is more constructive to define target CCF structures that a forest enterprise or district wishes to adopt first and only then to search for suitable sites. Digital information such as sub-compartment data bases and Geographical Information Systems (GIS) can help there.

There are many indicators of species and size diversity in plant size (Magurran, 2004). All of them can be used to quantify species and size diversity of forest stands or of larger forest entities, see Section 4.4. Skovsgaard (2000), for example, suggested a simple but effective complexity index C combining age (or size) and species diversity in a simple quantity:

$$C = \sum_{i=1}^{n} \frac{a_i}{A} \left(1 - \frac{1}{\frac{R_i + m}{m} \times s_i}\right) \tag{2.1}$$

The index was originally termed UMF (short for **u**neven-aged **m**ixed **f**orest). n is the total number of stands or forest entities considered, A is the total area of the forest and a_i is the area of stand or forest entity i, e.g. a forest stand. R_i is the range of ages (measured in years) or of tree stem diameters (measured in cm), and s_i is the number of tree species present. Integer number m balances different magnitudes of numbers of R_i and s_i. When using age for R_i, Skovsgaard (2000) suggested setting $m = 10$. Using tree size may be a better choice, as age is often unknown or cannot easily be measured. R_i can also be replaced by the coefficient of variation of stem diameter or by the Gini coefficient (Hui and Pommerening, 2014, see Section 4.4.2.1) and the addition of and division by m could then be omitted. Given Skovsgaard's (2000) original index using age, a monospecific and even-aged stand would contribute a value of 0 to C, whilst very diverse stands with large age differences and many species approach a value of 1. For example, a diverse selection forest in Switzerland including five species and tree ages ranging from 0 to 90 years would have an index of $C = 0.980$ (Skovsgaard, 2000). The purpose of index C is to give a rough measure of complexity and requires input data that can easily be acquired or even estimated without sampling. The complexity index or any similar index can be implemented in data bases and GIS for providing criteria for initial land searches. The larger the index value, the shorter the period of transformation to CCF and the more likely it is that transformation is successful. Naturally, high index values do not necessarily guarantee a good potential for CCF, and there are many successful examples of CCF in monospecies woodlands. As always, other information should be considered as well and a detailed field evaluation is always advisable.

In Britain, potential CCF sites were often selected that were located in lowlands or in sheltered valley positions (Mason and Kerr, 2004). This is related to the low overall woodland cover in the country which is coupled with high wind speeds and high precipitation particularly in the west of the country where conifer management dominates, see Figure 2.13. This choice is contrasted by the situation in Central Europe where selection forests are often maintained above 1000 m asl (Schütz, 2001b). Here, overall woodland cover is comparatively high and forests with high structural diversity are considered and have proved to be very resilient (Dvorak and Bachmann, 2001; Hanewinkel et al., 2014). British silviculturists also discourage CCF on fertile sites with a high risk of competing ground vegetation (Mason and Kerr, 2004; Forestry Commission, 2008). Forestry Commission (2008) recommended the following general criteria for selecting CCF candidate stands:

- Strand structure,
- Good thinning track record (low h/d and large c/h ratios),
- Advance regeneration,
- Ground vegetation,
- Litter layer,
- Deer density (judging by browsing, fraying, bark stripping),
- Access and topography.

Existing diverse horizontal and vertical stand structure often indicates that transformation will not be too difficult and can be considered an asset that forest managers do not need to achieve themselves. Since homogeneous, dense forest stands and stands where advance regeneration is absent, prolong the transformation process, it is good advice not to select such stands for transformation in the first instance, see also Section 2.2.1. Good thinning track records suggest that individual trees are resilient which is a pre-requisite for transformation. Even without proper documentation thinning histories can be concluded from individual-tree morphology and resilience indicators.

Tree morphology and resilience indicators

In silviculture, a distinction is made between *collective stand resilience* and *individual-tree resilience* (Pommerening and Grabarnik, 2019). Both resilience types determine overall stand resilience or stability. Collective stand resilience is the result of an unmanaged and undisturbed comparatively dense stand matrix that is resilient to wind and snow as long as larger gaps in the main canopy or density reductions do not occur. Individual-tree resilience on the other hand is the bottom-up result of the resilience of individual trees, particularly of the main-canopy trees. Thinning operations temporarily reduce collective stand resilience. However, if they occur regularly, the residual trees build up individual-tree resilience through gradual morphological adaptation that helps them withstand increased loads of wind and snow. Individual-tree resilience is a necessary pre-requisite of CCF in general and can be fostered by methods of individual-tree silviculture (cf. Chapter 3). Individual-tree resilience is particularly important for trees with shallow root systems, e.g. *Picea* spp. and *Betula* spp. Good forest management usually attempts to balance individual-tree and collective stand resilience (Pommerening and Grabarnik, 2019).

Several tree size ratios have been proposed to quantify and monitor individual-tree resilience. One of these ratios is the *height–diameter ratio* or *slenderness* ratio (also referred to as h/d ratio, Eq. (2.2), Table 2.1). The more growing space a tree is granted, the larger the stem diameter (dbh) relative to the total height tends to be and, therefore, the smaller the resulting height-diameter ratio:

$$h/d = \frac{\text{Tree height}}{\text{dbh}} \qquad (2.2)$$

Table 2.1 Interpreting height–diameter ratios.

h/d value	>100	80 – 100	<80	<45
Interpretation	Very unstable	Unstable	Stable	Open-grown tree

Source: Adapted from Burschel and Huss (1997).

The more slender the trees are, the more they are prone to wind and snow damage. Open-grown trees (Hasenauer, 1997; Smith et al., 1992), which have grown up in the absence of interaction with other trees throughout their life, per definition have developed the highest possible degree of individual-tree resilience. Forest-grown trees usually have much larger height–diameter ratios than open-grown trees, since the former are commonly surrounded by neighbouring trees influencing their morphology. In conifers, snow damage is rare for trees with $h/d < 80$. h/d ratios that often preserve the impact of past density patterns as a legacy effect: With increasing (local) density, h/d ratios increase and decrease with decreasing (local) density. As such slenderness is also a useful indicator of thinning urgency and hints at past management (Abetz and Klädtke, 2002; Pretzsch, 1996). The h/d ratio is site-dependent, i.e. on more fertile sites, there tend to be larger h/d values than on less fertile sites (Figure 2.3). As an adaptation to wind, trees maintain a lower h/d ratio on exposed sites (Kramer, 1988; Pommerening and Grabarnik, 2019). The h/d ratio also decreases with increasing tree size.

The crown ratio (c/h ratio, Eq. (2.3)) is a crown form factor (Pommerening and Grabarnik, 2019). It is calculated as the ratio of crown length and total tree height, i.e. it can be interpreted as the percentage of tree crown relative to total tree height. Crown length is usually calculated as the difference between total tree height and height to base of (live) crown. For height to base of crown, there are various definitions in the literature (cf. Pretzsch, 2009). A sensible definition is 'the first living branch (broadleaves), the first whorl with at least three living branches (conifers) from the base of the tree that is contiguous with the rest of the crown'.

$$c/h = \frac{\text{Crown length}}{\text{Tree height}} \tag{2.3}$$

The crown ratio is often used to assess individual-tree resilience particularly of coniferous trees in a similar way as the h/d ratio, although the first more likely is an indicator of resistance to stem breakage while the latter indicates resistance to windthrow. An interpretation guide for the c/h ratio is given in Table 2.2.

Table 2.2 Interpreting crown ratios.

c/h value	<0.30	0.30 – 0.50	>0.50	>0.62
Interpretation	Very unstable	Unstable	Stable	Open-grown tree

Source: Adapted from Schütz (2001b).

(Continued)

(Continued)

The crown ratio is also related to photosynthesis and assimilation and, therefore, used to answer the question of how much of a tree's crown can be reduced in prunings or is naturally reduced in dense forests without major assimilation losses. As such the crown ratio is also an indicator of tree vigour. A common silvicultural requirement is that frame trees (see Chapter 3) have a crown ratio of 0.50–0.66 (Bartsch et al., 2020; Pommerening and Grabarnik, 2019). Ratios c/h and h/d are correlated: With increasing c/h ratio, h/d values typically decrease. The relationship between both ratios is stronger for conifers than for broadleaves. Broadleaved species tend to have larger h/d values than conifers (Pretzsch, 2019).

Pretzsch (1996) suggested a simple classification system that involves both aforementioned ratios for judging whether a forest stand has high, medium or low resilience with regard to wind and snow (Figure 2.2). According to this system, any stands are likely to have high resilience that either have complex forest structure (like, for example, selection forests, see Figure 2.15 for a definition of complex structure) or overstorey trees with both $h/d < 0.8$ and $c/h > 0.4$. In situations where only one of the two criteria, $h/d < 0.8$ or $c/h > 0.4$, is met by the overstorey, stand resilience is said to be medium.

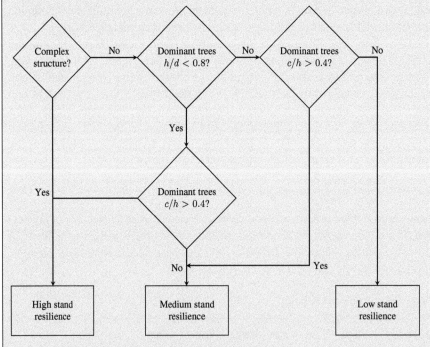

Figure 2.2 Flow chart for judging on stand resilience based on the mean h/d and c/h ratios of the main stand canopy (adapted from Pretzsch, 1996).

If the dominant trees of a forest stand do not fulfil any of the two aforementioned criteria, this stand is likely to have low resilience to wind and snow. Estimators of h/d and c/h are the mean ratios of the overstorey trees, i.e. of the dominant trees of a forest stand. With a little bit of experience, these mean ratios can usually be reliably estimated visually without measurement so that the assessment of stand resilience is a comparatively quick procedure. Since both ratios are measures of individual-tree resilience, they also indicate past disturbance and/or thinning history: Low stand resilience usually hints at poor past management with few interventions if any whilst high stand resilience is often the result of frequent interventions.

The natural trend of declining h/d ratio with increasing tree size is very clear in Figure 2.3a. The forest stand Clocaenog 2 is situated higher up on the slope of Cefn Du, i.e. the trees in Clocaenog 2 are more exposed than those in Clocaenog 1. This is indicated by lower h/d ratios in Clocaenog 2 across the whole size range. Presumably, better soil moisture and soil nutrient conditions at the lower slope may also have contributed to higher h/d ratios in Clocaenog 1. The h/d peak at lower stem diameters in Clocaenog 2 was formed by small groups of *Pinus contorta* DOUGL. EX LOUD. trees in a majority of *P. sitchensis*. The decline of h/d ratios at Artist's Wood (with predominantly *Pseudotsuga menziesii*) is not only more gradual than in the two plantation plots but also more varied, as can be seen from the noisy trend curve and the coefficient of variation. For the c/h ratio, no clear trend can be seen that would apply across all three forest stands (Figure 2.3b). However, there possibly is a U-shape pattern indicating that c/h ratios are high for small

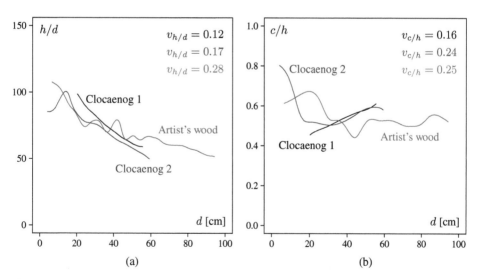

Figure 2.3 (a) h/d and (b) c/h ratios over stem diameter d in three forest plots in 2002/3. For the trend curves, non-parametric regression based on a Gaussian kernel, with bandwidths $w = 6$ cm and $w = 10$ cm was used. v denotes the coefficient of variation. See also the photos of the stands in Figures 1.11a and 2.4 for a visualisation.

trees, then considerably drop for the mid-diameter range and finally increase again towards the large-size range.

When it comes to identifying land suitable for CCF, the presence of advance regeneration is another asset, since competing ground vegetation is already inhibited by existing seedlings. Ground vegetation such as grass and shrubs, on the other hand, can potentially make transformation difficult. Soil scarification to expose the mineral soil can sometimes be an option, but this involves additional work and management input. Along similar lines, there should be a good, receptive seedbed for germination. This involves not only the absence of competing ground vegetation but also the litter layer. Too much raw humus accumulation will make seedlings germinate in litter and unless the seedlings get their roots quickly into mineral soil they may die when the litter dries out. Therefore, a litter layer of less than 5 cm is recommended (Forestry Commission, 2008). Deer density can be judged by the abundance of browsed seedlings, by bark-stripped tree stems, and by fraying evidence. If too many seedlings are browsed, it may be necessary to fence the forest stand. Access and topography are important criteria for management technology and logistics: Since thinnings are crucial to CCF, there need to be good forest roads and a network of extraction racks for harvester and forwarder machines to drive on (see Section 5.3.6). Topography should not be too varied, at least not when getting started with CCF. Following the principles of selecting CCF trial areas (see Section 1.10), it is good advice to identify sites for CCF that are promising, i.e. where transformation does not appear to be difficult and the above criteria can help in this decision process. It is also good advice not to select CCF candidate sites bordering on large clearfelling or other open areas, as wind turbulences can cause windthrows and crown breakage inside the CCF area.

2.2.1 'Accidents' that Turn into Flagships

As mentioned in Section 2.2, transformation to CCF can be considerably shortened when forest stands are selected that already have quite diverse forest structure. Such stands can help demonstrate visions of desired, future CCF stands and the road ahead towards them much faster than any plantation where transformation has only started recently. As discussed in conjunction with Skovsgaard's (2000) complexity index in Section 2.2, it is often quite likely that transformation of more diverse forests, in terms of tree size and species, is successful. There is a particularly interesting group of diverse forests meriting special attention. This is forest land that most likely was previously hit by natural disturbances or 'unorthodox' management involving heavy tree removals that dramatically lowered overall stand density. The spontaneous response of forestry staff and visitors to such recently managed stands – may they have been caused by nature or by humans – often is negative. This is because immediately after the impact, the forest in question looks somewhat devastated, and it is hard to imagine that anything good can come from the current situation. However, this impression is often deceiving and many of such sites later give rise to a very diverse development of tree sizes and species. Then early and untimely condemnation of the staff responsible can gradually give way to high praise. Incidentally, these examples often show that the thinning interventions we usually plan for are in fact too weak. When participating in field trips, it is a good idea to remember such accidents and to earmark them as potential CCF management demonstration sites.

Almost any forest district has its odd corners where something out of the ordinary happened that the staff involved feel they need to hide but later turns into treasure. A good example of a single-tree selection forest that had its origin in an accident caused by excessive logging is Artist's Wood in Gwydyr Forest just outside the village of Betws y Coed in the Snowdonia National Park (North Wales, UK). The stand was named after an artists' colony that established in Betws y Coed in Victorian times, see Figure 2.4. The soil type is iron pan soil, and the forest stand is located at an altitude of 80 m asl. The coastal climate is influenced by the Gulf Stream and annual precipitation is in excess of 1200 mm. The total area of the Artist's Wood stand is 3.7 ha (Schütz and Pommerening, 2013). Artist's Wood was planted by the Forestry Commission Wales in 1921 and established as a *P. menziesii* plantation with some *P. sitchensis* in wetter depressions. *Tsuga heterophylla* (Raf.) Sarg. and broadleaved species later colonised parts of the understorey. In 1997, there was a high demand for sizeable telegraph posts in Ireland and the Forestry Commission, attracted by decent prices for quality Douglas fir timber, carried out what initially were thought to

Figure 2.4 Artist's Wood – a lowland single tree selection system with mainly *P. menziesii* (Mirbel) Franco in the Snowdonia National Park (Wales, UK) starting off from a management 'accident'. The photo was taken in 2003, and there was approximately 10 m³ of volume in each of the large *P. menziesii* overstorey trees. Source: Courtesy of Stephen T. Murphy.

be excessively heavy selective fellings at Artist's Wood to meet this demand. Many large *P. menziesii* trees were removed in this operation. The fellings triggered several waves of *P. menziesii* regeneration and released parts of the mid-storey. The author of this book visited the site during a public field trip in 2000 for the first time and decided to include the stand as a management demonstration site of the Tyfiant Coed project (Bangor University, UK). Subsequently, Artist's Wood was earmarked as management-demonstration site and scheduled for a direct transformation to a selection system (see Section 2.4) carried out as part of a heavy local selection thinning in 2004 (see Section 5.2). In 2002, standing volume was $570\,m^3\,ha^{-1}$ and basal area at this time was $34\,m^2\,ha^{-1}$. In the treatment in 2004, 25% of basal area was removed to result in a residual basal area of $27.7\,m^2\,ha^{-1}$. For Britain, Artist's Wood has been considered a rare example of best CCF practice involving *P. menziesii* (Helliwell and Wilson, 2012). Also internationally, there are not many examples of single-tree selection systems based on species other than *Abies alba*, *Picea abies* and *Fagus sylvatica*. The achievements at Artist's Wood are also remarkable given that *P. menziesii* is a shade-intermediate (cf. Appendix B) though very productive species. As a result the main canopy needs to remain much more open than in the classic combination of *A. alba*, *P. abies* and *F. sylvatica* in continental Europe; however, this is compensated for by the large volume accumulation of individual trees.

Common examples of accidental CCF situations include stands with main canopies of *Pinus* spp. and *Larix* spp. Such overstoreys are typically very light. Whilst providing excellent shelter for new cohorts of trees, they do not cast excessive shade even if the main canopy is closed. *Pinus* spp. and *Larix* spp. therefore often act as nurses for shade-intermediate species such as *P. sitchensis* (see Figure 2.5). The management

Figure 2.5 *P. sitchensis* (Bong.) Carr. and *P. abies* (L.) H. Karst. natural regeneration colonised a *Larix kaempferi* (Lamb.) Carr. stand at Clocaenog Forest (plot 6 of the author's Tyfiant Coed project). Source: Courtesy of Gareth Johnson.

demonstration plot 6 in compartment 5224 c/d at Clocaenog Forest, one of the national CCF trial areas in Wales (cf. Section 1.10), is located at 350 m asl on podzolic brown earth. The stand was planted as a mixture of predominantly *Larix kaempferi* with very few *P. sitchensis* and *P. contorta*. With time, the latter species mostly disappeared. Annual precipitation is in excess of 1300 mm. Regeneration here largely did not originate from the parent trees, but from other forest stands in the area. If such colonisation does not include unwanted, e.g. invasive species, it is possible to take the natural regeneration forward in successive thinnings. In this particular case, regeneration was dense and structure – in terms of horizontal dispersal and vertical differentiation – was moderately developed. Selective thinnings in the under- and mid-storeys can then help increase size and dispersal diversification whilst keeping the overstorey largely intact. Although forest structure at Clocaenog 6 was less developed than at Artist's Wood, this example illustrates a situation where on a given site, there is not only advance regeneration present but also the new cohort of trees already has experienced many years of self-thinning and self-differentiation so that some size hierarchies are recognisable that can then be taken forward by forest management. Transformation to CCF in such situations therefore does not have to go to great lengths to establish a CCF woodland. Results are likely to be available soon, and risks are comparatively low.

Another common management accident relates to 'failed' *Picea* spp. replanting following clearfelling or a large-scale disturbance. Such replanting sites can face considerable mortality among *Picea* spp. trees due to unsheltered conditions and often are invaded by *Betula* spp. and other broadleaved species (Figure 2.6). This is a very common situation in Wales, and when the author started his work there, he was specifically asked to focus on this particular type of 'management accident'. Up to an age of approximately 30 years, *Betula* spp. dominate *Picea* spp. in Wales due to faster growth. After that the advantage turns in favour of *Picea* spp. Mixtures of *Picea* spp. and *Betula* spp. also occur naturally and appropriate management can modify competition between the species so that both of them survive for a large part of a forest generation. Management of competition is made easier when

(a)   (b)

Figure 2.6 Accidental mixed *P. sitchensis* (Bong.) Carr. – *Betula* spp. woodlands in North Wales. (a) Clocaenog forest, Tyfiant coed plot 7 planted with *P. sitchensis* in 1985. (b) Coed y Brenin, Tyfiant Coed plot 4, where *P. sitchensis* was planted in 1988. Source: Courtesy of Gareth Johnson.

both species are grown in separate groups rather than in intimate mixtures (cf. Section 4.3). Depending on the species abundances involved in such 'accidents', the sites can be used for forest restoration and conversion projects. Managing mixed *P. sitchensis* and *Betula* spp. stands were, for example, seen as a first step towards a restoration of the Atlantic rainforest woodland communities in Wales.

'Accidents' are ideal starting points for the transformation of plantations to CCF, taking them on saves much time, as they are ahead of the transformation schedule and deliver results fast while risks are usually low. Almost every forest district or forest estate has such odd corners, and although they may not appear typical of a country or a region, they usually provide many clues for successful CCF management elsewhere and grant forest managers the vital experience with this management type. Accidental management or disturbances can also be helpful to identify suitable forest plots for research projects that involve CCF, because parts of such forest stands resemble structures that are typical of well-established, advanced CCF, or of ongoing transformation.

2.3 Starting from Scratch – Instant New CCF

The silvicultural literature often suggests transformation as the main pathway to CCF (Schütz, 2001a). This strategy, however, requires an existing forest stand as the starting point. If the starting point is bare land, e.g. a clearfelling or a large windthrow site, it is often recommended to replant a plantation and then to transform that plantation to CCF. Although this is clearly a possibility, it is a lengthy and indirect way to go. There is always the alternative to plant or to sow a CCF woodland directly.

If bare land was left to natural devices, eventually diverse tree vegetation would establish through the gradual process of succession; however, this could take many decades. In fact, this is the route Swiss forestry decided to take in many areas after the devastating storms at the end of the 1990s. Any human intervention in such processes, e.g. planting can be interpreted as an accelerator of forest establishment (Heger, 1949). Theoretically, there are three main approaches to establishing diverse, mixed stands on bare land: Direct planting, direct seeding and the indirect establishment of mixed stands under the shelter of a *nurse crop*. Although they can be successful, methods of direct seeding and planting often fail or need to be complemented by repeated 'beating up' because the climatic and soil conditions on bare land are often unsuitable to many tree species (see Figure 2.7), let alone the interference of various wildlife feeding on seeds. Only pioneer species, species specialised on the colonisation of bare land, such as *Alnus*, *Betula* and *Larix* manage to persist in sufficient numbers, whilst the mortality of other species is usually great in such conditions (Bartsch et al., 2020). Therefore, in this section, the focus is on the nurse-crop concept. Two related but very different methods for starting from scratch, the Bradford–Hutt plan and the Anderson group selection system are discussed in Sections 2.3.1 and 2.3.2. Both methods were invented in countries new to CCF.

In the most general meaning, a *nurse crop* includes trees or shrubs that foster the development of other tree species (target trees) in terms of survival, growth, or certain desired traits such as timber quality. Nurse trees typically protect target trees in their juvenile phase from frost, drought or wind (Helms, 1998). They can also be employed to shade the stems of mature target trees to prevent epicormic growth, e.g. in *Quercus* spp., which would otherwise diminish timber quality.

Anderson (1960) applied the nurse-crop method to establish his trials in Glentress (see Section 2.3.2) and used the term 'advanced forest', while in the United States, this approach is sometimes called the 'one-cut' shelterwood method (Smith et al., 1997). Other authors referred to this method simply as 'shelter' (Klang and Ekö, 1999). A more appropriate synonym would be *artificial shelterwood system*, as the nurse-crop concept shares many similarities with the uniform shelterwood system, see Section 5.3.1. The difference is only that the sheltering canopy is planted more or less at the same time as the understorey trees but can play the same role due to the faster growth of the nurse species on bare land. Fiedler (1962) and Heger (1949) described the history of the nurse crop idea in a very detailed way. According to their findings, the corresponding German term *Vorwald* and its application has been in use for almost 200 years in Germany and Switzerland. The procedure is to establish a nurse crop of a pioneer species or early coloniser such as *Alnus glutinosa* (L.) GAERTN., *Betula* spp., *Larix* spp., *Populus tremula* L., *Salix* spp. or *Sorbus aucuparia* L. by planting or seeding at a comparatively wide spacing. Forest managers can also take advantage of nurse crops that have naturally established. *Betula* spp., for example, often naturally colonise recently replanted *Picea* spp. sites and can then provide shelter for *Picea* spp. due to the rapid early growth of *Betula* spp. A little later, or at the same time, the target species are introduced either by underplanting or from natural regeneration. Such stratified stand mixtures, composed of late successional species in the lower strata and light-demanding early successional species (cf. Appendix B) in the upper strata have been recommended as a means of gaining a higher timber volume yield compared to a monoculture (Assmann, 1970). It is important that the nurse species only moderately shelter the target species and only moderately take up nutrients so that the growth of the target species is not too much inhibited or suppressed (Bartsch et al., 2020). The method provides an open canopy which protects the target species against weather extremes such as excessive sunlight and evapotranspiration, high precipitation, frost and wind (Örlander and Karlsson, 2000; Langvall and Ottosson Löfvenius, 2002), see Figure 2.7. On damp sites, the nurse crop helps drain the soil through transpiration (Klang and Ekö, 1999). It improves soil conditions through the recycling of nutrient-rich litter from broadleaved nurse trees that is easily decomposed and suppresses competing ground vegetation. Finally, the light shelter promotes the development of certain desired properties (Bartsch et al., 2020), e.g. the temporary sheltering of the target species that can result in an improvement of stem quality, i.e. stem straightness and branchiness (Fiedler, 1962; Burschel and Huss, 1997). Another important benefit of the nurse crops is the retardation

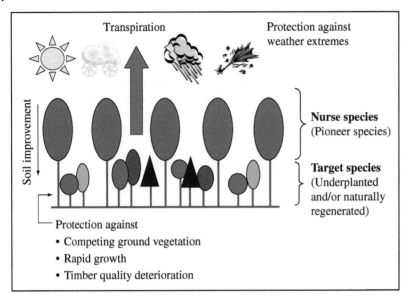

Figure 2.7 The concept and functions of a nurse crop/artificial shelterwood system used in afforestation. Source: Pommerening and Murphy (2004). Explanations can be found in the text.

of growth of the target species. Usually, forest managers working in commercial forestry cannot wait and are more interested in rapid tree growth, but from afforestation trials with *A. alba* it is known, that the lack of shelter leads to failure later on in stand development (Schmidt, 1951; Kramer, 1970). *A. alba* grown under canopy shows a slow but long-lasting growth with a late and slow culmination. In contrast to this, *A. alba* planted on bare land without shelter has a rapid growth when young which is followed by an early culmination at a lower level than under canopy with quickly decreasing volume production thereafter. It is believed that in many cases, the growth pattern of *A. alba* grown without canopy shelter is actually the reason for the increasing lack of resilience of *A. alba* stands around the age of 100 years (Kramer, 1970). Although not many studies have been carried out on this phenomenon with other tree species, it is possible that there is similar growth behaviour, with resulting greater instability in stands grown without canopy (Pommerening and Murphy, 2004). This is also supported by the continued, nearly constant growth of trees in selection forests (Schütz, 2001b). All these benefits are examples of what has been discussed in relation to the tenet 'Reliance on natural processes, promoting vertical and horizontal structure' in Section 1.5. The use of nurse crops can also be understood as a closer adherence to natural processes. In the sequence of natural succession from the initial herb and shrub stage to the climax stage, the nurse crop approach resembles a stage where pioneer tree species are gradually replaced by climax species. The only difference is that other stages are omitted in order to accelerate the process of producing timber or of achieving a mature forest.

There is a wide range of different nurse crop applications. Schmidt (1951) reviewed the growth of *A. alba* on poor sands in Eastern Frisia (Lower Saxony) near the North Sea coast of Germany, where afforestation with this species commenced in 1758, and presented a regional yield model. He stated that after the first systematic trials, it became clear that afforestation with *A. alba* would only be successful if a nurse crop of *Betula* spp. were used. A method was developed in which *Betula* spp. was planted first with an initial spacing of 2 × 2 m. When the *Betula* spp. trees reached a top height of 2 m, they were underplanted with *A. alba*. Schmidt (1951) suggested that, if available, plants should be taken from natural regeneration or other sheltered conditions nearby. They should be about 30 cm of height or 5–8 years old. Trials have shown that long-term mixtures of *A. alba*, *P. sitchensis*, *Abies grandis* (DOUGL.) LINDL. and *F. sylvatica* led to resilient and productive forests along the North Sea coast and *A. alba* then proved to be very resilient to windthrow (Pommerening and Murphy, 2004). Apart from *Betula* spp., Schmidt (1951) suggested *A. glutinosa* as a nurse species and mentioned that *Pinus mugo* TURRA is used in Denmark. When Anderson (1960) started his trials at Glentress in Scotland, he used a similar method to introduce *A. alba*. Another important target species that in Europe clearly benefits from the shelter of nurse crops is introduced *P. menziesii* (Bartsch et al., 2020; Stokes et al., 2020).

Siebenbaum (1965) discussed afforestation methods along the North Sea coast of the German state Schleswig–Holstein. The most successful method was the establishment of mixed *Abies-Picea*/broadleaved woodlands. *A. alba*, *A. grandis*, *Abies nordmanniana* (STEV.) SPACH), *Quercus petraea* (MATT.) LIEBL., *L. kaempferi*, *F. sylvatica*, *P. mugo* and *P. sitchensis* were planted at the same time in intimate mixtures. These diverse plantations differentiated into two strata comprising the pioneer conifers in the upper storey and the *Abies* spp. trees and the broadleaves in the lower storey. Afterwards, the mixed forest stands were then directly transformed to selection stands through subsequent thinnings. According to Siebenbaum (1965), this resulted in resilient stands which were far less prone to biotic and abiotic damage than any other plantation type (Pommerening and Murphy, 2004).

A very different understanding of nurse crops is evident in British and Irish forestry practice. Taylor (1985) stated that during the British afforestation campaigns in the 1920s and the 1950s/1960s *P. sitchensis* plantations suffered chronic nitrogen (N), phosphorus (P) and potassium (K) deficiency on many upland sites particularly where *Calluna* predominated. It was then found that by using nursing mixtures of *P. sitchensis* with *Pinus silvestris*, *L. kaempferi* and *P. contorta* DOUGL. ex LOUD. this effect could be eased. The nursing effect is not simply through the suppression of *Calluna*, but involves a mycorrhizal relationship between the nurse and *P. sitchensis* which increases nitrogen availability (O'Carroll, 1978). All three species proved to be suitable nurses, although *P. sitchensis* in mixture with *P. contorta* achieved slightly better results in Scotland (Taylor, 1985). Thelin et al. (2002) reported similar effects in mixed *P. abies*/broadleaved woodlands where the admixtures compensated for poor and acidified soil on sites with low fertility. The long-term objective of all these trials, however, was primarily to produce a final monospecies plantation of a desired target species such as *P. sitchensis* (Taylor, 1985).

Evans (1984) mentioned that using conifer nurses to aid the establishment of broadleaves was by 1984 almost universal in Britain, but suggested that the nurse species should not be allowed to overtop the target species. An ecological study presented by Pigott (1990) highlighted potential problems with the technique. He looked at the influence of evergreen coniferous nurse crops on the field layer in two woodland communities at Leith Hill Place Wood in Surrey (England) where a clearfell site was replanted with a mixture of *Q. petraea* and *P. abies* or *P. menziesii*. The nurse crop failed to stabilise the final oak stand, led to the accumulation of coniferous litter and reduced the diversity of ground vegetation significantly (Pommerening and Murphy, 2004).

Klang and Ekö (1999) reported on the use of nurse crops in Sweden. The driving factor for this practice is the frequent occurrence of early summer frosts which cause considerable damage to *P. abies* plantations. In these areas, seedlings were either planted under a shelter of trees from the previous stand or a shelter was formed by naturally regenerated *Betula* spp., established after clearcutting. Tham (1988, 1994) and Mård (1996) studied the influence of a birch shelter (*Betula pendula* ROTH and *Betula pubescens* EHR.) on the growth of *P. abies*. Although the nurse crop reduced the yield of *P. abies*, the total yield, including birch, was greater than that of monospecific *P. abies* stands. Bergqvist (1999) came to similar results in his investigation of *Betula* spp./*P. abies* nurse crop systems. *P. abies* growing under a birch shelter is a common type of two-storied stand in the Scandinavian boreal forest. In young conifer stands, there is often an abundance of naturally regenerated birch overtopping the conifers and the number of *Betula* spp. is usually reduced in pre-commercial thinnings.

Seitschek (1991) and Rittershofer (1999) summarised Bavarian examples on sites where early and late frost events suggest that the use of nurse crops would be beneficial. The initial spacing they quoted as a result of their experience is similar to the one stated by Schmidt (1951): 2 × 2 – 3 × 3 m. They came up with the interesting observation that the target species should be introduced when the stand height exceeds the spacing of the plants. But in any case, the authors emphasised the need for an early underplanting in order to keep the amount of competing ground vegetation low. In many situations, it is therefore advisable to plant nurse crop and target species at the same time (Seitschek, 1991). The same author also stated that the establishment of forest margins (cf. Section 7.3.3) can contribute to the resilience of such new CCF stands. *Alnus incana* (L.) MOENCH is recommended as nurse species only for dry sites and should be used in rare instances because of its tendency to develop root suckers. Also, in other studies *A. incana* has not performed well. *P. tremula* and *Betula* spp. usually provide less shelter than *A. glutinosa*. Perala and Alm (1990) reviewed the role of birch as a nurse crop worldwide. Seitschek (1991) preferred *A. glutinosa* as nurse crop species. Its success may be due to the fact that it forms a symbiotic relationship with the bacterium *Frankia alni* which belongs to the genus of the *Actinomycetes* and assimilates atmospheric nitrogen and therefore contributes to soil fertility (Savill, 2019). Moreover, common alder shows a monopodial branching system and has a monocormicly growing continuous crown which is very similar to that of coniferous trees (Bartels, 1993). These desirable properties make the author suggest that individual *A. glutinosa* trees should be allowed to develop in the target stand (Pommerening and Murphy, 2004). Greger (1998) described the afforestation of riparian zones along the River Elbe in the Stendal Forest District. There the local forest authorities established a nurse crop to rapidly cover degraded soils and improve soil properties. Shrub species such as *Rosa canina* L., *Crataegus* spp., *Prunus spinosa* L. and early coloniser tree species such as *Ulmus* spp., *Acer campestre* L. and wild fruit trees were used. The initial spacing for most nurse crop species was 2 × 2 m, though *Populus nigra* L., *Salix fragilis* L. and *Salix alba* L. were planted at 4 × 5 m. In contrast to Seitschek, the author suggested a comparatively late introduction of the two target species *Q. robur* and *Carpinus betulus* L. after 15–20 years. Elsewhere Heuer (1996) reported his experience concerning the establishment of mixed broadleaved and coniferous species under a *Pinus sylvestris* canopy in the borough of Berlin. One result of his study was that *Quercus* spp. tolerates lower light levels than commonly believed.

In recent years, several research groups studied the possibility to utilise biomass produced by nurse crops as a side product for energy wood production (Stark et al., 2013; Nord-Larsen and Meilby, 2016), see also Section 7.3.6.

The underplanting of nurse crops is not fundamentally different from the underplanting of mature forest cover. Also, some of the methods originally employed for afforestation with mixed species on bare ground have later been transformed to underplanting techniques by allowing wider spacings. General underplanting techniques are therefore discussed in Section 2.4.1.

Mixed species CCF afforestation on poor sandy soils in the Nordfriesland Forest District (Germany).

On sandy soils dominated by podzols and in an environment which is somewhat hostile to trees (frequent high winds and early and late frosts) the forestry staff of the Nordfriesland Forest District apply an afforestation method based on the nurse-crop or artificial shelterwood system. The method had gradually evolved along the North Sea coast in an area notoriously low in woodland cover. Much former agricultural land is converted to forest there. The regional PEFC (Programme for the Endorsement of Forest Certification) certification scheme requires that no herbicides are used for controlling competing ground vegetation. Therefore, alternative methods had to be invented to ensure the safe development of tree seedlings with as little tree mortality as possible.

As part of the afforestation method *A. glutinosa* trees are planted at the same time as the target species, such as *A. alba*, *Acer pseudoplatanus* L., *F. sylvatica*, *L. kaempferi* (LAMB.) CARR., *P. sylvestris*, *Q. robur* and *Tilia cordata* MILL. *A. glutinosa* typically grows much faster than these target species and thus provides cover against the severe environmental conditions for the understorey consisting of the target species. As a nurse species *A. glutinosa* provides not only shelter but also exerts vertical pressure to encourage monocormic growth and low taper. *A. glutinosa* also is a nitrogen-fixing species and produces prolific litter, which as it decomposes, increases the fertility of the soil and reduces its acidity. It is also very monocormically growing, and for this reason often being referred to as the 'conifer of broadleaves'. As a consequence, *A. glutinosa* offers the potential for a few individuals to be taken on into the main canopy and harvested for commercial use (Figure 2.8 and 2.9).

The target species *A. alba* often requires additional cover and protection which in the Nordfriesland Forest District is provided by *P. sitchensis* and planted either around it or even in the same planting holes. *P. sitchensis* trees are then removed a few years later once the *A. alba* trees are secure to ensure *P. sitchensis* does not begin to suppress

---

(Continued)

(Continued)

Figure 2.8 Early stage of an instant CCF mixed broadleaved woodland (mainly *F. sylvatica* L. and *Quercus robur* L. as target species, seen in the foreground) in the Nordfriesland Forest District (Germany). The nurse species *A. glutinosa* (L.) GAERTN. was planted at the same time as the target species but is now starting to form a light overstorey (tall straight saplings in the background) due to faster growth. Source: Courtesy of Sharron Bosley.

and compete with *A. alba*. *P. sitchensis* is also instrumental in deterring deer browsing due to its prickly foliage.

CCF afforestation projects in the Nordfriesland Forest District proceed in distinctive steps:

- *Consultation*. Prior to the afforestation, there is a public consultation meeting with all the stakeholders to introduce the project and to decide the details. All bodies affected have an input, such as the agencies dealing with water management.

Figure 2.9 Late stage of an instant CCF mixed broadleaved and conifer woodland in the Nordfriesland Forest District (Germany) with district staff, Bangor University students, and UK forest practitioners in the foreground. Source: Courtesy of Mike Hale.

- *Deep ploughing*. Breaking up the soil down to 80 cm and lifting the lower, deep layers upwards into a tilted position in order to make nutrients more accessible. At the same time, the surface seed bank and competing ground vegetation are buried in the lower soil profile.
- *Application* of 20–30 kg ha^{-1} of a special variant of winter rye to prevent the growth of competing vegetation, to create a microclimate for tree seedlings, to minimise rodent attack and for protection from drought and frost. The rye is mulched the following year and left on site along with any regrowth of rye.
- *Planting* of 3000–5000 tree seedlings per hectare at the same time as the winter rye is sown.
- A total of 10–30% of the total area is left to completely natural development (no ploughing or planting).
- *New forest margins* are created to provide conservation corridors between the forest edge and the surrounding fields, see Section 7.3.3.

2.3.1 Bradford–Hutt Plan

The Bradford–Hutt plan is a special variant of instant CCF developed in Britain. In the 1950s, the sixth Earl of Bradford and his forester Phil Hutt devised a practical experiment with a view to achieve CCF (Hart, 1991; Kerr et al., 2017). The main driver was a concern about rotation forest management and the desire to manage forests in an environmentally more balanced way. After unsuccessfully trying single-tree selection systems (see Section 5.4), they decided to devise a completely new CCF strategy. When the Earl purchased the Tavistock Estate in 1959 in South Devonshire (approximately 10 miles north of Plymouth), an opportunity arose to try new ideas at large scale. The area is one of high rainfall (1125 mm year^{-1}) with brown earth soils mostly of medium-to-high fertility and a maximum elevation of 150 m asl. The steep valley sides present an erosion danger, and there is risk of late frosts (Hart, 1991).

Although resolved to move towards CCF, Bradford and Hutt took the main elements of their new silvicultural plan from RFM, i.e. planting, thinning and clearfelling. These elements along with principles of the normal forest they decided to implement on a small-scale grid pattern, thus relying on the area-control method, see Section 1.8. Bradford and Hutt settled for the area-control method, because they were concerned about harvesting damage on regeneration trees and about operational aspects in forest stands where mature, harvestable trees could be anywhere (Kerr et al., 2017).

Since the Bradford–Hutt plan starts with planting in small gaps (square plots in this case) on bare land, it is a method suitable for both instant new CCF and transformation to CCF. The crown projection area of the main tree species at maturity, particularly *P. menziesii*, was expected to be close to 36 m^2. Based on this crown projection area, it was decided to adopt a size of 6 × 6 m as smallest unit (referred to as 'plot') set out in a geometrical pattern (Hart, 1991).

Taking local growth conditions into account, a rotation of 54 years was set. Nine plots were organised to form a timber-sustainable unit which would implement the full rotation from planting to harvestable mature tree (Figure 2.10; Table 2.3). A total of 30 such

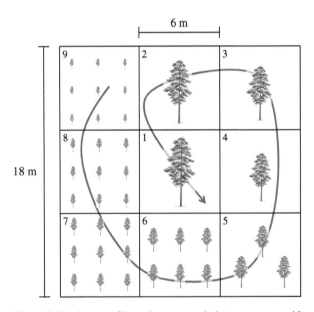

Figure 2.10 Layout of how the trees and plots are arranged in units according to the Bradford–Hutt plan. Source: Inspired by Hart (1991) and Kerr et al. (2017), see also Table 2.3.

Table 2.3 Tree numbers and ages in the nine plots of a unit according to the Bradford–Hutt plan, see Figure 2.10.

Plot	Age	Number of trees
1	54	1
2	48	1
3	42	1
4	36	1
5	30	3–4
6	24	6–7
7	18	9
8	12	9
9	6	9

units can be accommodated on a hectare. The organisation of the units suggests that the Bradford–Hutt plan is a group system of sorts (Hart, 1991; Matthews, 1991).

Extraction racks of 4 m width run on either side of the units with a distance of 18 m between them, i.e. the side length of a unit. In the vertical direction, four units are seamlessly stacked with rides following every fourth unit (Figure 2.11). Although the Bradford–Hutt plan can start either on bare land or in an existing forest, it is probably

Figure 2.11 Schematic representation of units, plots, extraction racks and rides in the Bradford–Hutt plan at Tavistock.
Source: Adapted from Hart (1991).

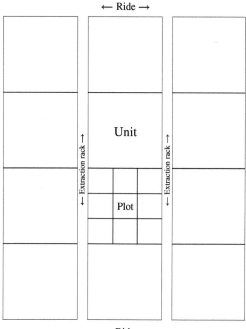

easier to start with bare land, since the design does not allow for natural regeneration and has a strict progression of age classes. Either the existing trees in the central plot are removed, and then plot 1 is replanted or nine trees are planted on bare ground straight away. Initial spacing is 2.0 × 2.0 m. Six years later, this process is repeated in plot 2 and then again repeated every six years in a typical spiral pattern (Figure 2.10). After 18–20 years, the nine trees are thinned so that by age 36 there is only one tree left to be felled at age 54 (Hart, 1991). Once the sequence has been completed after 54 years, one mature tree is harvested from each unit every six years (Kerr et al., 2017).

In summary, the Bradford–Hutt plan established small groups on a geometric pattern in such a way that a range of nine 'age classes' from 6 to 54 occur in a spatially predictable way (Hart, 1991). At the same time extraction of timber is straightforward, and timber sustainability is assured by the design of the plan.

At Tavistock, the Bradford–Hutt plan started in 1959 in existing stands of *P. sylvestris*, *Larix* spp. and *P. menziesii* in a total area of 350 ha (Hart, 1991). Shade-tolerant North American species such as *Thuja plicata* D. DON., *T. heterophylla*, *Sequoia sempervirens* (D. DON.) END. and *Nothofagus* spp. were planted later as part of the replanting (plot 9 in Figure 2.10). Because of the small size of the plots, shade-tolerant species had greater chances to survive. However, *P. menziesii* was increasingly favoured from plot 7 (year 18, Figure 2.10) onwards, as its greater light demand was more suited to the increased light associated with the available growing space at this stage (Kerr et al., 2017). The managers at Tavistock succeeded in maintaining diverse mixed-species woodlands that involved many North American and European species. Challenges for the application of the Bradford–Hutt plan included

- Defining plots and units in the field,
- Limited growth of plants in plot 1 (Figure 2.10) as canopy of neighbouring plots closes in,
- Concerns about timber quality.

The Bradford–Hutt plan has never been tried anywhere else in Britain nor anywhere else in the world (Kerr et al., 2017). After 54 years, Kerr et al. (2017) published an evaluation of the Bradford–Hutt plan. The authors compared the observed stem-diameter distribution of five woodlands in the estate with that of a q factor model with q varying between 1.24 and 1.36 (see Section 6.3.1) and found that the observed distributions only deviated slightly from the model. From this and other indicators, the authors concluded that the Bradford–Hutt plan was successful. Based on anecdotal evidence, the costs of managing the Bradford–Hutt plan were similar to the management costs on sites where conventional silviculture was applied on level ground, but higher for management and harvesting on steep ground. Timber quality has apparently always been moderate and pruning interventions have been necessary to correct that (Kerr et al., 2017).

The Bradford–Hutt plan is clearly an intriguing approach towards CCF. It can be applied not only as instant CCF on bare land, since the starting point is small-scaled clearfelling but also as gradual transformation. Even a combination of the Bradford–Hutt plan with a nurse crop is an option. A drawback is the rigid structure of the approach that leaves little options for unplanned events such as natural regeneration or disturbances. Several plots of a unit destroyed by a natural disturbance can seriously mess up the planned sequence of Figure 2.10. In a similar way, natural regeneration is not easy to accommodate. In fact, Kerr et al. (2017) even noted that management instructions included the recommendation to clear natural regeneration in all plots prior to replanting. This lack of flexibility is not unique to the Bradford–Hutt plan, but relates to all forms of the area-control method (Section 1.8). It would certainly be also possible to use the Bradford–Hutt plan to get started with CCF and later – when CCF management confidence has grown – to continue with free, individual-based size-control methods.

2.3.2 Anderson Group Selection System

Another area-control approach that is suitable both to instant CCF and gradual transformation is the Anderson group selection system. The system was devised by Prof. Mark Anderson of Edinburgh University and is a typical British interpretation of continuous cover forestry. In fact, the concept shares with the Bradford–Hutt Plan (see Section 2.3.1) and with the chequerboard method (see Figure 5.18) the reliance on cyclic planting, thinning, felling and normality. Note that *group selection system* here has a very specific meaning and is quite different from the more general use of the term elsewhere; hence, we refer to this system as the Anderson group selection system to avoid confusion.

Anderson was very interested in CCF management practised in continental Europe. He developed a particular interest in the single tree selection system (*Plenterwald* in German, see Section 5.4) including the species *P. abies*, *A. alba* and *F. sylvatica* as they had been perfected in Switzerland. Sadly Anderson's early attempts to mimic single tree selection systems he had visited in the Swiss Jura Mountains failed in Scotland (Anderson, 1960). One problem was heavy deer browsing on introduced *P. abies*, *A. alba* and *F. sylvatica*. As a result, the decision was made to replace the three European species by established North American species that were already present in the area (Wilson et al., 1999; Kerr et al., 2010). Discouraged by the browsing issue and other, mainly operational problems Anderson decided to take a step back from the sophisticated single-tree selection system and to combine the general concept of selection forests with something he knew to be safe, i.e. rotation forest management. Getting the British Forestry Commission on board, in 1952, Anderson established a trial area at Glentress Forest in Scotland (UK), an upland site, for what from now on he referred to as the group selection system. Glentress Forest is a state forest estate of 4400 ha located 3 km east of Peebles (50 km south of Edinburgh). The trial area comprised of 117 ha and was mostly planted between 1921 and 1946 at an altitude range from 240 to 560 m asl. Species were planted according to elevation and mainly included *P. menziesii* at lower elevations between 240 and 320 m asl; *L. kaempferi* and *Larix decidua* Mill. at mid-elevations (320–400 m asl); *P. sylvestris* and *Pinus nigra* var. *maritima* (Ait.) Melville on upper slopes (400–560 m asl). The upper sites were partially planted with *P. sitchensis* and *P. abies* (Wilson et al., 1999). The soils are generally well-drained and predominantly acid brown earths with surface water gleys in the main valleys. At higher elevations, there is a mixture of podzols and ironpans. Annual precipitation ranges between 1000 and 1500 mm year^{-1}. The Glentress trial is managed and monitored to the present day (Kerr et al., 2010); however, monitoring protocols have not always been consistent over the years.

The basic design of the Anderson group selection system is a rigid system where stands are subdivided into squares or other simple geometric shapes (Garfitt, 1995). At Glentress, square sizes between 0.1 and 0.2 ha were established with 0.1 ha favoured on exposed upper

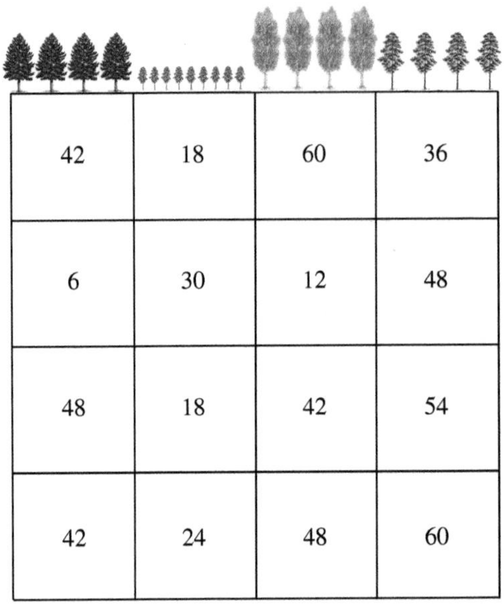

42	18	60	36
6	30	12	48
48	18	42	54
42	24	48	60

Figure 2.12 Layout of a hypothetical forest stand according to the Anderson group selection system operating on a six-year felling cycle. A rotation age of 60 years was assumed. The even-aged monospecies groups were assigned randomly for achieving good stand size and species diversity. Naturally, it is also possible to replicate some of the even-aged mono-species groups. Source: Adapted from Garfitt (1995).

slopes to provide more lateral shelter from adjacent, more mature groups and 0.2 ha on sheltered lower slopes (Wilson et al., 1999). Garfitt (1995) reported larger group sizes between 0.3 and 0.9 ha in other applications of this system (Figure 2.12).

The Anderson group selection system mimics the normal forest by planting monospecies tree cohorts of the same age in each square according to a six-year cutting cycle. At Glentress, the trial area was divided into six stands (between 15.0 and 21.5 ha) and approximately 12 groups in squares were accommodated in each stand (Wilson et al., 1999). Every six years, recently clearfelled areas were restocked by planting based on an initial spacing of 2×2 m and including natural regeneration where available. Anderson had proposed a rotation age of approximately 60 years. When fully established, each stand should include groups of varying ages ranging at six-year intervals from 1 to 60 years, and all trees in the oldest groups are clearfelled regardless of size and species (Kerr et al., 2010).

Unfortunately, costs and revenues have not been recorded accurately and fully mechanised tree removals proved difficult at Glentress, since no system of extraction racks and rides was previously put in place (Wilson et al., 1999). Species dynamics kept changing, *Picea* species tend to dominate the trial area now, but there appears to be an opportunity to replace planting by natural regeneration. Anecdotal evidence suggests good resilience to wind at Glentress (Hart, 1991). The Anderson group selection system is another perfect example of an area-control approach translated to CCF. Although for planning purposes, rotation periods are often discussed in CCF practice, strictly speaking they do not exist as far as CCF theory is concerned (see Section 1.3), and yet like the Bradford–Hutt plan (see Section 2.3.1), this approach heavily relies on this concept. The rigidity of the system initially was seen by Anderson and others as an advantage helping them to get started with CCF, and there was the hope that a greater variety of tree species could be maintained due to the group approach and the clearfelling/replanting. These anticipations and expectations were only partially fulfilled at the expense of little management flexibility, and the need to physically maintain the grid system in the forest. Smith et al. (1997) briefly mentioned a similar concept in their textbook.

2.4   The Mission of Transformation and Conversion

With both transformation and conversion an initial forest stand, usually a monospecies plantation or a woodland heavily colonised by a non-native invasive species, forms the starting point. Since transformation is a gradual process where the initial species composition is largely retained, the desired change in forest structure can be brought about by the application of thinnings, silvicultural systems and underplanting. Thinnings, particularly crown thinnings, can already be very effective in diversifying forest structure, see Section 5.2. They also promote individual-tree resilience, root development and seed production through crown release (Vítková and Ní Dhubháin, 2013). Transformation and conversion are among the most challenging tasks in silviculture, a mission associated with great uncertainty (Figure 2.13) and requiring both much experience and knowledge (Schütz, 2001b; Larsen and Nielsen, 2007). The contribution of silvicultural systems to transformation largely is the provision and the sheltering of natural and artificial regeneration. On first sight, this contribution does not seem large; however, forest structure can be much more effectively modified through guided regeneration processes that are often staggered in time and space than through thinnings alone. Silvicultural systems (cf. Section 5.3) therefore form an important part of transformation and to a large degree accelerate natural development (Malcolm et al., 2001; Vítková and Ní Dhubháin, 2013). However, they differ in terms of the structure of the new forest generation they produce. Uniform and strip shelterwood systems, for example, usually lead to single-storeyed woodlands, while group systems give rise to more complex forest structure. Artificial regeneration through underplanting naturally is also a possibility in transformation. Transformation therefore often mimics the understorey re-initiation phase as described by Oliver and Larson's (1996) model of stand development by encouraging new cohorts of trees (Vítková and Ní Dhubháin, 2013). Considering the principles of CCF as described in Section 1.5, it is usually the objective to apply silvicultural systems in such a way that diverse forest structure is achieved by allowing different cohorts of trees to establish at different times and locations in the forest stand (Schütz, 2001a). It is therefore not a problem, if regeneration does not cover the whole stand area at the first attempt. This is where novices to CCF sometimes become nervous and prematurely suspect failure; however, partial regeneration is only a first step and patience is required. Retarding and staggering the regeneration process in time usually lead to greater success but involve longer regeneration phases.

In plantations, transformation should start when they have reached mid-rotation age (Schütz, 2001a). The main-canopy trees should ideally also be able to produce seeds at that time or there should be a chance of colonisation from mature, adjacent stands. Regular thinnings in the past support the transformation process. If stands are dense and the last thinning has been years ago, a good way forward is to start with one or two moderate thinnings to prepare the residual trees before commencing transformation, see Figure 2.13. *Advance regeneration* is usually a reliable indicator that transformation is likely to be successful.

Figure 2.13 Transformations of plantations to CCF can go wrong, particularly when the first interventions are too heavy and the trees have not been previously prepared. When this is coupled with high precipitation and frequent, severe gales, disaster can strike. This happens and is not a shame. Who knows, perhaps this was just the beginning of a wonderful, diverse woodland to impress us 20 years later? Example from a *P. sitchensis* (Bong.) Carr. stand in the Wicklow Mountains (Ireland) with many broken stems and tree tops after initial transformation thinning. Source: © Arne Pommerening (Author).

> Advance regeneration is a general term for regeneration banks of tree seedlings (trees < 1.3 m) or saplings (trees ≥ 1.3 m with a stem diameter < 7 cm) that are present (sometimes in high densities) in the understorey before opening up the main canopy (Mason et al., 1999; Palik et al., 2021).

Any advance regeneration should be supported by careful canopy openings in the first transformation step. Mason et al. (1999) rightly emphasized that any canopy openings should aim at releasing advance regeneration rather than that they take place in anticipation of future regeneration. Thinnings are also used to implement the canopy openings of silvicultural systems and these cuttings are often heavier than previous interventions so that the necessary light levels at the forest floor are achieved. A reduction of 5–10 m² basal area is quite common in individual interventions for transformation. In the boreal but also in the temperate forest, tree removal also serves the purpose of releasing offspring from competition for nutrients (Petrițan et al., 2011; Högberg et al., 2021). When it comes to selecting trees for removal, it is important to consider multiple resources and their limitations and interactions (Binkley, 2021).

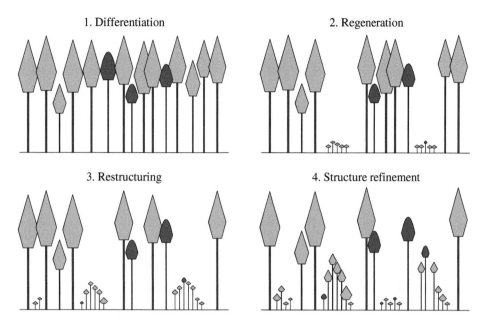

Figure 2.14 The four phases of transforming a mixed-species conifer RFM to a CCF forest stand. Source: Schütz (2001a). The figure was produced using the author's own R code based on the crown models published by Pretzsch (2009, p. 234ff.).

Schütz (2001b) defined the transformation process to selection systems by four distinctive stages, which can to a large degree be generalised for any transformation towards CCF (see Figure 2.14 for an illustration):

1. *Differentiation.* Using thinnings promote existing trees and groups of trees that are likely to contribute to structural differences and diversification.
2. *Regeneration.* Favour new regeneration groups that are randomly or irregularly dispersed.
3. *Restructuring.* Achieve a good horizontal and vertical dispersion of trees and tree groups contributing to diverse forest structure.
4. *Structure refinement.* Individualise tree groups vertically.

At the start of transformation, light thinnings are used to elaborate on existing trees and tree groups that markedly differ from the average tree in the stand. This is the first step towards diversification and at the same time, the resilience of individual trees is strengthened. At that stage, advance regeneration may already be in place, but only just survives in a state of stationary growth over a longer period. When these tree seedlings and saplings are gradually released in transformation, they can eventually emerge into the mid-storey and thus contribute to the differentiation of stand structure. Schütz (2001a) also referred to this process as 'verticalisation' of forest structure. The crown canopy should ideally only be opened where there is advance regeneration to avoid the invasion of shrubs and ground vegetation. Canopy openings should be gradual and not too large to begin with, as regeneration can occur very quickly and too abundantly so that the new forest generation becomes fairly uniform. This would somewhat defeat the objectives of CCF and can even cause operational problems later. Transformations can take between 20 and 80 years on average (Vítková and Ní Dhubháin, 2013), and the gradual nature of this process is illustrated in Figure 2.14. Hastily made and luxurious openings can lead to an overabundance of seedlings and consequently to costly respacing/thinning requirements. The successor forest stand then much resembles the plantation predecessor. The more homogeneous the initial forest stand, and the less advance regeneration exists, the longer the transformation usually takes.

Another good and rapid transformation method based on individual-tree size control (Section 1.8) is the structural thinning devised by Reininger (2001), see Sections 3.6 and 5.2.

Residual basal area can be interpreted as a surrogate measure of light and nutrient availability and the requirements vary with species and environmental conditions (Hale, 2003). In order to achieve satisfactory forest stand structure residual basal area should neither be too high nor too low. For example, in Britain and in Ireland, residual basal areas between 18 and 25 m^2 appear to work well for *Pinus* spp. and *Larix* spp., whilst 20–30 m^2 are more appropriate for *Picea* spp. and *P. menziesii* (Mason and Kerr, 2004; Vítková and Ní Dhubháin, 2013). Frequent and moderate interventions with basal-area reductions of no more than 20% is a good general guideline in absence of better, local information.

Completion of transformation generally depends on the degree of stand differentiation achieved and how this corresponds with the structure target (Schütz, 2001a). Various characteristics have been defined to be able to judge whether transformation is complete and what maintenance is necessary. These characteristics are discussed in Chapters 4 and 6. Mason and Kerr (2004) distinguished between *simple* and *complex* forest stand structure (see Figure 2.15). A simple structure would involve one or two canopy layers, whilst a complex structure would have three or more canopy layers on a permanent basis.

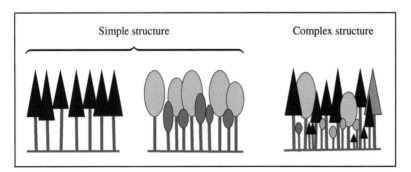

Figure 2.15 *Simple* and *complex* CCF *structure* according to Mason and Kerr (2004). Simple structure includes one or two permanent canopy layers (illustrated by the first two images from the left for a monospecies conifer and a mixed broadleaved stand with two species) and complex structure has three or more permanent canopy layers (illustrated by the image on the right involving five tree species).

Table 2.4 Attributes associated with simple and complex forest structures (Palik et al., 2021), see Figure 2.15.

Simple structure	Complex structure
1–2 canopy layers	3+ canopy layers
Single tree species in dominant canopy layer	Multiple species in dominant canopy layer
Single age cohort	Multiple age cohorts
Low tree size diversity	High tree size diversity
Low tree species diversity	High tree species diversity
Poor ground vegetation	Rich ground vegetation
Little deadwood	Abundant deadwood
Horizontal and vertical homogeneity	Horizontal and vertical heterogeneity

Source: Adapted from Palik et al. (2021).

Palik et al. (2021) pointed out that the terms simple and complex structure can also be associated with additional attributes or criteria (see Table 2.4). When applying silvicultural systems (see Section 5.3), forest structure initially often becomes more complex; however, this is only of a temporary nature, and the structure usually simplifies again once the regeneration process is complete. Such temporary changes are therefore not included in the description of CCF target structures.

The transformation process illustrated in Figure 2.14 aims at a complex structure, but simpler structure can be achieved using the same general framework. Conservative, cautious management behaviour aiming at a simple stand structure can be attractive when forest owners, managers or conservationists are CCF novices. The seemingly safer option may, however, later come with greater costs when too much tree regeneration established itself. Light-demanding tree species such as *Pinus* spp. and *Larix* spp. are easier to accommodate in simple stand structures (Mason and Kerr, 2004, cf. Appendix B).

Mason and Kerr (2004), Malcolm et al. (2001) and Vítková and Ní Dhubháin (2013) mentioned wind hazard and soil fertility as potential impediments to transformation. Whilst wind and snow hazard increase tree mortality, soil fertility can put competing ground vegetation at an advantage and hamper natural tree regeneration. Schütz (2001a) published a flowchart explaining the decision process forest managers should in his opinion go through when considering a transformation to CCF (Figure 2.16). *Forest structure, resilience* and *cover-building trees* play a key role in this rationale. Cover-building trees are a variant of frame trees (see Chapter 3) that support the transformation process as resilient trees by ensuring woodland climate and soil preservation. The role of cover-building trees, however, is also to retard the growth of some groups of trees whilst others can grow more freely and thus to ensure increasing irregularity of structure and spatial size inequality. Schütz (2001a) recommended 40–80 cover-building trees per hectare and most of them would exceed target diameters (cf. Section 3.4) in commercial scenarios.

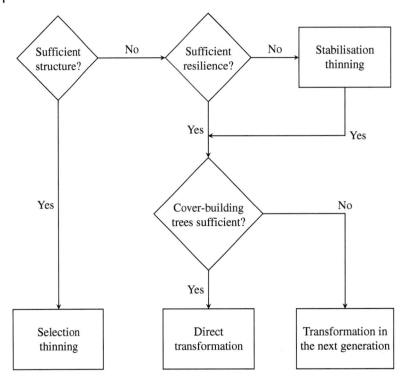

Figure 2.16   Flow chart illustrating the decision-making process when transforming RFM to CCF forest stands. Source: Adapted from Schütz (2001a).

Another important feature of Schütz' concept is to postpone transformation to the next forest generation (i.e. until after clearfelling and replanting), if it does not appear feasible to carry on with the current forest stand for lack of individual-tree resilience. This resilience particularly relates to the cover-building trees. Schütz (2001a) reckoned that the crown ratio (c/h, Eq. (2.3)) is a very useful characteristic to judge the resilience and longevity of cover-building trees. Large c/h values are correlated with high probabilities for both properties. The decision whether to proceed with the current or the next forest generation mainly depends on forest structure. If irregularity of forest structure in terms of horizontal and vertical tree size differentiation (or size inequality) is sufficiently developed, a direct transformation can be carried out by using *selection thinnings*. Another important consideration is whether a continued regeneration process and as a consequence a continued supply of seedlings is ensured (Schütz, 2001b). This question, however, is often related to forest structure. Selection thinnings should not be confused with *selective thinnings* (see Section 5.2). Whilst the former denote heavy crown thinnings targeting predominant trees, i.e. the typical thinning in selection forests, the latter refer to all thinnings where individual trees are taken out selectively rather than systematically in patches or rows of trees. There is also the option of improving structure and resilience by specialised *stabilisation thinnings*. These thinnings are based on crown thinnings and would favour not only individual trees that are likely to withstand wind and snow but also support trees that contribute to structural heterogeneity. Even in seemingly homogeneous plantations, there is always a number of trees that in terms of self-differentiation markedly differs from the average tree size. These often need little help to achieve good development. Even slightly suppressed trees in the mid-storey can make important structural contributions to the envisaged future stand structure. The objective of stabilisation thinnings should be to achieve irregularly dispersed groups of existing, formerly suppressed and new regeneration trees (see Figure 2.14). At this stage, large, uniform canopy openings would counteract any attempt to increase structural diversity. Once differentiation and size diversity have increased, the question comes up whether there is a sufficient number of cover-building trees and whether they will last long enough to support the transformation. If this question can be answered with 'yes', direct transformation to CCF can proceed through selection thinnings. The purpose of selection thinnings is to improve self-organisation processes in stands with sufficient structural differentiation, see the example of Artist's Wood in Section 2.2.1. Otherwise, Schütz (2001a) recommended transforming the next forest generation to CCF after clearfelling and replanting. The cover-building trees are crucial to the entire transformation process, as they provide seeds, shelter and ensure the desired differentiation of tree sizes and patchiness of regeneration. These cover-building trees need to be maintained until the first trees of the next forest generation can take over the role of the cover-building trees. The success of regeneration is crucial to transformation, but initiating regeneration should only be considered after checking stand stability and resilience as well as the state of the potential cover-building trees. Regeneration should then be encouraged by small openings staggered in time to promote diversity of structure.

The principles of forest conversion are not too different from those of transformation. Conversion is often applied not only in the context of conservation or forest restoration but also in situations where the current species composition is utterly unsuitable (see Section 2.4.3). The intended species change is commonly required quickly so that the speed of change is an important difference between transformation and conversion. The (partial) planting of native trees may act as an important accelerator. In this context, it is wise to take advantage of the natural benefits provided by overlapping current and future forest generations. This can be achieved by underplanting, see Section 2.4.1. Underplanting has proven to be a successful component of forest conversion. One mistake often made in conservation and forest restoration was that mixed woodlands of native broadleaves and non-native conifers were completely stripped of the latter. The few remaining broadleaves were unable to uphold woodland conditions, many slender *Betula* spp. trees heavily bent over, so that their tips nearly touched the ground, as they very suddenly lost their support, and many of them died only a short time afterwards. The general intention of such restoration is certainly laudable, since it follows the concept of passive restoration (Newton, 2007), i.e. the forest site is allowed to recover naturally through the process of seed dispersal and succession. However, a slow gradual approach, where in each thinning much more conifers are removed than broadleaves, is sometimes a better strategy, since all trees, even the non-native ones, help maintain woodland conditions and provide shelter (Forestry Commission, 2008). In some cases, it may be necessary to remove abundant regeneration of non-native trees. This may, for example, be partly achieved by extracting felled trees through clusters of such regeneration and thus reducing their numbers.

2.4.1 Underplanting

In silviculture and forest management, underplanting is a general term and can be defined mainly in two different ways:

> **1.** Underplanting can imply planting an understorey of trees supporting the overstorey (sometimes referred to as *stand improvement underplanting;* Stokes et al., 2020), e.g. underplanting a 40-year-old stand of *Quercus* spp. with *F. sylvatica* or *C. betulus* acting as nurse or trainer to shade the stem axes of the oak trees and thus to suppress epicormic growth which could devalue timber quality.
> **2.** The underplanting of an existing tree canopy with another species with the objective to provide artificial regeneration, for example, for achieving a mixed-species stand (sometimes referred to as *regeneration or enrichment underplanting;* Stokes et al., 2020), e.g. the underplanting of *P. abies* stands with *F. sylvatica*, a common procedure of forest restoration in Central Europe (Bartsch et al., 2020).

In this chapter on CCF starting points, the focus is mainly on definition 2, i.e. on underplanting thus establishing artificial regeneration. Underplanting techniques can be used in conjunction with instant CCF on bare ground (Section 2.3) and as part of transformation/conversion (Section 2.4). Underplanting relates to planting seedlings or saplings under a light canopy of mature trees but also to planting small trees in canopy gaps (Bartsch et al., 2020). Underplanting is an important tool of transformation and conversion, and there can be many reasons for underplanting: Natural regeneration can fail or not be sufficient, the parent tree species or provenances may not be suitable for a given site, or there may be a desire to enrich a site with other species in order to achieve a mixed-species forest stand. Underplanting can also help (re-)introduce light-demanding tree species in a stand of otherwise shade-casting species. In a restoration context, it is often the objective to achieve a complete species change and underplanting is the first step towards that. This incomplete list of reasons may demonstrate that underplanting is an important tool not only in forest conservation/restoration but also in forest management attempting to mitigate climate change (Figure 2.17).

Theoretically silvicultural systems do not necessarily need to deliver natural regeneration, although this is certainly the main purpose of most of them, cf. Section 5.3. It is also possible to combine, for example, shelterwood systems with artificial regeneration through underplanting, hence, the aforementioned term 'artificial shelterwood system' for such

(a)   (b)

Figure 2.17 Forest restoration through conversion based on the concept of an artificial shelterwood system: (a) Underplanting of *F. sylvatica* L. under a dense canopy of *P. abies* (L.) H. KARST. (Neuhaus Forest District, Germany) and (b) *F. sylvatica* under a light canopy of mature *P. sylvestris* L. trees (Sellhorn Forest District, Germany). Source: Courtesy of Tim Summers.

combinations (Figure 2.18). It is also an option to combine sparse natural regeneration and partial underplanting. In partial underplanting, one obvious strategy could be to plant in canopy gaps with spacings of 1.5 × 1.5 m or 2.0 × 2.0 m. Normal nursery stock is usually sufficient for underplanting, and the use of larger plants (saplings) has so far not shown any advantage (Bartsch et al., 2020). Underplanting is also a good way to respond to sudden disturbances that have started to disrupt the crown canopy whilst not totally destroying it yet. Rapid planting can then prevent the invasion by grass species and unsuitable tree species (Bartsch et al., 2020). Similar to the situation in uniform shelterwood systems (see Section 5.3.1) it is important to prepare the residual canopy trees and to foster their resilience through appropriate thinnings and preparatory fellings.

Monitoring the well-being of understorey seedlings and saplings

Managing light conditions is an important concern of forest management. The concept of CCF heavily relies on natural but also on artificial forest renewal. New cohorts of tree seedlings and saplings can only survive and eventually emerge into the main canopy, if the light conditions are adequate. For the assessment of sufficient light conditions, light measurements are possible; however, there are also good correlations between more readily accessible population characteristics such as stand basal area (Hale, 2003) and tree structure (Metslaid et al., 2007). An informative characteristic of the latter group is known as the *leader-to-lateral branch* or *apical dominance ratio* (ADR). There is positive correlation between this ratio and the amount of light received (Duchesneau et al., 2001; Grassi and Giannini, 2005; Metslaid et al., 2007). Leader-to-lateral branch or ADR can generally be defined as

$$ADR = \frac{\text{Length of terminal leader}}{\text{Length of lateral shoots}}. \qquad (2.4)$$

(Continued)

(Continued)

Apical dominance in this context relates to the ability of the terminal shoot to have a growth advantage over lateral branches generally attributed to the plant hormone *auxin*. The length of lateral shoots in the denominator is measured at the last node before the terminal leader and can either be calculated as the maximum (Duchesneau et al., 2001) or the mean length (Grassi and Giannini, 2005) of all healthy, undamaged shoots at that point. Trees with ADR < 1, i.e. the terminal leader is shorter than the lateral shoots, are light stressed. Trees with ADR > 1 have sufficient light for a successful development resulting in normal height growth and ADR $= 1$ marks the threshold or compensation point. To obtain a better idea of trends it is recommended to measure ADR several years in sequence (Grassi and Giannini, 2005; Pommerening and Grabarnik, 2019).

Worldwide there is much experience with underplanting involving a wide range of species, sites and climatic zones (e.g. Emmingham et al., 1989; Smidt and Puettmann, 1998; Kenk and Guehne, 2001).

Figure 2.18 Diversifying *Picea* spp. stands and enhancing resilience through underplanting: *A. alba* Mill. planted under a dense canopy of *P. sitchensis* (Bong.) Carr. as part of the author's Tyfiant Coed project at Clocaenog Forest (Wales, UK). The underplanting was carried out in spring 2007 and the photograph shows the situation in February 2011. Note the abundance of *P. sitchensis* seedlings on the forest floor that regenerated naturally. See also Stokes et al. (2020) who analysed this underplanting experiment. Source: © Arne Pommerening (Author).

Most authors prefer the establishment of monospecies groups in patches of, for example, 30 × 30 m alternating each other or groups with only two different species rather than an intimate mixture of species over the whole afforestation area (Rittershofer, 1999; Burschel and Huss, 1997), see Section 4.3. This facilitates the organisation of plant cover according to small-scale differences in site conditions and also makes the management of the young growth considerably easier. Intimate mixture of species will eventually result naturally through the reduction of stem numbers in each group as a consequence of thinnings and natural mortality. The initial spacing of underplanted trees can be quite wide, e.g. 2–3 × 2–3 m, since the overstorey provides sufficient shelter and shade to achieve good tree form even for economic objectives. It is not necessary to plant up the whole stand area. Therefore, compared to conventional replantings on clearfell sites savings can still be achieved, even if natural regeneration is only sparse. To create more structural differentiation or to meet the light requirements of different species, it is also possible to stagger the underplanting of patches in time. Careful spatial planning is required to ensure that thinnings in the overstorey do not damage any of the underplanted trees.

2.4.1.1 Nest Planting

A quite sophisticated but efficient design, '*(oak)-nest-planting*' or *(oak) cluster planting* (Saha et al., 2017), was originally devised in Russia where it is known as the '*Ogijewski method*' or '*regeneration in small places*' (Podkopaev, 1961; Tarasenko, 1962). The method has been successfully applied to the establishment of mixed oak in Russia, Ukraine, Belorussia and Poland on fertile sites with competing ground vegetation (Szymański, 1986, 1994; Saha et al., 2017). The approach owes its name to the fact that initially it was predominantly used to establish oak (*Quercus* spp.) forest stands; however, it is not limited to this species group. The invention of the silvicultural technique was inspired by the observation of natural seed dispersal and germination of *Quercus* spp. and of other species with heavy seeds in native and semi-natural woodlands, where early stages of stand development are often characterised by seedling and sapling clusters (Anderson, 1951; Saha et al., 2017; Wilhelm and Rieger, 2018). As in natural forests, it was originally assumed that the cluster pattern would later largely disappear as trees grow and self-thinning commences, but remaining clusters are also welcome, as they contribute to a heterogeneity of structure desired in CCF. The technique has repeatedly been used for underplanting. Nest or cluster planting was also developed and promoted by Mark Anderson in Britain (Anderson, 1930, 1951) and at some time was even referred to as 'Anderson's spaced group planting'.

Szymański (1994) described the successful establishment of oak-nests under a *P. sylvestris* canopy. This author also applied the method as an enrichment underplanting technique for beating up of natural regeneration. Szymański's lectures and publications inspired researchers in France, Germany, Austria and Switzerland to try the nest planting method and to set up experiments. Most of these new plantings were carried out on bare land caused by clearfelling and catastrophic windthrow or on abandoned agricultural land (Saha et al., 2017). The basic design presented in Figure 2.19 suggests 200 nests per hectare with 21 target plants, e.g. *Quercus* spp. plants, and 16 shade casting *guard* or *trainer* plants, e.g. *C. betulus* plants, in each nest. Guard trees are commonly planted on the perimeter of the nests, and their role is to provide lateral shade and shelter for the target trees and to suppress ground vegetation. Lateral shade can significantly improve timber quality and

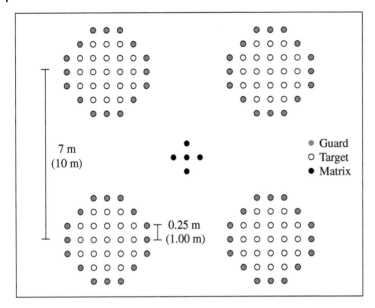

Figure 2.19 The basic principle of the nest or cluster planting (Ogijewski method). Modified from Pommerening and Murphy (2004). Both the original measures proposed by Szymański (1994) and the measures revised by Gockel (1995) are given, the latter in brackets. *Guard*, *target* and *matrix* denote tree species fulfilling different functions as explained in the text.

reduces the growth of species that require slow juvenile development such as *A. alba*. Guard trees eventually becoming competitors of target trees may need to be removed or coppiced/pollarded (Saha et al., 2012). The basic design in Figure 2.19 results in 4200 target and 3200 guard plants per hectare. *T. cordata* and *F. sylvatica* have frequently been used as alternative guard tree species. Saha et al. (2012) reported the use of *P. abies* as guard trees and from North Frisia (Germany) the use of *P. sitchensis* as guard trees for *A. alba* is known. An approximately circular nest shape is achieved by omitting the corner plants in what otherwise would result in a quadrate nest design (Bartsch et al., 2020).

The anticipated advantages of nest planting include (i) a reduction in site preparation and establishment costs compared to conventional row planting and (ii) the promotion of natural regeneration of other species and hence greater diversity (Saha et al., 2012). Three years after planting competing ground vegetation is largely naturally suppressed by the close initial spacing (Bartsch et al., 2020). In each nest, a group with a positive microclimate was established preventing frost and sun damage. Browsing by deer was hoped to be concentrated only on the boundary trees of the nests so that the target trees in the centre are scarcely affected, but experiments could not confirm this (Pommerening and Murphy, 2004; Saha et al., 2012). Self-differentiation within the nests as a consequence of the close spacing is supposed to enhance timber quality. Most silvicultural attention is therefore paid to the central target tree of each nest which is anticipated to become a frame tree. Irrespective whether this strategy really works out or not, the expectation is that in the long run, each nest produces one frame tree (see Chapter 3) which in commercial scenarios eventually reaches target diameter. The space or *matrix* between the nests can be planted with short-rotation tree species such as *Prunus avium*, *Salix* spp., *S. aucuparia*, *P. tremula* L. and *Betula* spp., but it is also possible to leave it to natural seeding from adjacent forest stands (Szymański, 1994; Burschel and Huss, 1997). In Figure 2.19, these trees are referred to as *matrix trees*. However, it is not recommended that this space be left without tree cover for too long (Guericke, 1996; Strobel, 1986). It has been suggested that the first pre-commercial thinning be carried out at a stand age of 20 years. The number of trees per nest is then reduced to the best five. The next intervention is then due at a stand age of 30 years (Szymański, 1994), but these numbers much depend on species and environmental conditions.

High mortality of planted seedlings in German nest planting trials of the 1980s encouraged Gockel (1995) to revise the nest planting design. Based on the results from general spacing trials he introduced wider initial spacings between the seedlings and also increased the distances between nests (see Figure 2.19, numbers in brackets) resulting in 100 nests per hectare. Gockel (1995) also recommended using taller plants (saplings) of 80–150 cm in height to foster survival given the high mortality in early trials. He also proposed a truly circular planting design where 19 saplings were planted on concentric circular rings, see Figure 2.20. Naturally, it is possible to include additional target-species rings in each nest or to make other modifications. This revised nest planting design is sometimes referred to as *group planting* in the literature (Saha et al., 2012, 2017); however, we retained the original name to avoid confusion with other group designs.

Generally the wider initial spacing of nest plantings proposed by Gockel (1995) across many different sites proved to be superior to the original design in terms of survival, growth, resilience and timber quality of the target trees (Saha et al., 2017). This design leads to results comparable with traditional row planting, whilst saving costs. Natural pruning or self-pruning, i.e. the natural process of losing branches along the lower part of the stem axis due to a lack of light, even turned out to be more effective in nest plantings than in comparable row plantings. Fencing of the planted area additionally increases the survival probability of the saplings (Saha et al., 2012). Leaving the matrix, i.e. the space between nests, to natural regeneration usually leads to high tree species diversity. Diversity can be further increased by varying the number and size of nests as well as the distance between them (Saha et al., 2017). This suggests that nest planting is particularly ideal for conversion and forest restoration. Saha et al. (2017) reported nest plantings with *Quercus ilex* L. in

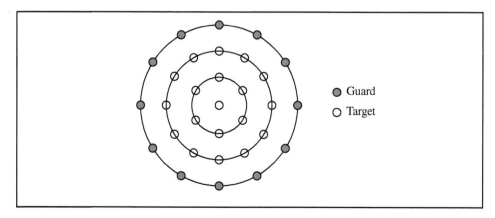

Figure 2.20 The circular nest or cluster planting design suggested by Gockel (1995). *Guard* and *target* denote trees fulfilling different functions as explained in the text.

Spain, *A. alba* in Italy, *Pilgerodendron uviferum* (D. Don) Florin in Chile and *A. pseudopla-tanus* L. in Switzerland. When using other species and sites, initial spacing and distances between nests may need to be adapted.

As with natural regeneration, also in underplanting, it is crucially important to have felling directions of mature overstorey trees and extraction racks organised before regeneration materialises or underplanting commences. This ensures that any felling of large trees or extraction of stems does not damage natural or artificial regeneration (cf. Section 5.3.6).

2.4.2 Graduated-Density Thinning (GDT)

Graduated-density thinning (GDT), a regional method from Wales (UK), has been proposed for transforming plantations to CCF. For readers trained in Central European silviculture, this may seem odd on first sight, since transformation usually involves a concept larger than just a series of thinnings, e.g. a silvicultural system that includes a number of thinnings or some greater plan like the concept proposed by Schütz (2001a). However, labels can be deceiving. Graduated density thinning is not a thinning method in the classic sense of the term and can be considered a transformation method for young conifer plantations with a top height between 12 and 15 m (i.e. approximately 20 years after planting). Similar to the Bradford–Hutt plan (Section 2.3.1) and the Anderson group selection system (Section 2.3.2), GDT is largely based on the area-control method, whilst most methods discussed in Section 2.4 are related to the size-control method which dominates CCF in Central Europe (see Section 1.8).

Graduated-density thinning was originally proposed by Latvian forester Talis Kalnars and applied by him in a number of woodland estates in Wales that were undergoing transformation in the 1980s (Vítková and Ní Dhubháin, 2013). The method was taken up widely in Wales and recently also in Ireland. GDT was involved in silvicultural experiments set up in the Irish counties Laois and Wicklow in approximately 20-year-old plantations of

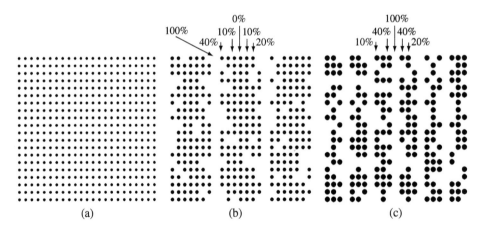

Figure 2.21 Graduated density thinning (GDT) simulated for a hypothetical plantation. Source: Adapted from Vítková and Ní Dhubháin (2013). (a) Initial plantation. (b) First thinning. (c) Second thinning. Trees are indicated by circular objects and their changing sizes suggest time and growth. For better clarity, natural mortality or colonisation was not considered here.

P. sitchensis. The starting point of this method is a conifer plantation with regular plant locations arranged in rows where the first thinning is due. During the first thinning, extraction racks are cut in every eighth row (R1 in Figure 2.23) and, for example, 40% of trees are selectively or randomly removed from the rows immediately adjacent to an extraction rack. In the second row from the rack, 20% of the trees are selectively or randomly removed followed by 10% in the third row. The fourth row is situated halfway between two extraction racks and is left unthinned (Figure 2.21). In the opinion of practitioners in Wales, the retention of the entire fourth row ensures collective stand resilience. The visual information in Figure 2.21 can also be summarised as shown in Figure 2.23 (Morgan, 2012, pers. comm.). The selective tree removal is a negative selection, i.e. trees with seemingly bad timber quality are removed (cf. Section 5.2), a legacy of RFM. This strategy leads to global thinnings from below.

In a subsequent thinning, the unthinned rows of trees are completely removed to make new extraction racks (R2 in Figure 2.23), whilst the previous racks are abandoned, with further selective thinnings on either side of the new racks using the same or modified graduated thinning intensities (Vítková and Ní Dhubháin, 2013). Naturally, any full row removal does not necessarily need to be used as extraction racks. The numbers in Figure 2.21 are based on empirical experimenting by practitioners and can, of course, be modified. Figure 2.22 shows an application of GDT to *P. sitchensis* in Wales.

Figure 2.22 An example of a first graduated density thinning (GDT) in an approximately 35-year-old *P. sitchensis* (Bong.) Carr. stand in the upland Bryn Arau Duon forest estate in South Wales (UK). Source: © Arne Pommerening (Author).

The procedure becomes clearer when examining R code designed to simulate GDT:

```
1    > graduatedThinning <- function(r, ranInd, colInd, myX,
2    + myY) {
3    + add <- rep(x = c(1, -1), times = 3)
4    + for (i in 1 : length(ranInd)) {
5    + for (j in 1 : length(colInd)) {
6    + rowInd <- sample(x = r, size = ranInd[i], replace = F)
7    + myY[rowInd, colInd[j]] <- rep(NA, ranInd[i])
8    + myX[rowInd, colInd[j]] <- rep(NA, ranInd[i])
9    + }
10   + colInd <- colInd + add
11   + }
12   + return(list(myX, myY))
13   + }
```

The listing above gives the key function of the simulation, graduatedThinning().
As the name suggests, this function handles graduated thinnings that are carried out
symmetrically on either side of the most recently cut extraction racks. Variable add in
lines 3 and 10 in particular ensure the symmetric cutting. The trees to cut are selected
randomly, and this is taken care of by the sample() function in line 6. Following this, the
simulated forest stand is initialised including two matrices for storing tree coordinates in
the listing below (line 3).

```
1    > rows <- 20
2    > columns <- 23
3    > myPlotX <- myPlotY <- matrix(NA, nrow = rows,
4    + ncol = columns)
5    > spacing <- 1
6    > for (i in 1 : columns)
7    > myPlotY[,i] <- seq(from = 1, to = rows, by = spacing)
8    > for (i in 1 : rows)
9    + myPlotX[i,] <- seq(from = 1, to = columns,
10   + by = spacing)
```

Following the initialisation, the coordinates are simulated using the seq() function in
lines 6–10. For understanding the code, it is helpful to think of the forest stand as a matrix
with rows and columns. Now, we are in a position to simulate the first graduated thinning:

```
1    > rowRemovals <- c(8, 16)
2    > myPlotY[,rowRemovals] <- rep(x = NA, times = rows)
3    > myPlotX[,rowRemovals] <- rep(x = NA, times = rows)
4    > graduatedNumbers <- c(40, 20, 10)
5    > rnd <- rows * graduatedNumbers / 100
6    > columnIndex <- sort(c(1, cbind(rowRemovals - 1,
7    + rowRemovals + 1), columns))
8    > nm <- graduatedThinning(r = rows, ranInd = rnd,
9    + colInd = columnIndex, myX = myPlotX, myY = myPlotY)
10   > myPlotX <- nm[[1]]
11   > myPlotY <- nm[[2]]
12   > rm(nm)
13   > par(mar = c(0.5, 0.5, 0.5, 0.5))
14   > plot(x = myPlotX, y = myPlotY, axes = F, xlab = "",
15   + ylab = "", pch = 16, cex = 2)
```

In lines 1–3 of the above listing, extraction racks are cut in columns 8 and 16. Afterwards the percentages for the graduated density thinning are defined in line 4 and then converted to absolute numbers in line 5. Following this, the column indices are defined and along with other settings inputted in the graduatedThinning() function in line 8. Now the simulation results are retrieved in lines 10–11 and visualised in lines 13–15. The second intervention is coded in a very similar way and again uses function graduatedThinning(). However, the initialisation differs a bit from the first intervention:

```
1  > rowRemovals <- which(!is.na(colSums(myPlotX)))
2  > myPlotY[,rowRemovals] <- rep(NA, rows)
3  > myPlotX[,rowRemovals] <- rep(NA, rows)
4  > columnIndex <- sort(cbind(rowRemovals - 1,
5  + rowRemovals + 1))
```

In line 1, the columns are identified that are removed entirely from the plantation, i.e. these form the new extraction racks (R2 in Figure 2.23). The selected columns are those where up to this time no tree has been removed. The removals are carried out in lines 2 and 3 and then the column indices are re-arranged in a vector (lines 4–5). Afterwards, lines 8–15 from the previous listing can be applied. The code is useful for simulation studies based on forest growth models where it can be integrated but also for visualisation as shown in Figure 2.21. The R code can also be used for preparing the implementation of thinning trials in the field.

Permanent extraction racks are chosen as part of the third intervention. The number of interventions that follow after the third thinning varies according to site and stand conditions. Selective, local crown thinnings and finally target-diameter harvesting (see Section 3.5) follow. The proponents of this method claim that it fosters stand resilience and structural diversity whilst providing higher earlier financial returns than conventional thinning (Vítková and Ní Dhubháin, 2013). Row removals and the selection of bad quality trees for fellings are typical elements of plantation forestry. The selective element of the method introduces a touch of size control; however, the intriguing method is predominantly based on the tradition of area control. The size control 'antithesis' of GDT is the structural thinning devised by Reininger (2001), see Section 3.6.

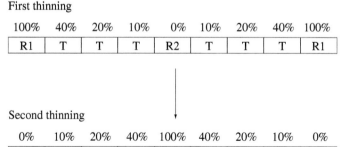

First thinning

100%	40%	20%	10%	0%	10%	20%	40%	100%
R1	T	T	T	R2	T	T	T	R1

Second thinning

0%	10%	20%	40%	100%	40%	20%	10%	0%
R1	T	T	T	R2	T	T	T	R1
(0%	30%	20%	0%	100%	0%	20%	30%	0%)

Figure 2.23 Alternative representation of Figure 2.21. At the bottom in brackets, another option for the second thinning is provided which is quite common in Wales, where the selective parts are carried out as a global low or crown thinning. T: selective thinning, R: extraction rack. (The numbers 1 and 2 refer to the time of first and second thinning).

2.4.3 Variable-Density Thinning (VDT)

Variable-density thinning (VDT) has been proposed and developed in North America since the 1990s. The concept was originally devised in the context of conservation, more specifically in forest restoration, but can also include the moderate use of felled trees for commercial purposes. In the past, fire suppression and plantation forest management had given rise to large areas of uniform forest land in the United States that was subjected to similar treatment (Willis et al., 2018). This increase in uniformity decreased wildlife habitats and biodiversity. For example, in the Pacific Northwest of the United States, there was a shortage of late-successional forest structures and establishing them in the area was a restoration priority. Large trees with a stem diameter in excess of 80 cm were rare, diversity of tree sizes was low and coarse woody debris from fallen trees and snags were scarce (Willis et al., 2018). In addition, there was often little variation in crown structure and hardly any development of mid-storey. Therefore, the desire arose to move structurally simple forests towards greater structural heterogeneity (Palik et al., 2021). As often in conversion, there was a requirement to convert simple forest structures, often plantations, fast. VDT was designed with an explicit spatial distribution of interventions and resulting structures in mind (see Figure 2.24). As such the method largely follows the area-control method, see Section 1.8. Kimmins (2004) referred to VDT as *structure management silviculture* that is based on conceptual models of natural disturbance and

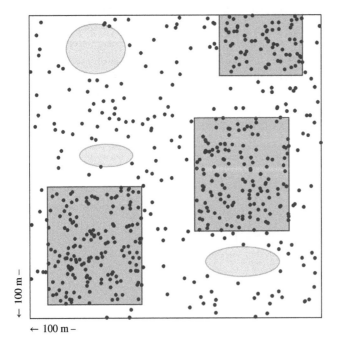

100 m →

← 100 m →

Figure 2.24 Tree locations (brown) of a simulated pattern after variable-density thinning inspired by Figure 2 in Willis et al. (2018). A global low thinning was carried out in areas shaded in white; skips are shaded in green and gaps in yellow. The shapes of skips and gaps can vary as can their sizes and depend on the objectives of forest restoration in a given region and on local stand conditions.

Table 2.5 Variable-density thinning patch types.

Patch type	Functions	Size/Density
No thinning (skips)	– Self-thinning, snags – Self-differentiation – Dark, moist habitats – Nature reserve	0.05 – 0.30+ ha
Moderate thinning	– Stem & crown growth – Reduce competition – Resilience – Understorey	25 – 45% relative density, any thinning type
Heavy thinning	– Big trees, large crowns – Dispersed retention – Epicormic growth – Understorey-mid-storey – Regeneration	Release of individual trees or tree groups, 0.2 – 1.0+ ha, 10 – 12% relative density, any thinning type
Canopy gaps	– Remove overstorey – New species introduction – Regeneration – Understorey-mid-storey – Root rot pockets	0.04 – 0.40+ ha, larger for light demanders and stands with large top heights, 0-12 standards per ha in gaps

Source: Adapted from Palik et al. (2021).

stand development. VDT distributes trees and microclimatic conditions unevenly within a forest stand or larger entities of forest land. Targets for spatial patterns of the different patches as listed in Table 2.5 can be provided. Further details can be obtained from studies of natural old-growth or mature forests (Willis et al., 2018). Thinned patches are forest areas subjected to thinnings (cf. Section 5.2) with different intensities. Such areas can, for example, border on canopy gaps, which are next to a retention area with no interventions (see Figure 2.24). There are three main patch types in VDT, i.e. *skips*, *gaps* and *thinned areas*. Skips are areas skipped over in the thinning. Gaps are small openings in the main canopy to encourage the development of understorey vegetation where mature boundary trees gain additional growing space (Brodie and Harrington, 2020). All these different patches come with different tree densities; hence, the term 'variable-density' thinning, cf. Figure 2.24. Each patch type and associated density fulfils different roles in terms of ecology and conservation (see Table 2.5). Also, the horizontal and vertical heterogeneity in itself is anticipated to increase diversity and habitats. With time, forest stands treated in this way would become structurally more complex consistent with characteristics of late-successional forests (Willis et al., 2018). In contrast to GDT, variable-density thinnings were initially intended as one-off treatments and are usually not repeated; however, this is now being re-considered (Brodie and Harrington, 2020; Palik et al., 2021).

Additional gaps might be created or existing gaps expanded to release established regeneration similar to the group (shelterwood) system, see Section 5.3.2. As indicated in Table 2.5, thinnings are intended to accelerate tree growth processes of stem and crown according to different grades such as moderate and heavy. Thinnings in this context are often carried out as global low thinnings, but global crown thinnings are also possible, see Section 5.2.

Conversely, unthinned patches are retained (Figure 2.24) resulting in slower individual tree growth, crown contraction and higher mortality, thus encouraging the formation of deadwood (Willis et al., 2018). Standing and lying deadwood can be preserved here whilst protecting the safety of forest operators. Heavy thinning can be thought of as a heavy release thinning or retention/seed-tree system, where the number of residual trees is extremely low and spacing very wide. Gaps can be used to release advance regeneration, to encourage new regeneration or simply to leave these open areas to natural devices. Another option is to plant in gaps, e.g. a species that is native to the area but has recently much decreased in abundance (Palik et al., 2021).

There are many variants of VDT. A simplified version is *thinning with skips and gaps* (Willis et al., 2021). Using this method, only skips and gaps are allocated to certain locations in a forest stand whilst the area outside the skips and gaps, the *matrix*, is subjected to a light thinning from below (Figure 2.24). O'Hara et al. (2012) described four basic types of VDT that occurred in North American forestry practice: Randomised grid, *Dx* rule, spacing thinning and localised release. *Randomised grid* is a design where the whole forest stand is subdivided into quasi-rectangular thinning cells of more or less equal size. The number of residual trees including the option of no residual trees would be determined by throwing a dice or drawing integer random numbers, e.g. between 1 and 6, where 6 would denote no residual trees and numbers other than 6 would give the residual numbers of trees. According to the *Dx rule*, residual trees are freely identified throughout the forest stand and all other trees larger than a certain threshold diameter are cut within a radius (in m) of the stem diameter (denoted as *D* in O'Hara et al. (2012) and measured in cm) of the residual tree times a fixed multiplier *x* (usually an integer number). A maximum radius is defined to avoid unreasonably large radii. Retention trees are selected regardless of the distances between them. In *spacing thinnings*, a suitable grid for retention trees is defined, and the 'best' trees as close to the grid points as possible are selected. A certain radius of allowance around each grid point is specified which can vary between stands or even patches to achieve variability. 'Best' trees are defined in terms of size and species. All trees other than the retention trees are cut. A variant of spacing thinnings is *localised release*. As part of this VDT variant, residual trees are selected as for the spacing thinning. A fixed-radius circle is constructed around each residual tree and all other trees within the circular areas are removed in the spirit of a halo thinning, see Section 5.2. Outside the circular areas no thinning is applied where two or more circular areas are very close; otherwise, global thinnings from below are carried out. There is also the option to leave a few circular areas unthinned. Another VDT variant is the *individuals*, *clumps* and *openings* (ICO) method (Churchill et al., 2013). ICO prescriptions are even more strictly guided by targets for these three patch types derived from reference stands whilst also considering projected future climate. The spatial reference patterns are quantified and translated into efficient marking guidelines. Usually, individual trees and clumps are identified in the field, but computer algorithms, in conjunction with remote sensing, have recently been used efficiently to recognise tree clumps and to delineate treatments. Large openings are marked in the field whilst small openings often are the result of alternating individual trees and clumps (Palik et al., 2021).

In general, the thinned area comprises the majority of VDT treatment, while skips and gaps balance the thinnings (Palik et al., 2021). Determination of the amount and size of different patches should be based on consideration of the desired functions (cf. Table 2.5).

Since VDT only started in the 1990s, all results to date are still preliminary. Willis et al. (2018) reported that VDT accelerated the recruitment of shade-tolerant species into the mid-storey, which was a target in their study. The main stand canopy also became deeper and more heterogeneous, However, structural stand diversity targets were not met so far. There was particularly a deficiency in the number of large trees. Overall, the authors concluded that VDT had largely been successful in accelerating stand development towards providing late-successional habitat conditions within a 14-year period. O'Hara et al. (2012) noted that in their experiments more complicated VDT variants resulted in stands with greater heterogeneity at the expense of implementation ease. Those VDT variants that were easiest to implement were least effective at creating stand heterogeneity. As expected, gaps proved to be an excellent tool for encouraging new regeneration and promoting the growth of advance regeneration, thus increasing vertical tree size diversity. However, the authors saw a need for adjusting gap size in order to accommodate light-demanding species (Willis et al., 2021, cf. Appendix B). Essentially, VDT is a combination of thinnings and silvicultural systems designed to initiate a rapid conversion of plantations to some form of CCF that promises conservation advantages.

2.5 Keeping it Going: The Maintenance of CCF

Once a CCF woodland has been established, it needs to be maintained. Maintenance of CCF does not involve less work and thought process than the establishment of CCF on bare ground or transformation/conversion. Maintenance of CCF is related to sustainability and to the question of how the specific forest structure required for processes of biological automation and rationalisation can be perpetuated (see Section 2.6).

Most thoroughly, maintenance has been considered in the context of the single tree selection system (Schütz, 2001b), see Section 5.4. The concepts and models developed for this silvicultural system are described in Chapter 6, and some of them are also relevant to other types of CCF. More general methods of maintaining CCF forest stands are described in Chapters 5 and 7. Methods specific to particular management objectives are given in Section 7.3.

2.6 Biological Automation and Rationalisation

The recognition that processes of self-differentiation and self-organisation of trees in forest ecosystems can be successfully integrated in forest management has been an important change of paradigm in forestry and marked the beginning of the emancipation of forestry from agriculture (Pommerening and Sánchez Meador, 2018). This process probably started with the discovery that the use of natural regeneration can markedly improve forest management results and with the subsequent invention of silvicultural systems (cf. Section 5.3) in Central Europe. In Germany, early initiatives to retain a small number of seed trees in each harvesting operation go as far back as the fifteenth century (Mantel, 1990). The need to invent silvicultural systems arose when early forest managers realised that certain tree species responded badly to being planted on bare clearfelling sites and large-scale, costly failures of such replanting attempts were the consequence. One species for which this was discovered already in the eighteenth century was *F. sylvatica*. Seedlings and saplings of this species are highly susceptible to early and late frost events and G. L. Hartig recommended the uniform shelterwood system in 1791 for improving the success of *F. sylvatica* regeneration (Hasel and Schwartz, 2006; Bartsch et al., 2020).

Self-organisation: Irreversible process in nonlinear systems, which create complex structures of the total system as a result of interactions between parts of the system (complex interacting systems) (Wolfram, 2002; Puettmann et al., 2009).

In the beginning, the understanding dawned on forestry staff that working with natural processes in the long run provided more success and less disappointment than working against them (Morosov, 1959). Peterken (1996) reported instances where British forestry staff in the 1970s repeatedly tried to eradicate the prolific regeneration of native broadleaves at great financial expense in order to force a non-native conifer plantation upon nature on a particular site in England:

'I well remember the first meeting I attended (in 1970) of what is now the Institute of Chartered Foresters. At the very first stop, we were shown a 35-year-old oak stand which had been killed with herbicides and underplanted with Corsican pine. By the time of the meeting the transplants had been so heavily grazed by deer that they were barely visible. I never decided what worried me more: the reek of chemicals; the forest of dead oaks; the fact that the only problem under discussion was how to establish Corsican pine, or the suspicion that this might just be a practical joke, perpetrated on a new member as an initiative test'.

From both an economic and an ecological point of view, any forest manager today would refrain from such actions and rather take the broadleaved regeneration trees forward that nature offered for free and that were most likely well-adapted to the given site. The principle of working with nature rather than against it and integrating natural processes even in commercial forestry is a key principle of CCF (see Section 1.5). Smith et al. (1997) stated that in silviculture natural processes are deliberately guided to produce forests that are more useful than those of nature and to do so in less time. The processes involved in this principle are often labelled *biological* or *nature automation* (Schütz, 2005) and includes processes such as natural regeneration, natural pruning, size differentiation and self-thinning. In this endeavour, forest structure plays a key role. Forest structure is manipulated and tailored in such a way that it ensures biological automation. Even silvicultural systems can be interpreted as instances of forest structure optimised to deliver natural regeneration.

Along with decreasing revenues from timber production, the increase in frequency and intensity of disturbances in recent decades, particularly storms and fire, has provided further incentive for making use of biological automation. In Switzerland, for example, large areas of forest land that were destroyed by severe gales at the end of the 1990s were not replanted, which was a considerable rift in the forest tradition. In many cases, even the woody debris was not removed after the gales. This land, however, was not taken out of production, but still is very much part of commercial forestry in the country today. Surprisingly, within 20 years, this non-intervention treatment gave rise to interesting mixed-species forest stands. The decision was made to try new avenues that would make greater use of *biological rationalisation* and save costs whilst advancing conservation through carrying out less intensive forestry. This concept has considerable potential, since it is based on the well-known CCF principle of biological automation (Schütz, 2005). Biological rationalisation is a general, cross-cutting theme that heavily relies on the principles of individual-based forest management, see Chapter 3.

According to Schütz (2005) it is the combination of two principles that lead to biological rationalisation, i.e. the achievement of intended management objectives with a minimum of management input:

1. Biological (nature) automation,
2. Concentration.

Biological or nature automation implies leaving as much as possible to natural forest development, see also Section 1.4. This involves the *opportunistic* strategy to accept what grows naturally in terms of species composition. Interventions should be limited to the bare minimum where natural processes conflict with management objectives, and there is no other way to solve this problem with less input. Note that this principle is echoed by the British CCF semi-synonym 'low-impact silvicultural systems' (LISS, see Section 1.5). Although the word 'impact' here mostly relates to visual and environmental impact, LISS has also been interpreted as 'low-input silviculture'. Low input or low impact reduces the effect of humans on forest ecosystems whilst reducing management costs at the same time. As this principle is commonly applied in CCF carried out for commercial purposes, with biological rationalisation in mind, any intervention is supposed to produce revenue and to provide benefit to the residual trees at the same time. Tight spacing maintained for more than two decades up to the first thinning boosts early height growth, improves stem straightness and facilitates self-pruning. In a commercial context and particularly in broadleaved forest stands, high tree density is maintained until the desired length of clean stem without branches has been achieved.

The concentration principle denotes the idea of achieving management objectives through limiting management inputs to a minimum of trees that matter most for the development of a forest stand and the generic methods for achieving this concentration are described in Chapter 3.

Traditional early thinnings (sometimes referred to as pre-commercial thinnings and respacing) often did not produce any or only little revenue in commercial forestry, and it is the strategy of biological rationalisation to avoid them altogether. Instead, it is intended to achieve the required tree size, quality (in a commercial context) and individual-tree resilience mainly through self-differentiation and self-thinning, i.e. through natural processes. Wilhelm and Rieger (2018) referred to this stage of stand development as *quality-achievement stage*, abbreviated and labelled as Q. This has proven to be possible, since only a comparatively small number of dominant frame trees (see Chapter 3), e.g. $100–150\,ha^{-1}$, is required. In his practical experiments, Ammann (2005a) found that the omission of early interventions indeed leads to a strong self-differentiation process. This process later helps identify the most dominant trees in the stand that persisted and flourished despite high initial tree densities. Those trees that naturally managed to dominate a forest stand turned out to be ideal choices for frame trees. Particularly in the context of biological rationalisation, it is important to recruit frame trees from the most dominant trees in the forest stand. This reduces the risk of both failure and costs considerably. Moderate numbers of frame trees per hectare have also proved crucial to maintain the process of biological automation and reduce management effort as well as costs. In the rare event of frame trees, differentiating unfavourably it is possible to replace them by suitable matrix trees in the vicinity of the deteriorated frame trees (Schütz, 2005). An important part of biological rationalisation is the strategy to concentrate growth on a few important trees with desirable properties, i.e. the frame trees, rather than to distribute growth evenly over the whole forest stand.

Later these frame trees are repeatedly released by removing some of their most dominant neighbours so that the frame trees can continue to act as dominant trees in the main canopy until the next intervention is due. This approach is intended to promote rapid crown and stem-diameter growth. In these release interventions, crown thinnings are applied which further support self-differentiation. Wilhelm and Rieger (2018) referred to this stage of stand development as *size development stage*, abbreviated and labelled as D (for dimension). Using the abbreviations of the two stand development stages, Wilhelm and Rieger (2018) have coined the term *Q/D concept* or *Q/D strategy*. The Q/D concept is another variant of biological rationalisation and is in fact very similar to the concepts suggested by Schütz (2005) and Ammann (2005a). They were independently developed by two groups in different countries, Switzerland and Germany.

In comparatively young and dense forest stands where the first or second frame-tree release thinning is scheduled, Ammann (2005a) recommended pollarding or girdling frame-tree competitors marked for removal rather than felling them at ground level. According to his experience both approaches are faster and more cost-effective than conventional fellings whilst the treated competitor trees still act as nurses for the frame trees, since they are usually recovering or dying gradually.

Using a chainsaw, pollarding can be achieved by (i) roughly removing all coarse branches of a competitor tree up to the operator's shoulder height. In a second step, (ii) approximately at the operator's breast height, the blade of the chainsaw cuts diagonally through the stem (cf. Figure 2.25). The felling direction of the tree is opposite the direction of this slanted cut.

Alternatively girdling can be carried out by removing a broad band of bark around the stem at breast height using a chainsaw or some specialised blade or steel brush (cf. Section 5.2).

Frame trees that resulted from self-differentiation processes were found to be dynamic and responded well to release thinnings, which again fuelled the self-differentiation process.

Example: Rationalisation management concept for *A. pseudoplatanus* L. and *Fraxinus excelsior* L. in Switzerland

Production objectives:

1. Target diameter 50–60 cm in 80–100 years,
2. 80–100 frame trees per hectare (corresponding to frame-tree distances between 8 and 10 m),

Prescriptions:

- No management while top height < 15 m,
- Selection of 80–100 dominant frame trees with seemingly good timber quality,
- Inter-frame tree distance: > 10 m, at least 7 m,
- Removal of 2–4 frame-tree competitors in first thinning at top height 15 m,
- In all thinnings, no removal of matrix trees,
- 2–3 further frame-tree release thinnings until only the frame trees remain in the overstorey,
- Last frame-tree thinning between 50 and 70 years (mean stand age),
- No further management until target-diameter harvesting commences.

Modified from Ammann (2005a)

Ammann (2005a) also developed prescriptions for other species such *F. sylvatica*, *P. abies* and *A. alba*. Top height was chosen as a criterion in order to make prescriptions more general and independent of specific environmental conditions and changes in growth patterns. The top-height threshold of 15 m has worked well in Ammann's practical experiments but is certainly not set in stone. The author, for example, suggested 17–32 m for *F. sylvatica* depending on site quality. Wilhelm and Rieger (2018) did not use top height thresholds but instead suggested that frame-tree selection and release thinnings should commence when the height to base of live crown (see Section 2.2) reached 25% of the achievable final tree height on a given site. The authors stated that if absolutely necessary, the quality of frame trees or of some of them can be enhanced by partial pruning. From this point on until target diameter harvesting the height to base of live crown should remain constant in all frame trees. Wilhelm and Rieger (2018) recommended a minimum distance of 12–15 m between frame trees regardless of species.

Any matrix tree that is not a potential competitor of frame trees is strictly ignored in active management to reduce costs but most importantly in order not to disturb the ongoing self-differentiation process (Ammann, 2005a). The unmanaged co-dominant and suppressed trees in the matrix occupy the space between frame trees without competing with them. They also contribute to collective stand resilience. In a commercial context, matrix trees can produce a range of by-products such as fuel and pulp. According to Swiss experience, leaving woody debris after storms, fires or bark beetle calamities behind improves soil nutrient and soil moisture regimes. The debris of stems and tree crowns can also be effective in deterring browsing animals (Ammann, 2020).

Schütz (2005) estimated that principles of biological rationalisation can lead to a ten-fold reduction of costs of current management practices in Switzerland. Biological rationalisation is similar to the principle of passive restoration described by Newton (2007). According to this principle, a forest ecosystem is largely allowed to recover or to develop naturally through the process of succession and direct input is limited. The concept of biological rationalisation has been tested mostly in the context of quality-timber production for a range of conifer and broadleaved tree species typical of Switzerland and Germany (Wilhelm and Rieger, 2018). Limitations and compromises include stands with *Picea* spp. that have a flat rooting system. In such stands, interventions potentially need to start earlier to foster individual-tree stability, specifically individual-tree resilience to wind and snow. Another limitation of this principle can arise when natural regeneration only includes the most dominant tree species on a given site and as a result tree diversity decreases from forest generation to forest generation. For example, it is potentially possible that shade tolerant tree species gradually outcompete more light demanding species and these can have difficulties to return under the conditions of biological rationalisation.

(a) (b)

Figure 2.25 *Juglans regia* L. frame trees (trees with light bark) in a matrix of *F. sylvatica* L. (trees with dark bark) from natural regeneration at a forest site near Brugg in Switzerland. (a) Released and well developed *Juglans regia* frame trees at the appropriate inter-frame-tree distance, note the next frame tree in the background. (b) *Juglans regia* frame tree (light bark) and *F. sylvatica* competitors (dark bark) that have just been pollarded to reduce competition gradually and in a cost-effective way. Source: © Arne Pommerening (Author).

For example, species like *P. avium*, *Juglans regia* L., *Sorbus torminalis*, *Larix* spp. and to some degree even *P. sylvestris* face considerable difficulties to survive in dense unmanaged stands that are left without intervention for more than two decades until a top height of 15 m has been reached.

A selected number of representatives of important minority species can, of course, be released before a top height of 15 m has been reached with a view to train them to become prospective frame trees (Ammann, 2005b). In that situation limited enrichment under-planting in large gaps can also help improve the situation (Schütz, 2005). It is always possible to include rare, disadvantaged tree species in the frame-tree selection so that they are promoted in subsequent thinnings and can contribute regeneration to the next forest generation. To give an innovative example, a limited number of large *Juglans regia* saplings, for example, has been trained as prospective frame trees in small artificial canopy gaps of an otherwise unmanaged stand at a forest site near Brugg in Switzerland (Figure 2.25). This illustrates the importance of thinking creatively and outside the traditional box of silviculture. Another option for maintaining tree species diversity is to choose silvicultural systems (Chapter 5.3) wisely, e.g. a group shelterwood system rather than a uniform shelterwood system to encourage or to plant light-demanding species in gaps (Ammann, 2005b, cf. Appendix B).

3

Individual-Based Forest Management

Abstract

Local or individual-based forest management has emerged in Western and Central Europe in the nineteenth century as an alternative to traditional global management methods. The concept includes breaking large forest stands down into smaller neighbourhood-based units that are easier to perceive by the human mind. The centre of each of these neighbourhood-based units is a frame tree (also referred to as final crop tree, elite tree or target tree) with clearly defined properties that depend on the management objectives. The number of frame trees is comparatively low and they are supposed to form the backbone or scaffold of a forest stand. In contrast to global methods, all management effort is directed towards frame trees only which makes woodland management both easier and more efficient. Local or individual-based forest management methods were first introduced in a commercial forestry context, but rather constitute generic principles that can be efficiently applied in any management situation, for example, in management for conservation, carbon sequestration and recreation. Originally an independent development, local forest management methods are now increasingly applied in the context of CCF, and the concept is also very helpful in CCF training.

3.1 Introduction

The introduction of *local* or *individual-based forest management* was a fundamental change in paradigm because this concept includes the strategy of breaking a large forest stand down into smaller neighbourhood-based units that Schädelin (1934) referred to as *thinning cells*, see Figure 3.6. A small number of clearly defined individual trees form the centres of these small, localised management units and all efforts are exclusively directed towards these trees rather than towards the forest stand as a whole, as this is typically done in traditional, *global* forest management (Pommerening and Grabarnik, 2019), cf. Section 5.2. In this book, these special trees are referred to as *frame trees*. For a forest operator marking trees or for a forest harvester driver, given the limited visibility in forest stands, the small management units or thinning cells are more natural to perceive and easier to work through one by one (Figure 3.1).

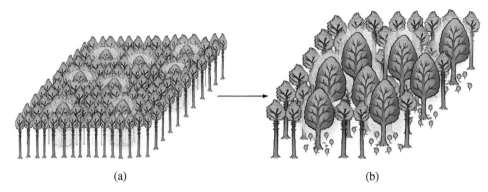

(a)   (b)

Figure 3.1 Envisaged dispersion and development of frame trees shortly after selection (a) and shortly before target-diameter harvesting (b). The frame trees are highlighted by yellow 'halos'. Source: Courtesy of Zeliang Han and inspired by Wilhelm and Rieger (2018).

Frame-tree management is commonly associated with continuous cover forestry (CCF) and is often inappropriately marginalised as a technicality assisting forest managers to put CCF into practice (Wilhelm and Rieger, 2018; Bartsch et al., 2020). Attempts to properly explain and to generalise the concept of individual-based forest management beyond yield and production scenarios are largely absent in the literature. Even in the countries where these concepts were first devised, forest scientists and practitioners appear not to be fully aware of the enormous difference that their introduction has made to forestry and conservation and of the great potential individual-based forest management has for delivering ecosystem goods and services (Pommerening et al., 2021a).

3.2 Definition and Terms of Individual-Based Forest Management

Individual-based forest management – sometimes also referred to as free-style or crop-tree release silviculture (Smith et al., 1997) – is a bottom-up approach and a size-control method (cf. Section 1.8) par excellence where in a first step a number of trees with certain characteristics matching the forest-management objectives are selected and often are also visibly marked in the field. In this book, following British convention, these trees are referred to as frame trees (also termed *final-crop trees*, *target trees*, *plus trees* and *elite trees*) for the appealing metaphor of a framework of special trees that form the backbone or resilient scaffold of a forest stand or a tree population (Pommerening and Grabarnik, 2019). Frame trees are supposed to stay on in the forest stand long-term and most importantly to benefit from any future forest operation. To achieve this, in a second step, based on the nearest-neighbour principle, trees in close proximity of frame trees are selected that are likely to influence the frame trees negatively in the next 5–10 years (Figure 3.2). For instance, in temperate climates, the most important criterion is to ensure that the crown growth of frame trees is not obstructed by neighbours. In arid and boreal climates, where tree densities can be low and competition for light is less important, other environmental criteria must be used

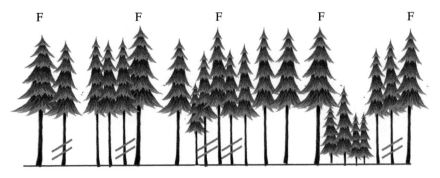

Figure 3.2 Sketch of an imaginary single-species conifer forest. The frame trees are indicated by the letter 'F' and the stems of neighbours selected as perceived frame-tree competitors to be thinned are crossed by double red lines. Modified from Pommerening et al. (2021a).

(see, for example, Figure 3.3). Forest managers identify competitive neighbours as an interpretation of the individual frame tree's 'eye view'. Whilst lacking more direct measures, competitiveness is usually defined by a combination of size (crown and total height) and inter-tree distance (Klädtke, 1993; Pommerening and Grabarnik, 2019).

Individual-based forest management has been influenced by the concept of natural competition and facilitation processes (see Section 4.1), as they were proposed in forest ecology and perceived by forest managers in their daily fieldwork (Pommerening and Sánchez Meador, 2018). This perception of how neighbours of frame trees acted as competitors may initially have been intuitive and even naïve, but it has certainly influenced individual-based forest management considerably. Field observations and ecophysiological theory (Göttlicher et al., 2008) appeared to support the view that only the immediate surroundings of important trees mattered. This has given rise to the idea of local thinnings, where only the neighbourhood of frame trees is considered; hence, the term *local thinning method*, cf. Section 5.2. Schütz (2003) used the term *situative thinnings* instead (Pommerening and Grabarnik, 2019). This is a selective process of low-impact forest management in the sense that neighbours of frame trees are considered for removal only, if absolutely necessary so that in the same intervention with some frame trees no neighbour, with others one or two and again with others perhaps three neighbours are selected for removal (Figure 3.2). This decision is a compromise between promoting the frame trees' growth and promoting their functional properties, e.g. timber quality or habitat value. Everything else in the forest stand, particularly where there is no frame tree, is strictly left to natural processes without human interference. All non-frame trees can be referred to as *matrix trees* and an important part of their role is to serve frame trees as nurses through 'mild, healthy competition' and potentially provide by-products such as energy wood or pulp (Pommerening and Grabarnik, 2019). As mentioned earlier, frame trees and their immediate neighbours form a mosaic of small local management units (Schädelin, 1934), and it is important to consider neighbouring units when selecting new frame trees so that the distances between them are not too short and the removal of competitors is coordinated between units (Figures 3.6 and 3.7). Because of the nearest-neighbour principle, this type of thinning is referred to as local thinning type in contrast to global thinnings that are

Figure 3.3 A mixed-species and low-density woodland with *Pinus pinea* L. in the overstorey and *Quercus ilex* L. in the understorey near Évora in Portugal. In arid environments like this, judging competition by a combination of the sizes of crown and total height as applied in temperate climates is usually not appropriate. Source: © Arne Pommerening (Author).

traditional methods and aim at affecting the whole forest stand in a uniform fashion (see Section 5.2).

In a way, frame-tree selection as part of individual-based forest management also serves as a didactic aid that helps field staff to separate 'important' from 'unimportant' trees regardless of the definition of importance. In our experience, it frequently happens that people unfamiliar with individual-based forest management are at a loss when asked to select trees for thinnings. They simply 'cannot see the trees for the wood' and the frame-tree method literally is an eye-opener fostering their observation skills and perception (Pommerening and Grabarnik, 2019). However, frame-tree based forest management does not come easily to everybody and therefore needs to be included in appropriate training (see Chapter 8).

The frame-tree method can even be applied in intensively mechanised forest management. It suffices that a specialist or experienced forest manager marks frame trees permanently and in a clearly visible way. It is perceivable that after initial training, harvester drivers can then be trusted to select frame-tree competitors whilst driving through a forest stand and carrying out the thinning (Eberhard and Hasenauer, 2021). Selecting and visibly marking frame trees is an effort that is required only once in the lifetime of a forest stand (Pommerening et al., 2021a). This has proved to be a good compromise between no tree marking at all and intensive tree marking that includes both frame trees and competitors prior to actual thinnings (Pommerening and Grabarnik, 2019).

The fact that all thinnings are oriented towards frame trees and occur in their neighbourhood leads to greater efficiency of management operations through a rationalisation of efforts, see Section 2.6. Only frame trees receive management attention and the success of thinnings can be more efficiently checked in subsequent years because only the frame trees need to assessed. This rationalisation advantage of local thinnings clearly supports the concept of permanent frame trees, since only then management success or failure can be reliably identified and thus treatments become more consistent. This markedly simplifies silvicultural monitoring. If maintaining or increasing species diversity is an objective of stand management, it is easy to select frame trees in such a way that the desired long-term species composition is ensured including rare tree species (Pommerening and Grabarnik, 2019; Bartsch et al., 2020). At the same time, harvesting, extraction or bark-stripping damage (caused by animals such as deer and grey squirrel) only matters, if frame trees are affected (Klädtke, 1993). The method also helps to efficiently communicate forest management approaches through quantitative descriptions and this contributes to higher transparency (Pommerening et al., 2021a).

The frame-tree concept is very suitable for a diverse range of management objectives. For example, in forest cemeteries that have recently become very popular across the world (Quinton et al., 2020), trees typically act as grave markers and therefore it is important to keep these trees in good health, cf. Section 7.3.7. Consequently, they can act as frame trees and individual-based forest management ensures the vigour of these living grave markers. When it comes to climate-change mitigation, individual-based forest management provides useful methods for carbon forestry (Pukkala, 2018): Individual trees are selected and all subsequent management aims at maximising their carbon sequestration. Once these trees have reached natural life expectancy or need to be removed for whatever reason, their timber can be processed to produce furniture or construction material to ensure that the carbon they have stored is not released any time soon. Alternatively, it is possible to put large tree stems harvested from frame trees into long-term storage (Zeng, 2008), see Section 7.3.2. Frame-tree management has also been suggested for reducing drought stress in conifer stands (Kohler et al., 2010; Gebhardt et al., 2014). The authors found that *Picea abies* trees could better cope with drought effects, if they were granted more growing space, i.e. if they grew in a neighbourhood with low-local tree density. This is the typical structural context of frame trees and individual-based forest management. Repeated moderate reductions of local tree density around frame trees can enhance the stand-level capacity for plant-available water. In a conservation scenario, native trees can act as frame trees, whilst their competitors are invasive neophytes forming the matrix.

3.3   History of Individual-Based Forest Management

Early forest-management methods had their origin in agriculture, and it seemed 'natural' to consider whole forest stands or even larger units of forest land and their global characteristics when assessing timber sustainability rather than individual trees. Towards the end of the nineteenth century, management systems based on forest stands such as the normal forest were devised to ensure timber sustainability (Gadow and Bredenkamp, 1992; Hasel and Schwartz, 2006; Bettinger et al., 2017), see Section 1.8. For some time alternative approaches based on individual trees such as the single tree selection system (cf. Section 5.4) were even legally banned in various European countries because it was believed that timber sustainability could not be guaranteed in highly diverse selection forests (Schütz, 2001b). The stand approach also influenced research and supported a focus on unit-growth responses to different management strategies including yield tables as forecasting and planning tools (Pommerening et al., 2021a).

The fundamental idea of individual-based forest management has roots that apparently go as far back as 1763 when Duhamel du Monceau mentioned the use of frame trees for management of *Quercus* spp. in France (Klädtke, 1993; Schütz, 2003). In the nineteenth century and at the beginning of the twentieth century, the concept was gradually introduced to commercial forestry applications in Austria, Germany and Switzerland, initially with little success and limited uptake. Then Schädelin (1926, 1934) and his successor Leibundgut (1966) systematically developed and promoted the concept in Switzerland, which they referred to as '*Auslesedurchforstung*' (selective thinning). The rationale of this method was to concentrate commercially valuable timber on just a few trees, i.e. the frame trees as opposed to maximising overall volume production of a forest stand, which was the traditional objective of forestry at the time. This implied that a lower overall yield was accepted provided that the frame trees and their management in the long term produced an economic surplus (Knoke, 1998). Schädelin's original idea was to work with non-permanent frame trees, i.e. in each intervention, new frame trees would be selected, which could, of course, largely overlap with the frame trees selected in the previous intervention, but did not have to. Schädelin also proposed to initially appoint a large number of frame tree candidates (also referred to as *potential crop trees*) in early stand-development phases, which were later reduced to a smaller, definite number. The definite frame trees would then be promoted in later stand development phases through heavy release thinnings, i.e. by providing ample space around the crowns of each frame tree. Along with a number of precursors and colleagues, Abetz (1975, 1976) in Germany and Pollanschütz (1971, 1981, 1983) in Austria modified this concept by advocating a permanent selection of frame trees, which they referred to as '*Z-Bäume*' ('future' or 'target' trees from German 'Zukunft' and 'Ziel'). Permanence of frame trees is often ensured by visibly marking frame trees with ribbons or spray paint in the field, and this has proved to be good practice. Abetz did not suggest heavy release thinnings in later stand development phases, but recommended the continuation of frame-tree based thinnings. He also suggested selecting right from the start only such a number of frame trees that corresponded to the definite (final) number of frame trees. Theoretically, the maximum definite number of frame trees is the number that can be supported by a given mature forest stand shortly before harvesting would commence. To mark the differences to Schädelin's original idea, the Abetz/Pollanschütz approach was referred to as '*Z-Baum-Durchforstung*', i.e. frame-tree thinning, however, this separation of terms is artificial, as from a theoretical point of view both approaches are essentially just two variants of the same general individual-based concept (Pommerening et al., 2021a).

Abetz' and Pollanschütz' approach of selecting only a definite number of permanent frame trees triggered a long and sometimes heated debate that took place among practitioners and researchers about the risks associated with relying on frame trees that potentially would die, get damaged or differentiate in unfavourable ways (e.g. becoming less dominant, getting exposed to bark stripping by deer or grey squirrel, or deteriorating in timber quality) (Abetz, 1990; Dittmar, 1991; Klädtke, 1990; Schober, 1990; Spellmann and Diest, 1990). The debate reached its zenith around 1990. Part of this debate included Schädelin's original idea of, as briefly mentioned before, initially appointing a larger number of frame tree candidates from which later, when eventually it has become clearer, if the trees were suitable, a smaller number of definite and permanent frame trees is recruited. However, research and field experience suggest that the risk of losing frame trees is low, if (i) they are not selected too early (Mosandl et al., 1991), (ii) they are recruited from the most dominant trees of the forest stand, (iii) their number is comparatively low, i.e. ideally not more than approximately 150 trees per hectare, and (iv) they are consistently favoured in regularly recurring interventions (Klädtke, 1997). Initially working with candidate frame trees has turned out to be an unnecessary complication (particularly with CCF beginners) and usually leads to confusion and inconsistent management (Bartsch et al., 2020), where the fundamental idea of focussing on individual trees is gradually abandoned in favour of global stand management. Along similar lines, it has also been found to be more efficient to appoint permanent frame trees that are visibly marked in the field (Pommerening et al., 2021a).

The trend towards individual-based forest management was supported by the advance in computer technology. Towards the late 1970s, the increasing availability of individual-tree data and computing resources gave rise to new analysis and modelling methods that made individual-tree approaches even more feasible and particularly helped with checking up on timber and other forms of sustainability. Numerous variants of the local-thinning and frame-tree concept are now applied in many European countries and occasionally also in North America and more recently, in China (Pommerening and Sánchez Meador, 2018; Pommerening et al., 2021a).

3.4 How and When Frame Trees Are Selected

Frame trees are essentially 'trees of interest' or 'trees of importance' that can be flexibly defined in various ways, e.g. in terms of economic value, habitat value, stand resilience, spiritual or aesthetic value. This implies that individual-based forest management can be applied in commercial forestry as well as in conservation and recreation management, since it is a general concept. Even in plantation management, individual-based methods can be implemented and make sense. That is why a dedicated chapter on individual-based forest management was included in this book. The number of frame trees per hectare often varies with species and site and in accordance with silvicultural objectives.

> In the spirit of the size-control method (cf. Section 1.8), *target diameters* specify the diameter when frame trees have sufficiently matured so that they can be considered for individual-tree, selective harvesting.

Table 3.1   Regional guidelines for target diameters of important species in North Germany.

Tree species	Target diameter [cm]	Corresponding age [years]	F-tree distance [m]
Acer pseudoplatanus	60	90–110	12
Alnus glutinosa	35–50	80–100	12
Fagus sylvatica	60–70	100–110	10
Fraxinus excelsior	60	90–110	12
Larix spp.	65	70–80	12
Picea abies	45	60–80	8
Pinus sylvestris	45–50	80–110	10
Pseutosuga menziesii	60–70	70–100	12
Quercus spp.	70–80	150–180	15

Excellent/good growth conditions and regular heavy thinnings that start early are assumed. F-tree distance is the minimum distance between frame trees.
Source: Bartsch et al. (2020) and Codoc (2020).

Target diameters are usually determined by the regional timber-processing industry and associated markets. In conservation scenarios, they typically depend on size–habitat relationships. Therefore, target diameters can greatly vary from stand to stand and from region to region. However, after a number of years during which experience with target diameter harvesting has accumulated, regional guidelines usually develop (see Table 3.1 for an example). In a commercial context, target diameters can additionally be defined in terms of expected timber quality with decreasing target diameters from high to low timber quality (Susse et al., 2011).

Assuming light to be the most limiting environmental factor, the maximum number of frame trees per hectare is determined by the crown size of the frame trees at the time when target diameter harvesting commences. Since the question of how many frame trees should be selected depends on species-specific crown sizes, and these again partly depend on environmental conditions, it is theoretically possible to calculate a suitable maximum number of frame trees for a given target diameter. From relationships between stem and crown diameter, the required growing space for each tree of a certain size can be approximated. To simplify calculations square spacing and uniform growth rates can be assumed for determining the maximum number of trees per unit area, i.e. number of frame trees $= 10\,000/\text{crown diameter}^2$ (Abetz and Klädtke, 2002). Generally speaking, the number of frame trees decreases with increasing target diameter (cf. Figure 3.4). Broadleaved trees tend to have larger crowns and therefore fewer broadleaved frame trees can be accommodated than conifer frame trees. Assuming a target diameter of 60 cm, the given site can only support a maximum number of 93 *Fagus sylvatica* frame trees. For *Picea sitchensis* with an average commercial target diameter of 50 cm, this maximum number is 175 trees ha^{-1}. The *Pseudotsuga menziesii* curve is close to that of *Picea sitchensis*, however, for the former species, a target diameter of 60 cm is quite common. As a result the maximum number of frame trees per hectare is 153 trees ha^{-1} for the given site (Pommerening and Grabarnik, 2019, Figure 3.4).

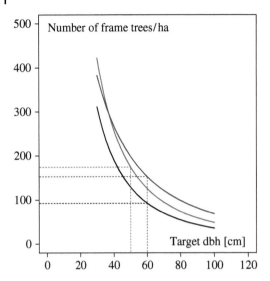

Figure 3.4 Number of frame trees per hectare approximated from crown diameter models published by Savill (1991) for British tree species over target stem diameter (dbh). Square spacing was assumed. Black: *Fagus sylvatica* L., red: *Picea sitchensis* (Bong.) Carr., blue: *Pseudotsuga menziesii* (Mirb.) Franco. The dashed lines indicate the number of frame trees per hectare given a target diameter of 50 cm for *Picea sitchensis* and of 60 cm for both *Fagus sylvatica* and *Pseudotsuga menziesii*. Source: Pommerening and Grabarnik (2019).

To achieve the structure that CCF requires in most applications, it is generally advisable to stay well below these maximum or potential frame tree numbers in order to achieve a more diverse horizontal and vertical woodland structure (Weihs, 1990). In environments where water or nutrient supply is low, the number of frame trees need to be even lower.

In the absence of better information, a good 'rule of thumb' and default is to select approximately 100 frame trees per hectare. In broadleaved forests, where trees usually have larger crowns than conifers, 60–80 trees per hectare are appropriate whilst the number can range between 100 and 150 in conifer forests. Alternatively, the corresponding distances between frame trees of 10–15 m can be used as a reference in practical implementation (Wilhelm and Rieger, 2018). These general defaults apply to most CCF management situations. In special situations such as in management for conservation and for the transformation of plantations to CCF (see Section 2.4) or when the selection process is delayed, 50 frame trees per hectare suffice (Schütz, 2001a).

Much larger frame-tree numbers were recommended in the past (Schädelin, 1934; Abetz, 1975; Pollanschütz, 1981), but these original numbers created problems and have largely been abandoned. Too many frame trees usually lead to homogeneous stand structures and leave too few options for identifying trees to be thinned and consequently crown releases of frame trees are too weak. On the contrary, selecting too few frame trees may not use the full potential of a forest stand and leaves large parts of the forest virtually unmanaged with no density reductions other than by natural mortality. In mixed-species stands, a larger number of frame trees is temporarily possible, when the frame trees of one or several component species are harvested much earlier than those of others (Wilhelm and Rieger, 2018). Otherwise, the number of species-specific frame trees in mixed forests should be a reflection of the desired species composition and the sum of frame trees across all species should not exceed 80–150 frame trees (Pommerening and Grabarnik, 2019). In Switzerland, frame-tree distances are more commonly in use than frame-tree numbers, cf. Table 3.1 and Figure 3.7. For deriving suitable distances between frame trees of different species, the maximum or the mean of the two species-specific distances can be used.

Frame trees are selected at a comparatively early stage of stand development, e.g. when a top height of 12–15 m is reached. In conifer forests, the selection of frame trees usually starts at lower top-height values than in broadleaved forests (Klädtke, 1993). If the necessary data are available, a more refined criterion for the appropriate timing of selecting frame trees is the species-specific growth pattern on a given site: Once the culmination of height growth has occurred, the selection of frame trees can begin; otherwise, it has to be postponed (Pretzsch, 2009). Wilhelm and Rieger (2018) proposed to select frame trees when their height to base of (live) crown has reached 25% of expected final total height. Frame trees are selected according to clearly defined criteria and these depend on the forest management objectives. Therefore, they can differ markedly, for example, in commercial scenarios, typical criteria are

- Vigorous (no diseases, no physical damage, high growth rate, large stem diameter and total height, crown class 1–3, D and CD, see Chapter 4),
- Non-deciduous trees particularly with shallow root systems, e.g. *Picea* spp., should have a h/d ratio (Eq. 2.2) which does not exceed a value of 80,
- Good timber quality (straight stems, low branchiness, little taper),
- Dispersion (as uniform as possible or clustered).

The criteria vigour, timber quality and dispersion are also priorities and should be observed in this order. Individual-tree resilience, cf. Section 2.2, is usually given the highest priority, although dominant trees, if chosen, usually are very resilient. In a conservation or in a recreation setting, some of these are replaced or complemented by other criteria such as habitat value, aesthetic value and species:

- Morphology similar to open-grown trees (e.g. large stem diameter with considerable taper; long, wide crowns, large root plate, buttresses and forked stems, i.e. 'fairy-tale trees'),
- Non-deciduous trees particularly with shallow root systems, *Picea* spp., should have a h/d ratio (Eq. 2.2) which does not exceed a value of 80,
- Leaves, bark, etc. (easy to decompose, scenic, presence of lichens, aesthetic value),
- Species (rare or offering many spacial habitats),
- Dispersion (anything but uniform).

Such criteria can be flexibly amended according to changing management objectives, changing climate and shifting societal expectations. However, vigour is likely to be the most important and generic criterion across a wide range of management objectives, as the frame trees are otherwise difficult to maintain. Frame trees are therefore always likely to be the most dominant and prolifically growing trees of a forest stand or at least dominant within the respective species population in mixed-species stands.

Dispersion and distances between frame trees are often referred to as the least important criterion; however, when appointing frame trees, it is crucial that they never compete with each other. A good spread of frame trees across the whole stand area also avoids creating patches that are never thinned, although this is not necessarily a bad thing. Often, unthinned parts of a forest stand are not a problem or in a conservation context are even desired (see Section 2.4.3), but on upland sites and with species that are susceptible to windthrow they can amplify the negative effects of disturbances (Pommerening and Grabarnik, 2019). A selection of frame trees according to vigour as the most important criterion often helps to establish a regular dispersion of frame trees at the same time, since competition processes often naturally enforce larger distances between dominant trees (Figure 3.5). Frame trees should not be selected near roads, rides, extraction racks, or forest margins, as the risk of damage is very high at these locations.

Frame trees then become cell nuclei of small local management units (Schädelin, 1934) within a forest stand that include frame trees and their nearest tree neighbours (Figure 3.6). Following the idea of a bottom-up approach, all local management units aggregate to a forest stand or tree population. This implies that all local management units are connected with neighbouring units they share boundaries with and thus interact with them. In the process of tree marking, this connectivity needs to be considered. For example in Figure 3.6, frame tree no. 63 shares boundaries with frame trees 26, 75, 89, 93 and 106. For five frame trees (no. 8, 16, 40, 48, 106 and 109), the forest manager has only selected one nearest neighbour or competitor for removal. For trees no. 26, 46, 63, 75, 89 and 111, two neighbours were selected for removal. For frame tree no. 93, the respective forest manager did not select any competitor for removal because he did not consider this necessary for the next five years to come. Note that in agreement with CCF theory small, close neighbours of frame trees

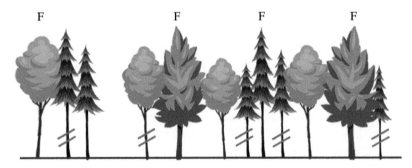

Figure 3.5 Sketch of an imaginary mixed-species forest where frame trees were recruited from different species populations to support tree species diversity. The frame trees are indicated by the letter 'F', and the stems of neighbours selected as perceived frame-tree competitors are crossed by double red lines. Source: Modified from Pommerening et al. (2021a).

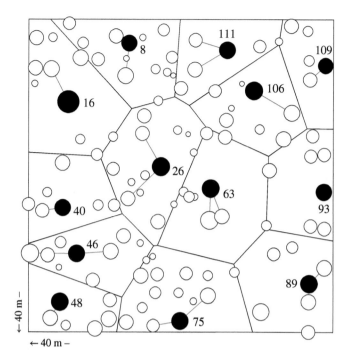

Figure 3.6 Schematic representation of how frame trees (filled circles) form local management units or frame-tree influence zones within a forest stand that includes frame trees and their nearest matrix neighbours (open circles). The red lines lead to frame-tree neighbours selected for removal, whilst, based on Dirichlet tessellation, the black lines denote imaginary boundaries of the local management units (frame-tree influence zones). The numbers next to the frame trees are identification numbers and the circle diameters indicate tree size.

were simply ignored as unlikely competitors (unless they were invading the frame tree's crown from below) and that thus the eviction of trees was limited to a minimum. To complete the concept of local management units, Figure 3.7 offers a 3d impression showing how neighbouring frame trees can be seen in the forest and emphasising the importance of appropriate distances.

Heterogeneous environmental conditions (e.g. soil depth and moisture), for example, in upland forest ecosystems, can make it sometimes necessary to take natural trends into account by allowing clustered arrangement of 2–3 (usually conspecific) frame trees in groups (cf. Figure 3.8b). This is also an attractive option for conservation, where habitat or remnant trees occur at close proximity or for some reason or other need to be situated close together. In such situations, it is necessary to provide large distances between one frame-tree group and another one so that the frame trees can tap into resources from outside the groups (Busse, 1935; Kató, 1973; Mülder, 1990). In mixed-species forest stands, the frame-tree groups can be recruited from different species populations to reflect a desired future species composition or to meet conservation targets (Figure 3.5). Groups are left intact for as long as possible and treated like a single frame tree (akin to a superorganism). Potential competitors are removed from outside the group.

Figure 3.7 Three-dimensional sketch of the local management units in an imaginary mixed-species forest, cf. Figure 3.6. Frame trees are shown in green, and matrix trees are given in grey and white. The red lines indicate the distances between frame trees, cf. Table 3.1. Inspired by Codoc (2020). Courtesy by Zeliang Han.

Inspired by Kató (1973), a simple way to adjust distances between frame tree groups (Figure 3.8b) could be as follows:

Assume 100 frame trees per hectare are to be selected. This results in an average distance $\overline{dist} = 10$ m assuming square spacing. This distance would apply to individual frame trees, but now we have organised them in groups, and we need to compensate for smaller inter-frame-tree distances within the groups by larger distances between the groups. One group of three frame trees has distances 3.3, 2.0 and 2.4 m. This results in an average deviation $W = \frac{(10-3.3)+(10-2.0)+(10-2.4)}{3}$ m, i.e. $W = 7.4$ m. A correction additive term a can be defined as the square root of W or 40% of W. In the former case, this would give $a = 2.7$ m, in the latter $a = 3.0$ m. This value a is then added to $\overline{dist}$ resulting in a distance of 12.7 or 13.0 m between the nearest trees of the current frame-tree group and the nearest frame-tree group yet to select.

For another frame-tree group in the same forest including three trees, the distances between them may be 6, 7 and 9 m. In that case $W = 2.6$ m. The two variants of a would be 1.6 and 1.0 m, respectively. This would result in adjusted distances 11.6 and 11.0 m between this and the nearest group. If according to this approach two neighbouring groups have different distance requirements, the maximum requirement is applied to define the distance between them.

The general idea of this simple approach is to apply a saturation relationship to the additive term so that adjusted distances between frame-tree groups do not become unreasonably large. This is justified given the high likelihood that the individuals of the group share resources and cooperate rather than compete. It is also possible to calculate $a = b \times \sqrt{W}$, where b is another factor, e.g. $b = 1.1$. The best procedure for calculating a (and possibly b) and identifying realistic values of these parameters for any given species and site needs to be verified in practical and research experiments.

A uniform dispersion of frame trees is usually preferred, since this spatial pattern is easier to implement and to maintain. A uniform dispersion does not necessarily imply that the structure of the whole forest stand will be homogeneous and simple. Provided the frame-tree numbers are reasonably small following the general recommendation in this book, it is quite possible that the overall forest stand structure can become heterogeneous as a result of local crown thinnings, self-differentiation in unthinned parts of the stand and advance regeneration. A clustered dispersion of frame trees is only desirable, if soil conditions are heterogeneous, the tree species involved have an inclination to form biological groups, or conservation objectives require frame trees to occur in small groups.

Some guidelines suggest first to mark frame trees and then to re-visit each frame tree by marking competitors for thinning among their nearest neighbours in a second step. This is probably the safest and most orthodox way to mark frame trees and frame-tree competitors. However, the author made the personal experience that it is easier and more natural for him

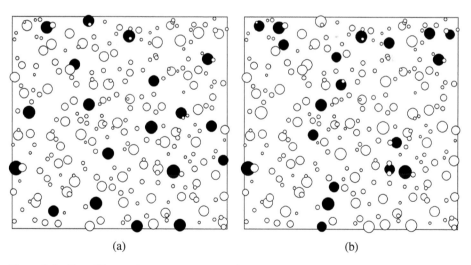

(a)   (b)

Figure 3.8 Two different simulated patterns of frame-tree dispersion at Embrach (Switzerland) in the same 91-year old predominantly *Fagus sylvatica* L. forest plot (51 × 49 m). (a) The most vigorous 76 frame trees ha^{-1} (filled circles) were selected with a view to achieve the largest distance between them resulting in a regular dispersion. (b) The most vigorous 76 frame trees ha^{-1} (filled circles) were selected with the objective to create spatially segregated clusters of 1–3 frame trees according to the group concept by Busse (1935), Kató (1973) and Mülder (1990).

to consider each management unit as a whole (see Figure 3.6), i.e. each frame tree and its neighbourhood simultaneously, and only then to proceed to the next. As a consequence frame trees and potentially competing neighbours are selected and marked at the same time. In my opinion, this also avoids falling back into the habit of global thinnings and it is easier to see how the different management units interact with each other (Pommerening and Grabarnik, 2019). However, eventually this is a matter of personal taste. When selecting potentially competing neighbours for felling, it is good practice to focus on the most urgent competitor first and only then to consider other trees that exert less competition pressure on the frame tree than the prime candidate. Rittershofer (1999) recommended marking trees in the field for no longer than half a working day. According to his experience, broadleaved stands should ideally be marked in winter when the leaves are off, since this state makes it easier to perceive forest structure. The trees should be marked in such a way that the marks can be seen from all directions so that it is possible to recognise frame trees already marked and their competitors from afar. For marking frame trees and competitors, it is useful to subdivide a stand into parallel strips and to mark trees within one strip before moving on to the adjacent one. This facilitates orientation. On slopes, the strips should be organised in a direction perpendicular to the slope direction. Then one should start marking trees at the top of the slope and gradually work downwards whilst looking back uphill towards the trees that have already been marked (Rittershofer, 1999). In Section 3.3, I have mentioned already that frame trees can be selected and marked *permanently* or *temporarily* and that there was a long-running debate on this in forestry journals. A temporary selection implies that new frame trees are selected prior to every new thinning intervention which retains flexibility and provides new opportunities but involves a series of additional, costly management inputs. Current understanding and best-practice advice is to mark frame trees permanently once and for all because management success or failure can then be more easily identified and as a consequence treatments become more consistent. This benefit is particularly important for CCF beginners. A temporary selection sometimes potentially appears to offer more flexibility, but in practice, this is often offset by management inconsistencies. Particularly, where possibilities for detailed tree marking in the field are limited (e.g. because of financial constraints), a permanent selection of frame trees is recommended. The debate in the early 1990s also included whether it is a good idea to appoint a large number (e.g. 200–300) of frame-tree candidates first and to reduce them to a smaller number (e.g. 100–150) of definite frame trees later, in case frame-tree candidates differentiate in unfavourable ways. As mentioned in Section 3.3, it is now understood, that such a strategy is an unnecessary complication and usually leads to more confusion and inconsistent management compared to the benefits it can offer. The strategy also creates additional work for reducing the number of frame trees in a second step. The same applies to keeping reserve frame trees (Bartsch et al., 2020; Pommerening and Grabarnik, 2019).

3.5 How Frame Trees Are Managed

In the past, it was recommended to carry out several density-reduction thinnings before commencing frame-tree management (e.g. in Abetz, 1975). However, as such thinnings (sometimes referred to as pre-commercial thinnings or re-spacings, see Section 5.2) are usually expensive, they have largely been abandoned as part of biological rationalisation (cf. Section 2.6), where such operations are left to self-thinnings, i.e. to processes involving natural mortality. Exceptions are made for species with flat root systems to avoid wind damage and insect calamities (Wilhelm and Rieger, 2018; Ammann, 2020).

For management purposes, the trees of a forest stand can be divided into frame trees and matrix (non-frame) trees. Normally, frame trees would not be felled in thinnings, and in commercial scenarios, the majority of them are finally removed as part of target-diameter harvesting (Pommerening and Grabarnik, 2019). Frame trees can be supported in local thinnings by removing all other trees within a certain radius around each frame tree, e.g. in young stands, where neighbouring trees have so far differentiated little, and it is hard to tell which neighbours matter more than others. The idea is to release frame-tree crowns on all sides (Smith et al., 1997). In conservation, this type of release thinning is known as *halo thinning* and often applied to old native remnant trees in non-native plantations (Short and Hawe, 2018). In middle-aged and older forest stands, it is more appropriate to select a small number of most competitive neighbours of a frame tree in the spirit of a crown thinning (cf. Section 5.2). In older stands, canopy gaps do not close so quickly, and it becomes impossible and undesirable to release frame trees using the pattern of a halo thinning (Smith et al., 1997). In that situation, forest managers identify competitors selectively as an interpretation of the individual frame tree's 'eye view'. Competitiveness is usually perceived as a combination of size (crown width and length; total height) and inter-tree distance. In some forest ecosystems, belowground competition for nutrients and water is more important than competition for light, and this needs to be taken into account when using tree size as a proxy (Petriṭan et al., 2011; Högberg et al., 2021). In that case crown measures may fail to be good indicators of competitiveness and frame-tree influence zones need to be much bigger, see Figure 3.3. In middle-aged forest stands, the selection of competitor trees around each frame tree automatically implies the crown thinning method. Later, when the frame trees are the most dominant trees of the forest stand and well looked after, frame-tree-based thinnings can become weaker or stop altogether at that stage (Pommerening and Grabarnik, 2019).

For commercial woodland management based on the frame-tree method Rittershofer (1999) designed a detailed functional tree stratification system. This classification is useful not necessarily as a literal set of rules but for understanding the mindset of individual-tree management. A similar system was also given by Assmann (1970) who referred to it as *Danish tree classification*. Such systems are helpful didactic aids and broad guidelines for modern individual-based tree management where trees need to be identified for release and thinning in the field. A classification system such as the one by Rittershofer (1999) may facilitate the decision-making process (Figure 3.9). In this and similar classifications, it is often recommended that rubbing trees and whips should be removed before competitors and wolf trees. The same priority applies to smaller neighbours growing into the crown space of frame trees from below. Hazard trees (including W, R, Wh and C) as well as nurse and indifferent trees together constitute the matrix, and all trees in these classes can be considered matrix trees (Pommerening and Grabarnik, 2019). It should be noted that in thinnings for nature conservation or in woodlands used predominantly for recreation, future veteran or habitat trees as well as aesthetically pleasing but dominant trees may qualify as wolf trees and need to be retained rather than removed.

Figure 3.9 Crown classes according to Rittershofer (1999) and result of a selective thinning in favour of frame trees followed by a growth process. Source: Pommerening and Grabarnik (2019).

Frame-tree based tree classification (Rittershofer, 1999)	
Frame trees (F)	Trees selected for their outstanding vigour, stem quality, productivity, resilience and crown morphology. Usually not more than ± 100 (80–150) trees ha^{-1}.
Hazard trees (H)	Diseased and damaged trees, trees that potentially can be a threat to frame trees:
Wolf trees (W)	Dominant, very competitive trees with exceptionally poor form that can outgrow and damage frame trees.
Rubbing trees (R)	Sub-dominant and suppressed trees rubbing frame-tree stems with larger branches or parts of their crown.
Whips (Wh)	Sub- or co-dominant trees with slender stems and very small brush-like crowns. Whips tend to move heavily in windy conditions and when they sway can seriously damage the crowns of frame trees.
Competitors (C)	Neighbours of frame trees that potentially have a negative effect on the growth and development of frame trees.

Nurse trees (N)	Trees of the over-, mid- and understorey that are beneficial to frame trees, e.g. by providing shelter against climatic extremes, by preventing epicormic growth or the growth of invasive ground vegetation (bramble, bracken).
Indifferent trees (I)	Dominant or sub-dominant trees are at sufficient distance to the frame trees or of small sizes so that they have no effect on them. Such trees can potentially replace damaged or diseased frame trees.

Crown thinnings (see Section 5.2) are instrumental in diversifying forest stand structure. Theoretically, the thinning type can also differ from frame tree to frame tree depending on tree size and dominance; however, frame-tree-based thinnings are usually carried out as local crown thinnings, i.e. dominant competitors of frame trees are removed. In local forest management methods, thinning intensity is defined by the number of competitors or by the sum of basal area of competitors to be removed in the vicinity of each frame tree. For traditional crown classes, see Section 4.2. In mature stands, where the frame trees approach their target diameters or even surpass them (e.g. in transformation scenarios when frame trees act as cover-building trees, see Section 2.4), it is also possible to apply local low thinnings rather than local crown thinnings. This is particularly recommended for mature *Picea* spp. and *Pinus* spp. frame trees.

In some studies, following the *competitor-for-frame tree rule* (Kruse et al., 2023), the sum of basal area is divided by the basal area of the frame tree (Schütz, 2000, see Figure 3.10). Although there is a considerable variance, the two examples from Ardross and Cannock Chase (UK) clearly show that the relative cumulative basal area of frame tree competitors decreases with increasing frame-tree size, i.e. larger more dominant frame trees need less removal of competitors (or smaller trees are removed) than less-dominant frame trees (Figure 3.10); hence, the recommendation to select dominant, vigorous frame trees. As frame trees become larger and more dominant, the trend curve approaches the horizontal line through 1 where the basal areas of competitor(s) and frame tree break even (Pommerening et al., 2021a). As Figure 3.10 has demonstrated, thinning intensity can obviously differ from frame tree to frame tree depending on the local release requirements of the tree in question, however, mean numbers can be defined as targets:

On average a removal of 0–2 frame-tree neighbours qualifies for a weak thinning, 2–3 neighbours for a moderate thinning and 3+ neighbours for a heavy thinning.

Local thinnings typically follow the growth dynamics of forest stands, particularly the tree crown-width development. Therefore, thinning intensity in local thinnings is heavy early in stand development and weak towards mature, old-growth stages when tree growth is much reduced (Bartsch et al., 2020). This temporal trend of thinning intensity can in principle also

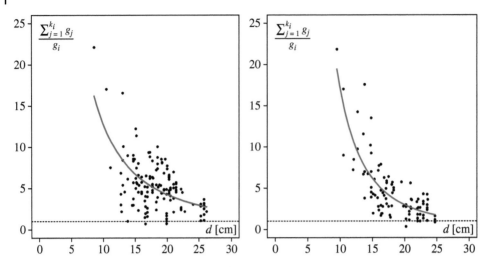

Figure 3.10 The relationship between the stem diameter at breast height, d, of frame trees and the ratio of sum of competitor basal area and frame-tree basal area for the marteloscope experiments Ardross 2013 and Cannock Chase 2012 (UK). g_i – frame-tree basal area, g_j – competitor basal area, k_i – number of competitors of frame tree i marked for thinning. The red trend line has been modelled using a simple power function. The dashed line marks the ratio value of 1.0, where cumulative competitor basal area and frame-tree basal area break even. Source: Modified from Pommerening et al. (2021a).

be concluded from Figure 3.10. Abetz and Klädtke (2002) recommended checking up on the performance of frame trees by monitoring the ratio of total height and stem diameter (h/d ratio, see Eq. 2.2) as a performance indicator: Increasing h/d ratios of frame trees imply that neighbouring tree crowns are closing in and as a consequence, the frame tree allocates more biomass to height than to diameter growth. The larger the h/d ratio, the more the stem diameters of frame-tree neighbours including potential competitors approach the size of the frame tree's stem diameter and reduce its diameter growth rate. A typical quantity illustrating the fundamental role of the h/d ratio in monitoring thinning requirements is the A thinning index that was suggested by Johann (1982):

$$dist_j^{\mathrm{crit}} = \frac{h_i}{A} \times \frac{d_j}{d_i}. \tag{3.1}$$

The index defines a critical distance, $dist_j^{\mathrm{crit}}$, between frame tree i and its nearest neighbours depending on the thinning-intensity parameter A. Any neighbouring tree j being located closer to tree i than the critical distance $dist_j^{\mathrm{crit}}$ needs to be removed according to this measure. Frame trees with a larger h/d ratio are relatively more heavily released than those with a smaller h/d ratio. Values of A are theoretically continuous; however, Johann (1982) defined discrete values from 4 to 8 with decreasing thinning intensity, i.e. 4 – very heavy and 8 – very weak thinning (Figure 3.11).

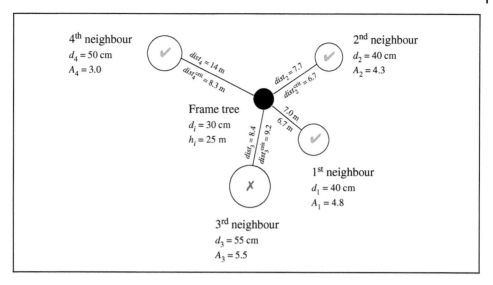

Figure 3.11 A worked example of the A-thinning index involving a frame tree, its four nearest neighbours and $A = 5$ (heavy thinning). The critical distances, $dist_j^{crit}$ and the A_j values were calculated based on Eq. (3.1), respectively. Ticked trees are not considered competitors and remain in the forest stand, whereas a cross means that the corresponding tree is removed. Source: Modified from Pommerening and Sánchez Meador (2018).

The thinning intensity parameter A can also be interpreted as a proportionality factor between the h/d ratio of frame tree i and the critical distance. Pretzsch (2009) pointed out that this proportional relationship does not hold for young forest stands with small diameters and older stands with larger diameters. In the first case, thinnings turn out to be too weak, and in the latter case, they become too heavy. His recommendation was to modify Eq. (3.1) in the following way:

$$
dist_j^{crit} = \begin{cases} 2 \cdot \dfrac{d_j}{d_i} & \text{for } \dfrac{h_i}{A} \leq 2 \text{ m}, \\[2ex] 6 \cdot \dfrac{d_j}{d_i} & \text{for } \dfrac{h_i}{A} \geq 6 \text{ m}, \\[2ex] \dfrac{h_i}{A} \cdot \dfrac{d_j}{d_i} & \text{otherwise}. \end{cases} \tag{3.2}
$$

However, in young forest stands, as already discussed, it may anyway be best not to remove competitors selectively but rather systematically by felling all other trees within a certain radius around each frame tree (halo thinning). The A-thinning index was originally developed for monospecies even-aged forests. For mixed species woodlands with different light requirements, different A-values can be applied to each species or species group.

It is also possible to re-arrange Eq. (3.1) to solve it for A whilst replacing $dist_i^{crit}$ by the observed distance between neighbour tree j and frame tree i:

$$A_j = \frac{h_i}{dist_j} \times \frac{d_j}{d_i}, \tag{3.3}$$

$\max(A_j)$ is then a frame-tree specific thinning intensity parameter (Hasenauer et al., 1996). The A thinning index also demonstrates the nearest-neighbour principle in local thinnings and is often used in computer simulations of individual-based forest management. The index of Eq. (3.3) has a high educational value when developing practical experience with selective thinnings. For analysing crown release, it is recommended to calculate the A thinning index before and after the thinning or marking of competitors (Pommerening and Grabarnik, 2019).

Assuming a commercial forest management scenario, for frame trees with $h/d > 80$ Abetz and Klädtke (2002) reckoned that it would take too long until trees reach their target diameters and individual-tree resilience to wind and snow would be quite low. For frame trees with $h/d < 40$, on the other hand, timber quality would be very low because of increased taper as a consequence of very open conditions. Therefore, $h/d = 80$ and $h/d = 40$ mark the boundaries of the realisation space of frame trees in the system devised by Abetz and Klädtke (2002). For example, the h/d development of *Picea sitchensis* tree no. 1947 in Artist's Wood (Gwydyr Forest, North Wales, UK) is well within the frame tree realisation space and close to the ideal line defined by $h/d = 60$ (Figure 3.12). This tree just happened to be a dominant tree in Artist's Wood without being particularly favoured in thinnings, which emphasises that frame trees are supposed to be dominant trees. Instead of plotting the h/d development of individual frame trees, it is also possible

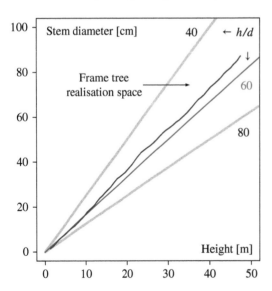

Figure 3.12 Frame tree realisation space (within the grey boundary lines) defined by $h/d = 40$ and $h/d = 80$. Ideal h/d is 60 and the blue line gives the observed h/d development of dominant *Picea sitchensis* (Bong.) Carr. tree no. 1947 in Artist's Wood (Gwydyr Forest, North Wales, UK). Source: Abetz and Klädtke (2002) and Pommerening et al. (2021a).

to monitor the mean h/d development of all frame trees of a given forest stand instead (cf. Figure 2.3). Naturally, in other climates, different numbers and relationships may apply. The monitoring of the h/d development also allows determining the thinning cycle, i.e. the time between two subsequent local thinnings (Pommerening and Grabarnik, 2019).

Based on growth trials and observational plots, Abetz and Klädtke (2002) also developed stem-diameter growth curves (so-called frame-tree norms) dependent on age or stand height for various species and sites to be used for comparison with field observations. The authors designed these frame-tree norms for comparatively slow (N norm) and for fast (S norm) stem-diameter growth. However, if data from long-term frame-tree management are not available, it is also possible to use the growth records of nearby dominant trees instead. Their growth rates would form an upper boundary of observational data and suitable growth curves for frame trees can then be defined through quantile regression (Cade and Noon, 2003; Pommerening and Grabarnik, 2019).

In commercial scenarios, the felling of individual frame trees can be considered once they have reached a previously specified target diameter. The frame trees are then not removed all at the same time, but staggered in time depending on their individual size maturity and on timber market conditions. Before their ultimate removal, frame trees are meant to contribute offspring to the next forest generation and some of them can continue to act as seed or habitat trees that are left in the forest until they die of natural causes; hence, this type of 'relaxed' target-diameter harvesting is sometimes referred to as *variable target-diameter harvesting* (Pommerening and Grabarnik, 2019).

3.6 Individual-Based Forest Management for Restructuring and Transforming Forests

In the past, comparisons between traditional global and local forest management methods were always based on global summary characteristics, particularly on stand growth performance. As a result, it was often found that certain variants of local thinning methods showed similarities with some global thinning methods, but discussions never went beyond such comparisons of stand yield characteristics (Bartsch et al., 2020). It has been a major misunderstanding and consequently an obstacle to a more universal uptake that the results of local forest management were mainly assessed in terms of how their stand yield characteristics compared with those of global forest management. These studies missed out on the fundamental difference of local forest management methods that centre on individual trees and therefore offer the opportunity to build spatial forest structure by assembling thinning cells or local management units (Figure 3.6) around each frame tree like tesserae in a mosaic. More than crown thinnings alone, individual-based thinnings are instrumental in increasing spatial tree size and species diversity because the localised approach prevents a homogenisation of forest structure provided the number of frame trees is moderate (Pommerening et al., 2021a).

Several researchers and practitioners have recognised this potential and used methods of individual-based forest management to increase size and species diversity and for restructuring forest stands. These methods also create a framework of resilient trees (Bartsch et al., 2020). The most prominent field of application of individual-based forest management is the transformation of plantations to continuous cover forestry (Schütz, 2001a), see Section 2.4.

To achieve transformation, Reininger (2001) developed and successfully tested a forest management method that he termed *structural thinning*. This is an important example illustrating how individual-based methods can be used to diversify and transform forests. The core of this method involves selecting frame trees in two different canopy strata allowing the maintenance of two-storeyed high forests on a continuous basis, cf. Section 5.5. He put his structural-thinning method to a long-term practical test in monospecific and mixed *Picea abies* woodlands at the Schlägl estate in the north-western corner of Austria at approximately 800 m asl (Figure 3.14). The method involves the simultaneous selection of permanent frame trees from upper (F_1) and lower canopies (F_2) at a fairly early stage of stand development (see Figure 3.13a). As part of this method and using modern standards, 100–150 F_1 trees are selected from the most dominant trees in the stand. All subsequent interventions strictly aim at releasing the crowns of F_1 trees from the perceived competition of matrix trees in the same canopy layer. During these operations, F_2 trees are only released to such a degree that they can survive in a shaded 'stand-by position' but do not emerge into the F_1 stratum just yet. However, in this concept, the main stand canopy is never fully closed at any time and the number of frame trees is moderate. Depending on initial stand conditions, 80–100 F_2 trees are recruited either from natural regeneration or suppressed trees of the same age as the main canopy trees or from both. F_2 trees are eventually released and allowed to progress into the main canopy when the target-diameter felling of the F_1 trees commences. The new F_2 trees are then recruited from natural regeneration. The selection and maintenance of F_2 trees diversify horizontal and particularly vertical forest structure.

The fundamental idea of this management method is that of a permanent two-storeyed forest (see Section 5.5), where target-diameter trees are not finally removed within a short period of time but are harvested individually as part of continued thinning operations

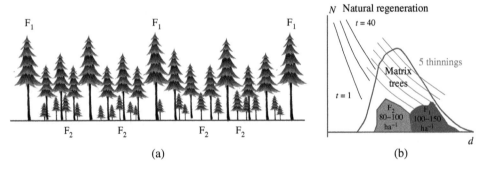

(a)   (b)

Figure 3.13 (a): Schematic illustration of the frame-tree method for promoting diverse forest structure described by Reininger (2001) aiming at a continuous two-storeyed conifer forest. F_1 – frame trees recruited from the upper canopy, F_2 – frame trees representing dominant trees in lower canopies. All other trees are matrix trees. (b): Empirical stem-diameter (*d*) distribution with modernised frame-tree numbers and indications of the transformation strategy (black and red curves). *N* – number of trees per hectare. Source: Adapted from Pommerening et al. (2021a) and Pommerening and Grabarnik (2019).

(a)   (b)

Figure 3.14 Impressions from the forest structure at the Schlägl forest estate in North-West Austria, specifically from the Hirschlatte monitoring site involving *Picea abies* (L.) H. KARST., *Fagus sylvatica* L. and *Abies alba* MILL. The forest stand has been monitored by Prof. Hubert Sterba of BOKU University Vienna since 1977 (Sterba and Zingg, 2001; Sterba and Sterba, 2018) whilst being transformed to a two-storeyed high forest by Reininger (2001). Source: Courtesy of Hubert Sterba.

(Weihs, 1990; Spiecker et al., 2004; Pommerening and Grabarnik, 2019). The most critical phase begins with the start of target-diameter harvesting. At that time, the F_1 trees have the greatest demand on space. Managers need to ensure that there are sufficient resources left for the survival of the F_2 trees. Research at the Schlägl estate has shown that the transformation of even-aged conifer plantations to two-storeyed high forests is possible within a comparatively short period of 40 years. The transformation strategy is indicated by the black and red curves in Figure 3.13b. Gradually, the bell-shaped stem-diameter distribution, which is typical of plantations, turns into a (negative) exponential one (see Chapter 6), which is typical of many but not all CCF woodlands, by the combined effects of natural regeneration and crown thinning of the main canopy (Reininger, 2001). The crown thinnings carried out at Schlägl essentially create conditions similar to a uniform shelterwood system in the overstorey (see Chapter 5.3.1) and thus keep the main canopy permanently open. Monitoring of the Hirschlacke stand at Schlägl has shown that the two-storeyed high forest gradually turns into a selection forest (see Chapter 5.4), when structural thinning is maintained (Sterba and Zingg, 2001). The author's Tyfiant Coed team of Bangor University (North Wales, UK) successfully applied this method to *Picea sitchensis* at Clocaenog Forest (plot 3 of the management demonstration area).

Li et al. (2014) proposed structure-based forest management as a method to modify spatial characteristics of forest stands. Here, the structure of the forest stand was modified by selecting trees for removal whose eviction would increase spatial species mingling, size differentiation and the diversity of tree locations, see Figure 4.2. In addition to the stem diameter distribution of residual trees, the authors also considered empirical distributions of nearest-neighbour structural indices.

The strength of frame-tree methods is in managing local neighbourhoods where the majority of important tree interactions takes place and thus in achieving structural differences and gradients throughout the forest stand. The real value of local forest management lies in a more efficient design of spatial forest structure and individual-tree resilience through a flexible and adaptive manipulation of neighbourhoods.

4

Forest Structure – The Key to CCF

Abstract

In forest practice, it is common knowledge that the understanding of tree and forest structure is central to any silvicultural activity. This is even more so in the case of CCF because of the explicit reliance on self-organisation processes that are meant to minimise the need for human interventions, the associated costs and the human impact on the forest ecosystem. Forest managers actively modify forest structure to achieve a certain, desired outcome, may the objective of forest management be traditional timber production, conservation, recreation or mental health therapy to name but a few. The ecological behaviour of forests as systems is much dependent on forest microstructure. All human interventions in silviculture and nature conservation are essentially goal-oriented modifications of woodland structure. Therefore, a good understanding of structural patterns helps predict the likely achievements of forest management. This chapter outlines and explains the significance and concepts of non-spatial woodland structure. This also includes the presentation of suitable criteria and summary characteristics for the monitoring of transformation of plantations to CCF and how these measures can be calculated using R.

4.1 Introduction

Structure is a fundamental notion referring to patterns and interactions between their components in more or less well-defined systems (Pukkala and Gadow, 2012).

> Characteristics of woodland structure provide information on the distribution of tree properties in space and time or in the population.

Typical of biological structures are repetitive patterns which are the result of complex interactions. They can be comparatively simple such as the structure of honeycombs or more complex. The fact that structures determine processes and that processes in return modify structures is well known in natural sciences. The patterns we observe and monitor in forests are typically the traces processes leave behind, and they allow us to develop

Continuous Cover Forestry: Theories, Concepts, and Implementation, First Edition. Arne Pommerening.
© 2024 John Wiley & Sons Ltd. Published 2024 by John Wiley & Sons Ltd.

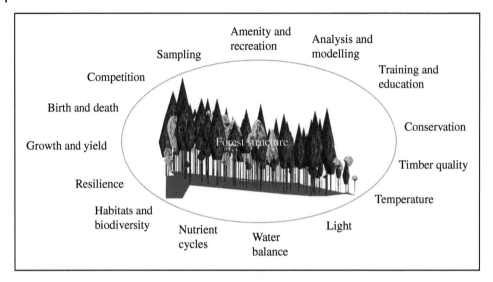

Figure 4.1 Selection of ecosystem functions, goods and services connected to forest structure. These can be interpreted as structure–property relationships. Source: Pommerening and Grabarnik (2019).

hypotheses that we can test (Gavrikov and Stoyan, 1995). However, structure is not only the result of past processes but also the starting point and often the cause for future developments (Pukkala and Gadow, 2012). Forest structure, for example has a strong influence on tree growth processes (see Figure 4.1). Every individual tree, however, also contributes to changes in forest structure, but this is a comparatively slow, long-term process (Pretzsch, 2009). Also, any impact on forests – whether natural or human-induced – is primarily a change of forest structure. In a second step, the modified structure then affects tree physiology and other processes in the forest ecosystem (Pommerening and Grabarnik, 2019).

There is a close relationship between structure and property, which is well known in materials science, physics, geology and the term *structure/property relationships* has been coined in those subject areas (Torquato, 2002). According to this term ecological processes not only leave traces as patterns but also the structure of materials or of a forest also determines to a large degree the properties of the system under study. The individual components of a material or the tree species or tree sizes of a forest arranged in various ways may result in very different properties for the material or forest as a whole. This suggests that the complex interactions between the components lead to a dependence of the effective properties on non-trivial details of the microstructure. In forestry, the principles of self-organisation and structure–property relations were first discovered in Central Europe in the eighteenth century when the first shelterwood systems (see Section 5.3) were designed and marked the beginning of the aforementioned emancipation of forestry from agriculture, see Section 2.6. At that time it was discovered that tree regeneration can be obtained naturally once the parent trees in the overstorey adhere to certain structural requirements. In the case of the uniform shelterwood system, density of the overstorey trees was reduced systematically so that a uniform pattern of released tree crowns and comparatively small gaps in between the tree crowns would be the result (see Section 5.3.1). This density reduction changed the light condition at the forest floor and soil moisture and nutrient regimes. After all, *density* is a fundamental characteristic of forest structure (Illian et al., 2008). Tree density management first began in the context of silvicultural systems and only much later was discovered as methods for stimulating tree growth and for improving timber quality (Hasel and Schwartz, 2006). Cotta (1816), for example, coined the thinning principle 'early, frequent but moderate'. However, in Central Europe, thinnings were only introduced late in the nineteenth century and systematic research on the long-term effects of thinnings on the residual forest stand started only around 1870. The long-term nature of growth and yield research is to blame for the fact that the first reliable results were only available in the 1950s. Thinnings are now a key element of forest management, particularly in continuous cover forestry (CCF). Thinnings are essentially tree removals (see Section 5.2) and an important tool for modifying forest structure. Silvicultural systems are also the results of very specific, localised thinnings. Along similar lines, planting, replanting and underplanting can also be interpreted as strategies with a view to modify existing structural patterns.

In forests, habitat functions, biodiversity, biomass production, recreation and even positive effects on human health are essentially system properties (see Figure 4.1). This implies that the results of forest management, e.g. forest products, but also conservation and woodland climate, are examples of such system properties, which are largely determined by forest structure.

Competition and survival of trees are such properties as well as the sampling error of forest resource inventories, which are strongly correlated with spatial forest structure. Spatial statistics and inventory research are therefore closely related. Pommerening and Stoyan (2008) and Nothdurft et al. (2010), for example, found that sampling results can be improved by reconstructing spatial forest structure and deriving estimators from the reconstruction. A proper understanding of woodland structure and its temporal evolution also plays a decisive role for forest modelling where the inclusion of spatial measures is often used to generalise models.

Landscape and forest structure determine the occurrence and population dynamics of brown bears, owls and woodpeckers to such a degree that direct conclusions concerning habitat and population development can be made from forest structure (Binkley, 2021). Similar results have been found for other birds, beetles, spiders and other animals living on and in forest trees. Pattern recognition helps identify distinctive spatial patterns and link them with the corresponding properties. This strategy has, for example been used when designing variable-density thinning (VDT), see Section 2.4.3.

In upland protection forests, the dispersion of trees and forest structure has an important influence on the ability of forests to intercept stones from rockfall, avalanches and landslides that threaten human settlements (Brang, 2001; Brang et al., 2006; Brauner et al., 2005; Dorren et al., 2005; Rammer et al., 2015).

In all these examples, the properties of the whole system, the forest, and the interaction between its components, the trees, depend on the microstructure of the system. In materials science, structure–property relationships are employed to predict the properties of a material from known details of the spatial microstructure. In a similar way, the projection of forest growth and yield has for a long time been the most important objective of tree and stand modelling in forestry (Pretzsch, 2009). Forest structural analyses form the basis for silvicultural strategies and pave the way to new ecological theories.

Naturally, there is a close link between structure and diversity. Measures of forest structure often are also good diversity indicators and vice versa. The term 'diversity' relates to the variability and unpredictability of living organisms and describes the variation at all levels of biological organisation (Gaston and Spicer, 2004; Dale and Fortin, 2014). Tree diversity is part of organismal diversity that relates to taxonomic scales, their differences and their components.

Commonly the diversity within a population of trees, i.e. a forest stand, is denoted as α diversity whilst β diversity describes diversity differences between populations/stands. Forest structure and α diversity can be subdivided into three different aspects, the tree *location*, *species* and *size* diversity (see Figure 4.2). It is recommended to address all three aspects in order to describe woodland structure as holistically as possible.

When statistically reconstructing forest structure, it helps to include summary characteristics representing all three aspects (Pommerening and Stoyan, 2008). In practical applications of the three aspects of forest structure, it is useful to think creatively. The examples in Figure 4.2 emphasise that almost any tree characteristic can and should be used to underpin these three basic categories as long as they fulfil the requirements of the corresponding definition.

Tree species and size diversity can be quantified in a non-spatially explicit or in a spatially explicit way. The diversity of tree locations is by definition always spatially explicit. Species diversity relates to a qualitative characteristic, i.e. species, whilst the other two aspects of α-diversity are quantitative. In Figure 4.2, other examples of qualitative information (central column) are given which can statistically be treated in a way similar

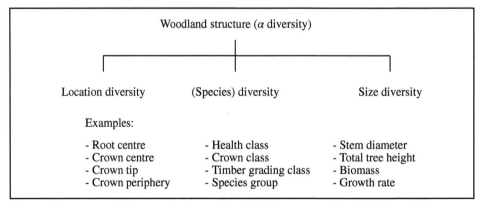

Figure 4.2 The three major characteristics of forest structure and α-diversity Source: Pommerening and Grabarnik (2019).

to species diversity (Pommerening and Grabarnik, 2019). Fichtner et al. (2018) defined the term *individual-based biodiversity effect* as the net effect of all intra- and interspecific interactions within the neighbourhood of a given tree and quantified it in terms of relative volume growth.

Another way of subdividing aspects of forest structure is to distinguish between horizontal and vertical elements. A sufficient degree of vertical forest structure is an important habitat requirement for many animal species such as the red squirrel (*Sciurus vulgaris* L.) (Flaherty et al., 2012). Vertical structure is also an important part of the definition of selection forests and of other forms of CCF (Schütz, 2001b), cf. Figure 2.15.

Many different concepts of diversity and structural measures have been developed to characterise and to analyse forest structure. They are mathematical expressions of species, size and location diversity of a population and an important pre-requisite for forest structure research (Magurran, 2004; Krebs, 1999). Such mathematical measures are usually more informative than just the number of species or the mean and variance of tree size. They also play a crucial role in monitoring the transformation of plantations to CCF.

Trees fill available space and their morphology reflects the geometry of that space. The morphology of open-grown trees is very different to the morphology of forest trees. This is the result of a range of complex interactions, one of which is competition. Interaction or interdependence implies that a change in one organism will result in a subsequent change of others (Kimmins, 2004). Competition for resources is a negative form of plant interaction. It occurs when two or more individuals attempt to utilise the same resource when that resource is in limited supply (Kimmins, 2004; Perry et al., 2008).

Competition sets in motion an interaction between individuals leading to a reduction of the performance (e.g. in terms of survival, growth and reproduction) of at least some of the competing individuals (Begon et al., 2006). Trees compete for light, water and nutrients. Competition pressure exerted by surrounding trees influences the growth of a tree. A tree at a location with few, small and distant neighbours likely captures more resources and grows better than the same-sized tree would grow in the presence of many neighbours that are close and large (Binkley, 2021). Too much competition pressure can be lethal and can cause what is referred to as natural mortality or self-thinning. In managed forests, the management of competition is one of the most important tasks of foresters: Through selective thinning and harvesting more resources are allocated to the remaining trees.

Thinnings (cf. Section 5.2) modify growing space and soon after release a tree puts energy into adapting its morphology and into occupying the newly available space (see Figure 4.3). The response to competition and other interaction processes can be understood as self-organisation. Forest structure is modified by interactions between individual trees, which to a large degree is influenced by the initial structure. Wind and snow are important environmental factors that influence tree growth dependent on growing space. This leads to biomechanical optimisation through goal-oriented biomass allocation. There is also positive interaction between trees that is often referred to as facilitation (Berkowitz et al., 1995; Dickie et al., 2005). Neighbouring trees often support other trees and shelter them from the forces of wind and snow. Also, mycorrhizal interactions between different species can have synergy effects in mixed species forest stands (Perry et al., 2008). All these interactions play important roles in CCF.

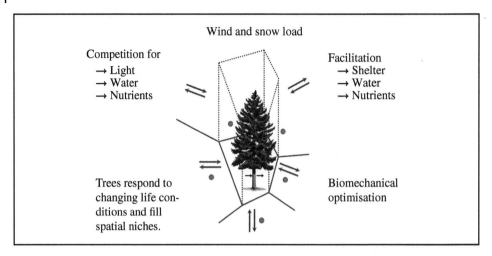

Figure 4.3 Tree reactions to changing growing space. The red dots symbolise the locations of surrounding trees in a forest. Source: Pommerening and Grabarnik (2019).

4.2 Crown Classes

An early though – from a modern perspective – somewhat crude approach to understanding forest structure was the development of crown classes. This was a qualitative system of classifying trees according to their sizes and morphological properties. It served and still serves to this day as the basis for distinguishing between seemingly similar trees for deciding which trees should stay and which should go in a scheduled management intervention. For inexperienced staff applying crown classes and selecting trees for removal is not an easy exercise in any forest and management type. However, crown classes were instrumental in making selective thinnings possible, since they were intended to guide forestry staff which trees to select for a given objective. They were also occasionally used in forest research for analysing tree growth and other properties by certain groups or clusters, i.e. the crown classes. Classifying trees according to 'social' or 'crown' classes goes as far back as 1844 (Hasel and Schwartz, 2006, Figure 4.4).

Crown or *canopy classes* (Hart, 1991) were originally devised for homogeneous even-aged forests and have traditionally been the basis for defining global thinning types (see Section 5.2) and intensities (Pretzsch, 2009). In CCF woodlands and particularly in mixed-species forest stands, differences in crown classes are usually much more distinctive than in Figure 4.4.

1. Predominant trees	Most dominant trees of a forest stand with exceptionally strongly developed crowns often above the level of the main canopy.
2. Dominant trees	Dominant trees forming the main canopy and having comparatively well-developed crowns.

Figure 4.4 Crown classes according to Kraft (1884). The numbers and letters are explained in the following box. Source: Kraft (1884).

3. Co-dominant trees	Crowns extend into main canopy, but comparatively weakly developed and narrow. Crowns start to degenerate. Lower limit of dominant trees.
4. Dominated trees	Dominated trees with heavily squeezed or one-sided crowns (flag-shaped).
	a. Free crowns tops in the mid-storey.
	b. Partly overtopped crowns in the understorey with beginning crown dieback.
5. Suppressed trees	Crowns completely overtopped
	a. but surviving (shade tolerant species only).
	b. dying or dead.

When applying the crown-class system, trees are classified according to their total heights, i.e. their relative position in the vertical forest stand profile, and their crown morphology, i.e. lateral and vertical crown dimensions. This visual assessment implicitly uses correlations between crown size and shape and competitive status of trees in a forest stand based on the assumption that light is the limiting environmental factor, which in many forest ecosystems such as the boreal forest or in arid zones is not true (Högberg et al., 2021, see Figure 3.3), but similar remedies as for light deficiency may apply. For practical and scientific use, it is also important to distinguish in this context between light demanding and shade-tolerant tree species, see Appendix B.

1. Dominant trees (D)	Most dominant trees of a forest stand with crowns above the level of the main canopy facing hardly any lateral competition.
2. Co-dominant trees (CD)	Dominant trees forming the main canopy with little lateral competition.
3. Sub-dominant trees (SD)	Crowns extend into main canopy, but face strong lateral competition. Crowns are therefore smaller and irregular in shape.
4. Suppressed trees (S)	Trees with overtopped crowns under the canopy of the main canopy.

For example, a mixed broadleaved forest stand with light demanding tree species in the overstorey can have overtopped *Fagus sylvatica*, *Carpinus betulus* and *Tilia spp.* trees with well-developed crowns (crown class 5a). Such trees are not necessarily in danger of dying any time soon.

There are different numbers of classes, names and coding systems in the literature (Assmann, 1970; Bartsch et al., 2020; Burschel and Huss, 1997; Köstler, 1956; Smith et al., 1997). Anglo-American sources often use letters (D, CD, SD, S) while numbers are common on the European continent (1–5). Also, continental European systems tend to assess both competitive status and crown shape, whilst Anglo-American consider the perceived competitive status only. Assmann (1970) associated classes 1–3 with the overstorey, class 4 with the middle and class 5 with the understorey.

Often, a more simplified version of Kraft's original system is used (Bartsch et al., 2020). This has, of course, limitations in more complex woodlands involving tree species with very different light demands (cf. Appendix B). Because of practical difficulties to decide which trees of a forest stand belong to which crown class and a certain degree of subjective judgement the original system by Kraft was refined and modified various times (Pretzsch, 2009). One could argue that class S in the Anglo-American system includes classes 4b–5 in the Kraft system, whilst class SD includes Kraft classes 3–4a. Class D perhaps matches class 1 and class CD is equivalent to Kraft's class 2. Crown classes clearly are blunt instruments, still they are helpful to learn to appreciate size and morphology differences of individual trees and therefore, pave the way for individual-based forest management (cf. Chapter 3). They have been and still are the basis of global selective thinning methods (Section 5.2). Crown classes also offer an easy approach to understanding the concept of competition.

The less dominant a tree, the smaller its absolute growth rate (AGR), the less resilient it is to wind or snow, and the smaller usually is its potential for benefitting from thinnings. The management effort required for maintaining sub-dominant and suppressed trees is high whilst the risk of failing is high, too. Frame trees (cf. Chapter 3) should therefore be at least co-dominant if not dominant trees (Codoc, 2020).

In commercial forestry, it has often been criticised that timber quality and other tree characteristics are not reflected by this system. An option to refine Kraft's original crown class system would be to stratify trees according to a number of height zones and to classify the trees whose crown tips fall into one or another of the height strata separately for each canopy stratum, see Table 4.3 (Oliver and Larson, 1996). In forest research, crown classes should only be applied with great care, as the classification is not based on quantitative criteria and can therefore be subjective (see Figure 8.1). A crown-class system for individual-based forest management is shown in Section 3.5. An alternative to crown classes is the use of the stem-diameter or basal-area percentile of each tree. This gives the relative position of each tree in the empirical stem–diameter distribution. A quantitative and spatial variant of crown classes is the size dominance nearest neighbour index, see Pommerening and Grabarnik (2019, Chapter 4).

4.3 Mixing Species – But How and When?

Managing mixed-species forest stands is an important tenet of CCF, see Section 1.5. In many parts of the world, diverse mixed-species woodlands are the normality in nature. In fact, mixed-species stands are more common in nature than single-species stands (Duchiron, 2000; Smith et al., 1997). Tendencies towards monospecies natural forests occur, for example, in lowland *F. sylvatica* woodlands in Central Europe; however, taking a worldwide perspective, such a tendency is not typical. A tree vegetation involving a diverse species composition usually leads to a better use of niches and higher efficiency of niche complementarity. For example, species with comparatively flat and deep rooting systems can facilitate each other by exchanging water and anions along the vertical soil profile. According to the *stress-gradient hypothesis*, this is particularly important with ongoing climate change and associated droughts, when facilitation effects are likely to increase (Bertness and Callaway, 1994). Such effects put mixed-species woodlands into a better position to buffer the impact of disturbances and sometimes even result in greater productivity compared to monospecies forests (Pretzsch, 2019). Mixed-species woodlands are also said to have good tree–soil interactions through exploring different species-specific soil depths and shedding leaves that come with different decomposition rates. More tree species involved usually also imply a greater choice of different habitats. Mixed-species woodlands are usually also more resistant to biotic and abiotic damage.

From the point of view of commercial forestry, the reduction and spreading of risks and the diversification of products that come with different species lead to better business results and increase economic flexibility (Duchiron, 2000). In mixed-species forests, there are also more options for the next forest generation, and such woodlands are usually popular with visitors who prefer seeing a mix of different species and sizes at close proximity (Jephcott, 2002; Petucco et al., 2018). Towards medium and advanced stand development stages, species diversity and structural heterogeneity can significantly increase stand productivity. The main reason for this is a better use of canopy space, light and nutrients by trees situated in different canopy layers (Pedro et al., 2015; Zeller et al., 2018; Zeller and Pretzsch, 2019).

Disadvantages or challenges associated with mixed-species forests may include potentially higher establishment and tending costs, as seedlings and saplings need to match microsites and specialised thinnings may be required to steer interspecific competition in a sustainable way. Also, forest managers need more silvicultural skills and ecophysiological knowledge to understand the interaction dynamics between different species for making appropriate management decisions. Frequent monitoring is necessary, particularly if many species are involved. Monitoring and intervention costs increase the more species co-exist that do not harmonise with each other on a given site. For mixed-species woodlands, yield and stand development are more challenging to anticipate and to project into the future, particularly if flexible, individual-tree spatially explicit simulation models are not available. Recently introduced tree species that are still comparatively rare in the area may attract deer (Partl et al., 2002) and tempt them to graze the seedlings (cf. Figure 4.5).

Figure 4.5 Native *Fagus sylvatica* L. saplings planted under non-native mature *Picea sitchensis* (Bong.) Carr. as part of conversion/forest restoration (see Section 2.4) in the Nordfriesland Forest District (Germany). School children on work experience in the district treated the lateral and terminal buds with sheep wool to deter roe deer (*Capreolus capreolus* L.) as an alternative to fencing. Source: Courtesy of Paul Cody.

There can be various kinds of motivation for introducing admixed species, and they can be prompted by problems in current forest stands (Bauer, 1962). Such reasons can include

- Improving economic stand value by introducing a more valuable species that is likely to fetch higher timber prices on the market than the current species alone,
- Playing a support role in terms of soil conservation (and conservation of soil biota) by producing litter with low total carbon to total nitrogen (C/N) ratio that easily decomposes and by controlling competing ground vegetation or nursing timber quality,
- Improving stand resilience to wind and snow disturbances,
- Averting bark beetle and fungi diseases through tree species diversification,
- Increasing the resilience to early and late frost events,
- Accelerating growth of the current species,
- Improving forest structure and forest microclimate through greater 'verticalisation' of canopies, i.e. through increasing the use of the vertical growing space.

Motivations for increasing species diversity vary much and cover the whole range from commercial to conservation forestry. Depending on the individual situation, there can also be a number of traditional, specific reasons for growing mixed-species forest stands. Most of these are based on local and regional experience and not all of them can be generalised. However, they may serve as good starting points for thinking about why a forest manager would want a mixed-species forest stand (see Table 4.1). Key to the example advice offered in Table 4.1 are the properties of the species suggested. For example, growing shade-tolerant species below light-demanding species can help not only to suppress competing ground vegetation but also to nurse commercially valuable stems of light-demanding species with a view to preventing epicormic growth. Care needs to be taken that the species used for such underplanting will not overtop and eventually outcompete the light demanders (cf. Appendix B). On the other hand, the amount of light available in lower canopy storeys and temperature can be increased by a mix of tree species that have shade casting and light canopies. Along similar lines, admixed tree species can improve litter decomposition at the forest floor. Naturally, there can also be aesthetic and commercial reasons for tree species mixtures. Mixing tree species has particularly two consequences (i) a change in the use of resources for tree growth and (ii) a diversification of growth conditions through a modification of canopy and soil space. This can potentially lead to spatio-temporal complementary effects resulting in a more effective use of resources and higher productivity (Pretzsch, 2019). Such effects can in turn increase the packing density of tree crowns.

Mixed-species forests can owe their existence to many different events. Natural regeneration where species from nearby stands colonise or even invade a stand is one possible origin. A frequent one is the colonisation of partially unsuccessful replantings or of mature stands with open canopies, recall Figures 2.5 and 2.6. For example, many seedlings in replanted *P. abies* plantations die and before forest managers can replant the gaps *Betula* spp. has invaded them. Naturally, it is also possible that the forest manager takes seedling mortality as an opportunity and deliberately refills mortality gaps on a replanting site with another species. The same opportunity arises after small-scaled disturbances in mature

Table 4.1 Examples of specific objectives for adding species to an existing forest stand.

Objective	Mixture type
Suppressing competing ground vegetation	Add *F. sylvatica* to *P. sylvestris*
Facilitating litter decomposition	Add *Tilia* spp. to *Larix* spp.
Promoting good soil structure	Add *C. betulus* to *Quercus* spp.
Increasing stand stability	Add *P. sylvestris* to *P. abies*
Promoting higher temperatures	Mix *A. alba* with *Larix* spp.
Improving light conditions	Mix *F. sylvatica* with *Quercus* spp., ~ *A. alba* with *P. sylvestris*
Improve forest structure	Add *Tsuga* spp. to *P. menziesii*, ~ *F. sylvatica* to *Larix* spp.
Good use of vertical root space	Mix *P. abies* with *A. alba*
For early returns	*P. abies* as temporary mixture in *F. sylvatica* and *Quercus* spp.
For final returns	*P. sylvestris* and *Larix* spp. as valuable timber in *A. alba*, *P. abies* or *F. sylvatica*

'Add' means adding a species to another one at a later stage and 'mix' implies mixing equivalent species from the start.
Source: Adapted from Bauer (1962).

forests. Increasingly, mixed-species woodlands are planted from scratch, either on clear-felling/windthrow sites or on former agricultural land, see Section 2.3. Another origin of mixed-species forests is underplanting, as detailed in Section 2.4.1.

For managing mixed-species woodlands, it is useful to distinguish different generic types of tree species mixtures (Duchiron, 2000). They come with certain properties, have different pre-requisites and support the understanding of species interactions. In *equivalent* or *genuine mixtures*, two or more tree species are equally important, have more or less the same function, the same economic or natural life expectancy, they are part of the same canopy layer(s) and no species assumes a support or nursing role (Figure 4.6a). Equivalent mixtures are usually *permanent* mixtures, i.e. the idea is to maintain the mixture at all times. There are two possibilities mixing species with (i) *similar* and (ii) *dissimilar* growth and competition patterns (Bartsch et al., 2020).

The tree species involved in equivalent mixtures typically have similar growth and competition patterns and are said to be *stable mixtures* because they can persist even without major human interventions. Such mixtures are naturally of particular interest in CCF, because keeping management input and impact to a minimum is usually part of the definition of CCF (biological rationalisation, see Section 2.6). Examples of species with similar growth and competition patterns include *P. abies – F. sylvatica*, *F. excelsior – A. pseudoplatanus* and *P. sylvestris – Q. robur*; however, their exact, local interaction also much depends on local environmental factors. The species involved can be mixed in

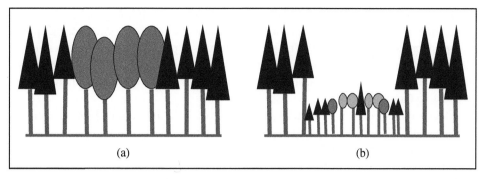

Figure 4.6 Equivalent (genuine) tree species mixture (a). Gap tree species mixture (b). Inspired by Otto (1994), Burschel and Huss (1997) and Schütz (2001b).

small-to-large groups or as individual trees (as shown in Figure 4.6a) and occupy the same canopy layer (Davies et al., 2008).

Where the species involved have slightly dissimilar growth and competition patterns, mixtures are potentially *unstable* in the long term, i.e. one or more of the component species may be outcompeted by others. Such species mixtures can only persist, if they are supported by costly human interventions or if they are spatially segregated. Corresponding forest structures have a natural tendency to lose the element of mixed species. In such cases, a comparatively gentle and inexpensive intervention involves arranging species with dissimilar growth and competition patterns in small-to-large species-specific groups with a view to spatially separate the species, at least for some time. In these groups, each species is protected from another, more competitive species and only faces intraspecific interactions. In addition it is possible to give the less-competitive species, a head start by planting it earlier or allowing it to establish earlier thus delaying the development of the more competitive species (Pretzsch, 2019). Inferior species (in terms of growth and competition) may still require support in subsequent thinnings for support.

Mixtures of advanced and delayed species include species with grossly dissimilar growth and competition patterns. Mostly, they are mixtures of light-demanding and shade-tolerant species, where the light-demanding species initially grow faster and subsequently the shade-tolerant species catch up with them and prove to be more competitive in the long-run (cf. Figure 4.10 and Appendix B). Examples of species with dissimilar growth and competition patterns include *P. abies – Betula* spp., *F. sylvatica – F. excelsior* and *F. sylvatica – Larix* spp. Despite dissimilar properties, these species often occur in mixtures, as in combination the species deliver highly desirable ecosystem goods and services. Naturally, such mixtures of advanced and delayed species require great silvicultural skills (Davies et al., 2008).

In *gap mixtures*, the main canopy may include only one or two species and species different from those in the main canopy exist in gaps (Figure 4.6b). The species not present in the overstorey can originate from advance regeneration colonising from adjacent forest stands or from artificial regeneration where different species were introduced by planting in canopy gaps. As there are abundant resources in larger canopy gaps, almost any tree species can grow here at least in parts of the gap area (cf. Section 5.3.2). The larger the gaps, the more light-demanding species are favoured. The species composition and horizontal dispersion of trees in the gaps much depend on the microsites in the gaps and their environmental conditions. The exposition and geometric shape of the gaps influence the species composition (Streit et al., 2009). The gaps can be cut as part of a silvicultural system (cf. Section 5.3.2) or stem from localised windthrows, snow damage, localised fires, or bark beetle infestations. As such, small localised disturbances can be an opportunity for diversifying an existing forest stand.

In *temporary species mixtures*, the species involved have different natural or economic lifetimes, i.e. the mixtures are not permanent or long-term. This can also include situations where one or more species fulfil a temporary support role. Temporary mixtures can be within the same canopy storey (Figure 4.8a) or in different ones (Figure 4.8b). Examples of the first situation can include *P. abies* in *F. sylvatica* or *Prunus avium* L. in *F. sylvatica* or *F. excelsior* (see Figure 4.7). The species with the shorter life expectancy can be mixed

Figure 4.7 A mixed-species woodland including *F. sylvatica* L. and *Fraxinus excelsior* L. on lime stone at Reinhausen Forest District (Lower Saxony, Germany). There is a large-sized *Prunus avium* L. frame tree (with dark bark and red ribbon) surrounded by much smaller *F. sylvatica* nurse trees in the centre of the photo. This is typical example of a temporary tree species mixture in a commercial scenario, as *P. avium* produces valuable timber and is harvested at an age between 70 and 90 years, whilst the other species remain standing for at least 150–180 years. Source: Courtesy of Keith Blacker.

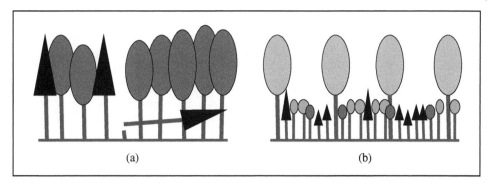

Figure 4.8 Temporary tree species mixture in the same canopy storey but different lifetimes (a). Temporary tree species mixture in different canopy storeys (b). Inspired by Otto (1994), Burschel and Huss (1997) and Schütz (2001b).

individually or in small- to medium-sized groups. Another variant of temporary species mixtures is implemented in nurse crops/artificial shelterwood systems (Figure 4.8b). We discussed this method in detail in the context of creating instant CCF woodlands on bare land (cf. Section 2.3). Here, it is the overstorey species that is usually removed earlier rather than the target species in the understorey.

Species mixtures can also occur between different canopy storeys (Figure 4.9a). Schütz (2001b) termed them *vertical tree species mixtures*. This mixture type can have various reasons, and most prominently is the consequence of applying shelterwood systems (cf. Section 5.3.1), the selection system (cf. Section 5.4) and the principles of the two-storeyed high forest (cf. Section 5.5). Due to a lack of natural regeneration from the main-canopy species, another species can colonise the understorey and eventually emerge into the mid-storey. This was, for example the case in the Pen yr Allt Ganol woodland described in Figure 4.25, where a second canopy of predominantly broadleaved species was established under the main canopy of *P. sylvestris* and *P. sitchensis*. Such mixtures can then develop towards a temporary or permanent two-storeyed high forest. One option is to keep two species even in the same canopy storey, but to maintain only co- or sub-dominant trees of the more aggressive species whilst cutting all dominant individuals of that species to avoid any imbalance between the two species.

Mixtures of main and supporting species commonly use a shade-tolerant species that is supposed to nurse the stems of commercially valuable overstorey trees for maintaining high timber quality, see Section 7.3.1 and Figure 4.7. This is often achieved by underplanting *Quercus* spp., *Larix* spp., *P. sylvestris*, or *Fraxinus/Acer* spp. with a shade tolerant species, i.e. *F. sylvatica* on poor soils and *Carpinus betulus* or *Tilia* spp. on richer sites, a few decades later.

A geometric variant of equivalent group mixtures are *strip mixtures* (Figure 4.9b). Strip mixtures are in principle just a variant of group mixtures, but they are quite common and therefore receive special attention here. The woodland Pen yr Allt Ganol is again an example where a large strip of *P. sitchensis* was planted within a wet depression in a matrix of *P. sylvestris*. Strip mixtures can also frequently occur along forest roads or at forest margins for conservation, aesthetic reasons or for preventing forest fires.

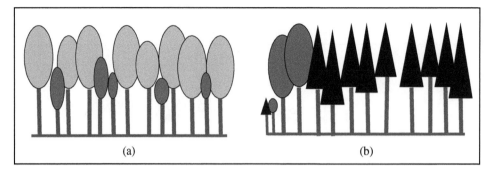

Figure 4.9 Permanent tree species between different canopy storeys (a). Permanent, equivalent (genuine) tree species mixture where one species is arranged in a strip (b). Inspired by Otto (1994), Burschel and Huss (1997) and Schütz (2001b).

When considering tree species mixtures it is important to check the compatibility of tree species on a given site in terms of growth patterns, life expectancy and light demand (cf. Appendix B). Here, it is particularly crucial to compare species-specific height growth patterns (Figure 4.10). According to the height-age curves in Figure 4.10, the species specific height-age curves intersect early, shortly before age 25. From that point on *P. abies* has a height advantage over all other species on the given site. If *F. sylvatica* were to be mixed with *P. abies* here, a head start of 10–15 years would be useful for *F. sylvatica* before *P. abies* is planted. The delay of introducing *P. abies* can perhaps also be shorter, as the two species otherwise harmonise quite well with one another.

The curves of *P. sylvestris* and *Quercus* spp. on the other hand are not so far apart. The fact that *P. sylvestris* overtops *Quercus* spp. can be tolerated, as *P. sylvestris* is very light demanding but also allows much light to penetrate its canopy (cf. Appendix B). Therefore, the relationship between *P. sylvestris* and *Quercus* spp. is not a problem on this site.

Where growth and competition patterns are similar, individual-tree mixtures and mixtures in small groups (up to 30 m in diameter) are possible. In situations where growth and competition patterns differ, it is more appropriate to spatially separate the species by arranging monospecific groups between 30 and 50 m in diameter.

In practice, to simplify underplanting and the management of natural regeneration, chequerboard patterns are often applied for organising monospecies groups in early stand development, which soon lose their rigid edges with ongoing natural mortality and thinnings. Some of the fields of such chequerboards are then deliberately left empty, i.e. unplanted, to allow for natural regeneration or to simplify the extraction of felled overstorey trees. In addition – as we have just discussed – there is the possibility to grant less-competitive species with slower growth a head start. The points of intersections of height-age curves of different species can give clues as to how large the head start needs to be (Pretzsch, 2019, Figure 4.10). Both the spatial separation of competing species and the temporal delay of the development of the more competitive species are crucial silvicultural tools for avoiding excessive intervention costs that would otherwise be necessary to maintain a chosen species composition. In arranging different species, e.g. in planting,

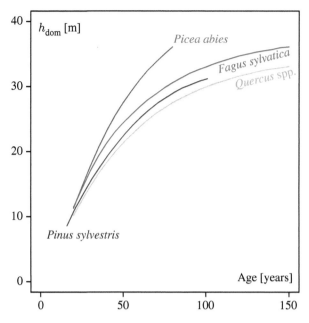

Figure 4.10 Top height development over age for four species taken from the British yield table system h_{dom} – top height. Source: Hamilton and Christie (1973).

they should be matched well with any microsites occurring, e.g. small depressions, slopes, dry outcrops and wet spots. As with any CCF, ideally species should be selected that are well adapted to the local site and are likely to survive the vicissitudes of ongoing climate change such as occasional summer droughts and strong winds (cf. Sects. 1.5 and 7.3.2).

Firmer clarifications of the suitability of tree species mixtures can be obtained through experimentation. Pretzsch et al. (2017) provided valuable details on experimental designs for studying the effect of tree species mixture.

4.4 Non-spatial Measures of Forest Structure

In silviculture and forest ecology, the term *structure* has for a long time been associated with *species* and *size* distributions of forest stands and tree populations. In the absence of efficient methods to survey and to process spatial information such distributions were the only sources of structural information. They are still of considerable importance, as despite advances in surveying and computing technology, data for non-spatial measures are much easier to collect and to process than those required for their spatial counterparts. Non-spatial measures of forest structure are also important pre-requisites for more precise, in-depth spatial analyses such as those carried out in point process statistics (Illian et al., 2008; Pommerening and Grabarnik, 2019).

Traditionally stem diameter and total tree height are the most common size variables used for quantifying non-spatial forest structure. More recently, biomass and carbon contents have become more important but are allometrically related to stem diameter and total tree height. Measures of non-spatial forest structure provide information on the distribution of tree properties in a tree population (Krebs, 1999). In this chapter, special attention is assigned to measures that can serve as indicators of structural changes in forest stands that are transformed to or maintained as CCF woodlands. The focus is on non-spatial measures of structure and diversity. Spatial indicators can be found in Pommerening and Grabarnik (2019, Chapter 4).

Pommerening et al. (2021b) discovered that spatial species and size diversity are often correlated in many woodlands throughout the world. This is related to the mechanisms of how different species communities with different size ranges spatially mix. Pommerening and Grabarnik (2019) used the verb *mingle* for the process of tree species mixing spatially in forest ecosystems. As part of this and following extensive studies, Pommerening and Uria-Diez (2017) proposed the *mingling-size hypothesis*. This hypothesis states that large trees have a tendency towards high species mingling, i.e. they tend to have many heterospecific nearest neighbours. This tendency enhances the conservation value of large, mature trees. From a forest management point of view, both findings imply that local size diversity often comes automatically with increasing species diversity and vice versa. This allows conservation or forest management to focus on maintaining or improving only one of these two diversity aspects, whilst obtaining the other as a by-product, thus providing even more options for biological rationalisation (cf. Section 2.6).

4.4.1 Species Diversity

4.4.1.1 Species Richness and Abundance

Species richness is the absolute number of species occurring in a certain region, whereas *abundance* is the absolute or relative number of a species per area or space unit (Gaston and Spicer, 2004). *Species diversity* is usually defined in terms of *species richness* and relative *abundance* (Newton, 2007). Species diversity is not the only aspect of biodiversity, but a rather important and the most commonly considered one (Kimmins, 2004). In the past, most importance has been assigned to species diversity, and there is a wide range of approaches to quantifying this aspect of diversity (Hui and Pommerening, 2014). The term 'species' can be flexibly extended to provenances or to species groups.

In the Allmitwald forest stand, *P. abies* dominates in terms of basal area, but the most abundant species is *F. sylvatica*. Both *A. alba* and *P. sylvestris* can be considered rare species (Figure 4.11a). Ranking species and plotting the empirical cumulative abundances in terms of number of trees and basal area are another way to establish differences in the number and size proportions (Figure 4.11b). The curves clearly show that the stand is dominated by one species with many but small individuals.

In species-rich Chinese ecosystems, Hui and Pommerening (2014) found that P_N and P_G often form an interesting saturation relationship (Figure 4.12). From rarest to commonest species, the abundances eventually increase more in P_G than in P_N, i.e. the most abundant species have more frequently individuals of larger size than the rarer species.

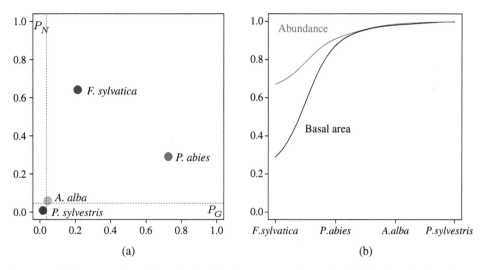

Figure 4.11 Visualisation of the relative abundance proportions of Table 4.2. Relative abundance, P_N, plotted over basal area proportions, P_G, with dashed lines showing the first quartiles (a) and empirical cumulative abundance (red) and basal-area proportions (black) over species ranked from most to least abundant (b).

There is often a gap between $P_G \approx 0.1$, i.e. the upper boundary of the other most abundant species in a forest stand, and the P_G values of 1–2 lead species that determine the asymptote (Figure 4.12).

This relationship is also a reflection of ecological carrying capacity and can be found in many species-rich forest ecosystems in different parts of the world (Pommerening and Grabarnik, 2019). Species richness is often interpreted as a surrogate measure for other kinds of biodiversity: More species usually lead to greater genetic variation, i.e. there is a greater diversity of genes in the population, and to greater size diversity. This implies greater ecological variation and a better exploitation of niches and habitats (Magurran, 2004; Gaston and Spicer, 2004; Krebs, 1999).

A species diversity index is a mathematical expression of species diversity in a community and provides important information on the occurrence and distribution of species in a community (Krebs, 1999). Such measures are more informative than just the number of species because relative abundance is considered as well. This is important for understanding the structure and processes of a community.

In the light of large-scale forest destruction and associated species loss, which may dramatically increase as a result of climate change and of the growth of the human population, the monitoring of species diversity assumes quite some importance. Forest management, silviculture and conservation have a particular influence on species diversity. In this context, the dynamics of forest succession and the effect of disturbances are crucial aspects of diversity research. Through niche complementarity species diversity can contribute to the resilience and growth of CCF woodlands (Binkley, 2021; Pommerening and Grabarnik, 2019). Processes of biological rationalisation (see Section 2.6) can also benefit from tree species diversity (Ammann, 2020).

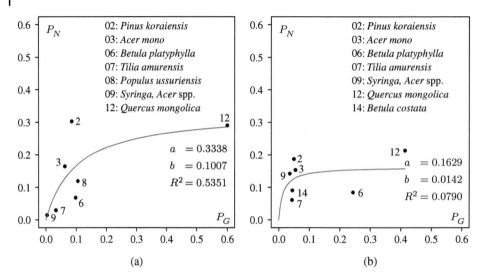

Figure 4.12 The relationship between the relative abundance proportions, P_N, and the basal area proportions, P_G, of the seven most abundant species in plots TFa (a) and TFk (b) in the Tazigou Experimental Forest Farm, Jilin Province, China (Wang et al., 2021). The numbers represent the original species codes and the parameters a and b are those of the Michaelis–Menten saturation function $P_N = \frac{a \times P_G}{b + P_G}$ used as trend line (red) (Hui and Pommerening, 2014). R^2 is the coefficient of determination. The authors of scientific tree names were omitted here for clarity.

4.4.1.2 Shannon and Simpson Indices

Examples of typical species diversity measures include the Shannon and Simpson indices. Both indices take the relative abundance of different species into account rather than simply expressing species richness. The Shannon index, also known as Shannon information index and related to the concept of entropy (Shannon and Weaver, 1949, see Eq. (4.1)), was independently derived by Shannon and Wiener and is an information theory index. It was originally proposed to quantify the uncertainty of information in a code or in a string of text (Krebs, 1999). The characteristic measured the uncertainty of the next letter in a coded message or the next species to be found in a community. The Shannon index assumes that individuals are randomly sampled from an infinitely large community and that all species are represented in the sample (Magurran, 2004). A monospecies forest would have no uncertainty and $H' = 0$. The Shannon index is affected by both the number of species and their equitability or evenness. The Shannon index was originally defined using the binary logarithm ($\log_2$), but natural ($\log_e$ or $\ln$) and decadic logarithms ($\log_{10}$) have been used as well. The evenness forms (first term of Eqs. (4.1) and (4.2)) are often used as standardisation to allow for comparisons between different monitoring sites (Pretzsch, 2009) and are constrained between 0 and 1. The terms $\log_e s$ in the denominator of Eq. (4.1) and s in the denominator of Eq. (4.2) represent maximum diversity. The transformation to evenness can, however, involve a certain loss of information.

$$E_{H'} = \frac{H'}{\log_e s} \quad \text{with} \quad H' = -\sum_{i=1}^{s} p_i \times \log_e (p_i) \tag{4.1}$$

p_i is the proportion of individuals found in the ith species, and s is the number of species.

Some ecologists consider $e^{H'}$, a transformation of H', also known as *equity*, see also Eq. (4.3) and the text following that equation. Equity gives the number of species that would have been found in a sample had all species been equally common. An evenness measure corresponding to $E_{H'}$ (Eq. (4.1)) can be calculated as $\frac{e^{H'}-1}{s-1}$ or $\frac{e^{H'}}{s}$. It is sometimes argued that the Shannon index confounds two aspects of diversity, i.e. species richness and evenness. An increase of the index may arise either as a result of greater richness or greater evenness or indeed both. Assuming the use of natural logarithms, Buzas and Hayek (1996) figured a way to decompose H' into its two components, i.e. $H' = \log_e S + \log_e E$ with $E = e^{H'}/s$. The decomposition allows a more refined interpretation of changes in diversity. For this purpose, the terms $\log_e S$ and $\log_e E$ can be made dependent on sample size or time or ratios such as $\frac{\log_e E}{\log_e S}$ can be considered (Magurran, 2004). Both species diversity measures can be calculated based on tree number proportions or size proportions. A common basis of size proportions is basal area, but volume and biomass are also possible. It is important to state what kind of proportions were used in the analysis. Basal area, g (measured in m²), relates to stem diameter, d through $g = \pi(d/2)^2$. In the case of size proportions, the conspecific tree sizes are taken into account and not only their numbers. This can make a marked difference, since unlike other vegetation trees can differ much in size. A comparison of number and basal-area proportions provides interesting information on the relationship between abundance and size dominance.

The Shannon index is easy to compute in R. Table 4.2 provides an example data set with absolute abundances, and these are used in the following listing:

```
> Example <- data.frame(
+ Ba = c(42.9, 12.6, 2.6, 1.0),
+ Sph = c(170, 375, 35, 5),
+ Species = c("NS", "BE", "SF", "SP")
+ )
```

The species coding is just an abbreviation of the English common names (**N**orway **s**pruce, **be**ech, **s**ilver **f**ir and **S**cots **p**ine) following the system of the British Forestry Commission. This information is just for convenience but not of importance in the following calculations.

Table 4.2 Species proportions (abundances) in terms of basal area and the number of individuals for trees with stem diameters larger than 5 cm in the Allmitwald, a mixed-species woodland with an overstorey age range between 90 and 110 years near Bern in Switzerland.

Tree species	Basal area [m² ha⁻¹]	Number of trees [ha⁻¹]
Picea abies	42.9	170
Fagus sylvatica	12.6	375
Abies alba	2.6	35
Pinus sylvestris	1.0	5
Total	59.1	585

In the following listing, the Shannon index is calculated in lines 1f. and 7f. for tree number and basal area proportions, respectively. These calculations are repeated by using the alternative decomposition method by Buzas and Hayek (1996) in lines 4f. and 10f. The corresponding evenness measures are calculated in lines 14 and 16.

```
1   > (ShannonSph <- sum(Example$Sph / sum(Example$Sph) *
2   + (-log(Example$Sph / sum(Example$Sph))))))
3   [1] 0.853377033981
4   > log(length(Example$Species)) + log(exp(ShannonSph) /
5   + length(Example$Species))
6   [1] 0.853377033981
7   > (ShannonBa <- sum(Example$Ba / sum(Example$Ba) *
8   + (-log(Example$Ba / sum(Example$Ba))))))
9   [1] 0.76849469859983
10  > log(length(Example$Species)) + log(exp(ShannonBa) /
11  + length(Example$Species))
12  [1] 0.76849469859983
13  > numberSpecies <- length(Example$Species)
14  > ShannonSph / log(numberSpecies)
15  [1] 0.61558140746646
16  > ShannonBa / log(numberSpecies)
17  [1] 0.55435174530972
```

In many data applications, the Shannon index is lower for basal area proportions than it is for tree number proportions. This is also the case in the Allmitwald woodland. Such an outcome implies that one or more species dominate the forest stand in terms of size, as can be clearly seen for *P. abies* in Table 4.2. In the Allmitwald woodland, *P. abies* dominates the stand in terms of size and *F. sylvatica* in terms of the number of trees per hectare, see Figure 4.11. The larger the difference in the results of the two variants of the Shannon index, the greater the size imbalance between species.

The index by Simpson (1949) gives the probability of any two individuals drawn at random from an infinitely large population belonging to different species (Magurran, 2004). As D increases, diversity decreases. Simpson therefore suggested that this probability was inversely related to diversity, but a complementary relationship implying $D = 1 - \sum_{i=1}^{S} p_i^2$ has also been put forward (Magurran, 2004; Sterba and Zingg, 2006).

$$E_D = \frac{D}{S} \quad \text{with} \quad D = \frac{1}{\sum_{i=1}^{S} p_i^2} \tag{4.2}$$

The Simpson index is often referred to as a dominance index which is weighted towards the abundance of the most common species whilst being less sensitive to species richness. In some publications, $\frac{n_i(n_i-1)}{N(N-1)}$ is calculated instead of p_i^2, where n_i is the absolute number of individuals of species i and N is the total number of trees (Magurran, 2004). Using Eq. (4.2), the R code for the Simpson index follows largely the same structure as the listing above for the Shannon index. Also, here the index is calculated both for tree number and basal area proportions.

```
1  > (SimpsonSph <- 1 / sum((Example$Sph /
2  + sum(Example$Sph))^2))
3  [1] 2.00395256917
4  > (SimpsonBa <- 1 / sum((Example$Ba / sum(Example$Ba))^2))
5  [1] 1.7403746020041
6  > SimpsonSph / length(Example$Species)
7  [1] 0.50098814229249
8  > SimpsonBa / length(Example$Species)
9  [1] 0.43509365050101
```

The Simpson index is calculated in lines 1–3 and 4f. for tree number and basal area proportions, respectively. The corresponding evenness measures are computed in lines 6 and 8. Whilst the Shannon measure emphasizes the species richness component of diversity the Simpson index is weighted by abundances of the commonest species. The Shannon and Simpson measures are among the most meaningful and robust diversity measures available (Krebs, 1999; Magurran, 2004).

Depending on their construction principles, the values of the Shannon and Simpson indices increase differently. The reciprocal variant of the Simpson index (Eq. (4.2)) apparently can take very large values with increasing number of species that markedly differ from those of the Simpson index. The complementary version of the Simpson index is always closer to the Shannon index than its reciprocal counterpart (Figure 4.13a). This potentially implies that the reciprocal Simpson index is able to differentiate different abundance

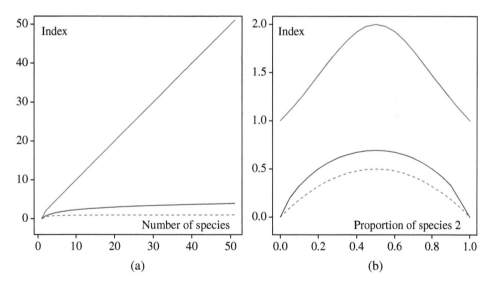

(a)   (b)

Figure 4.13 Shannon (blue, H' in Eq. (4.1)) and Simpson indices (red, D in Eq. (4.2)) before evenness calculation in comparison. (a): Index values depending on the number of species assuming that the species have equal proportions. The continuous red line gives the Simpson index for the reciprocal variant as in Eq. (4.2), the dashed red line shows the Simpson index for the complementary variant. (b): Dependence of the index values on the proportions of two species.

patterns better. For a fixed number of species, the index values are largest when both have equal proportions and this statement does not depend on the index type or construction principle (Figure 4.13b).

Jost (2006) demonstrated that both the Shannon and Simpson indices along with their variants and a number of other diversity indices are special cases of the general diversity index

$$^{q}D = \left(\sum_{i=1}^{s} p_i^{q} \right)^{\frac{1}{1-q}},$$ (4.3)

where superscript and exponent q may be referred to as the *order* of diversity. Dependent only on q and p_i, Eq. (4.3) can produce a whole family of diversity indices. For instance, ^{q}D gives species richness for $q = 0$. The reciprocal Simpson index (Eq. (4.2)) is derived from Eq. (4.3) with $q = 2$, i.e. a diversity index of order 2. Eq. (4.3) is undefined for $q = 1$; however, its limit exists and yields the exponential of the Shannon index (Eq. (4.1)), i.e. $e^{H'}$. The results computed from ^{q}D with varying q are also referred to as *Hill numbers* (Hill, 1973; Jost, 2006).

4.4.1.3 Species Profile Index

Based on the Shannon index H', Pretzsch (1995) proposed the *species profile index*. The index is the sum of the Shannon index values calculated separately for c relative height bands:

$$A' = -\sum_{i=1}^{s} \sum_{j=1}^{c} \begin{cases} p_{ij} \times \log_e (p_{ij}) & \text{if } p_{ij} > 0 \\ 0 & \text{otherwise} \end{cases}$$ (4.4)

In Eq. (4.4), s is the number of species, c the number of height bands, e.g. 3, and p_{ij} is the proportion of species i in band j. Naturally, it is also possible to use other indices such as the Simpson index instead of the Shannon index. For calculating the species profile index, the forest stand is subdivided into c height bands. Based on Assmann (1970), Pretzsch (1995) suggested $c = 3$ with definitions that depend on observed tree height, i.e. on the distance between forest floor and tree crown tip (Table 4.3). Naturally, any other height-band definition would also work and the species profile index is not conditional on the definition in Table 4.3. The trees of a forest stand are assigned to one of the c height bands according to their total height. Assuming the same result for the Shannon index, the species profile index is higher in stands with more vertical height differentiation compared to a single-canopy

Table 4.3   Possible tree height bands. The definition relates to individual-tree total height in percent of maximum tree total height of a forest stand.

Height band	Definition [%]
1	[0, 50)
2	[50, 80)
3	[80, 100]

Source: Pretzsch (1995) and Assmann (1970).

forest stand. Therefore, the index gives an idea of the vertical extent of species diversity (Pretzsch, 2009).

A'_{max} is given as $\log_e(s \times c)$ and therefore evenness can be calculated for the species profile index as

$$E_{A'} = \frac{A'}{\log_e(s \times c)}. \tag{4.5}$$

Other species diversity indices are described in detail in Krebs (1999), Magurran (2004) and Staudhammer and LeMay (2001). The general strategy is to monitor abundance and size-related species diversity in time and space, particularly during the process of transformation to CCF, but also as part of monitoring the routine maintenance of CCF stands.

4.4.2 Size Diversity

4.4.2.1 Size Diversity Based on Tree Stem Diameters

Stem diameters have traditionally been the most frequently used tree characteristics in silviculture and forest ecology. This is associated with the fact that stem diameters can be more efficiently measured in terrestrial surveys than, for example, total tree heights. In remotely sensed surveys based on airborne laser scanning (ALS), it may, however, just be the other way round, i.e. total tree heights can be more easily and more reliably measured (Robinson and Hamann, 2010). Most of the characteristics discussed in this section can also be applied to other tree variables such as height or biomass; however, they are most commonly applied to stem diameters.

Summarised as empirical diameter distributions, i.e. as histograms or density distributions, stem-diameter data often assume very characteristic shapes, which reflect and hint at stand type and silvicultural treatment and can be uni- or multi-modal. The shapes can be interpreted by silviculturists and ecologists and reveal much information about the structure of a forest ecosystem. Figure 4.14 visualises six common types of empirical stem diameter distributions. It is quite obvious that the shapes are very informative and with time and experience can be directly associated with visual impressions of stand structure obtained from forest visits or from digital photos (see also Figures 1.10 and 1.11). By studying the modes of the individual distributions hypotheses can be made as to which stand phase dominates and what are the most frequent tree sizes. Similar size distributions can be computed for total tree heights or any other quantitative tree variables, but these have not been considered much in the past.

That the stem diameters of a forest stand form a (negative) exponential or (reverse) j-shaped diameter distribution is an expectation often discussed in the context of CCF (Figure 4.14b), see also Chapter 6. Figure 4.14 demonstrates that this expectation cannot necessarily be used as a general criterion for all CCF woodlands. Diameter distributions (b)–(f) may be readily observed in various woodlands where CCF has been successfully established quite a while ago. On the other hand, it is possible to observe tendencies towards (negative) exponential stem-diameter distributions in even-aged plantations (Figure 4.15). For example, in plot 1 of the Tyfiant Coed project at Coed y Brenin, a *P. sitchensis* plantation was colonised by *Betula* spp. immediately after planting resulting in an even-aged mixed-species forest stand. Looking at the stem-diameter distribution by pulling all species together (Figure 4.15a) gives the impression of a trend towards a

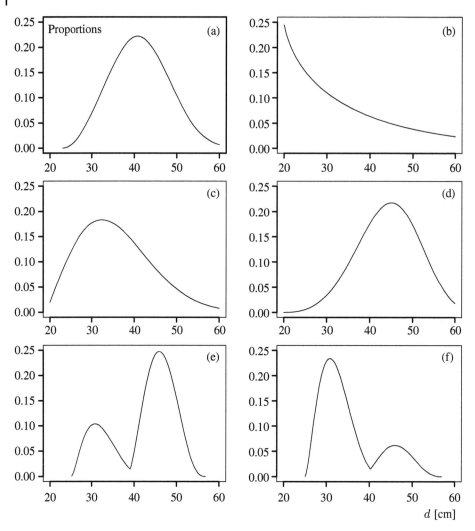

Figure 4.14 Typical stem-diameter (*d*) distributions of (a) an even-aged pure species stand (plantation), (b) an uneven-aged selection or natural forest (negative exponential), (c) an even-aged pure species stand (right skewed due to a few exceptionally big trees), (d) an even-aged pure species stand (left skewed due to a few exceptionally small trees), (e) a two-storeyed mixed forest and (f) a young (regenerated) stand with a few standards/seed trees. Source: Pommerening and Grabarnik (2019).

(negative) exponential diameter distribution. However, digging a bit deeper and analysing the species-specific diameter distributions separately (Figure 4.15b) reveals a glimpse of the true nature of the forest stand. The main stand component is apparently *P. sitchensis* forming a somewhat flat bell-shaped diameter distribution similar to Figure 4.14a. *Betula* spp. trees on the other hand form a diameter distribution more akin to that in Figure 4.14b and when both are added one to another the empirical distribution of the whole stand can convey the impression of a (negative) exponential diameter distribution, but this

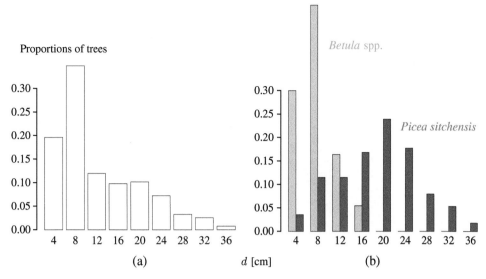

Proportions of trees

Betula spp.

Picea sitchensis

d [cm]

(a)   (b)

Figure 4.15   Total stand (a) and species-specific (b) stem-diameter distributions (*d*) of the 33-year old even-aged mixed-species woodland Coed y Brenin, Tyfiant Coed plot 1 (Wales, UK).

impression is deceiving, since the woodland is clearly even-aged with no CCF treatment prior to the time of survey in 2003.

In conclusion, stem-diameter distributions are very useful indicators of size structure in woodlands, and they have widely been used in forest ecology and forest management. However, the diameter distribution does not suffice as the sole indicator to judge forest structure or CCF progress. Stand history, site visits as well as other, more detailed analyses contribute important details to the full picture of stand appraisal.

It is also possible to use *population pyramids* for visualising the size structure of two important species in a forest stand. This is a visualisation and interpretation tool from demography often employed to highlight gender differences in society. Species-size differences are sometimes easier to see using population pyramids (Figure 4.16) than in standard bar charts (Figure 4.15) or histograms. In demography, certain shapes of population pyramids are distinguished and interpreted. For example at Coed y Brenin (plot 1), *Betula* spp. exhibits an *expansive* pattern typical of 'young' and dynamic populations. Such populations are often characterised by their typical pyramid shape with a broad base and a narrow top. *P. sitchensis* on the other hand shows a *constrictive* pattern typical of populations that are less dynamic and more stationary with small percentages of trees in the lower size classes. Constrictive pyramids often look like beehives and typically have a graph tapering in at the bottom and at the top (Wilson, 2016).

Non-spatial size diversity can be quantified by using common statistical dispersion characteristics such as the coefficient of variation, range and skewness (Table 4.4). The coefficient of variation is a measure involving the standard deviation expressed as a proportion of the arithmetic mean. This characteristic is particularly meaningful and can be compared between populations. The range is the difference between the maximum and minimum values of the data set. Skewness indicates asymmetry of a probability density

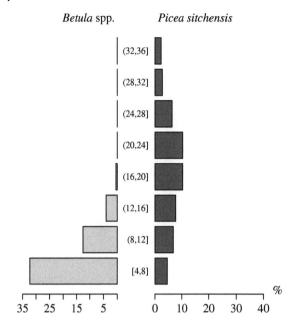

Figure 4.16 Population pyramids for visualising the stem-diameter distributions of the main species in the 33-year-old even-aged mixed-species woodland Coed y Brenin, Tyfiant Coed plot 1 (Wales, UK).

distribution. In asymmetric size distributions, there is usually a tail on the left or on the right side. These tails provide important information on an unusually wide range of stem-diameter values either towards very small or very large trees. In the mixed *Betula* spp. – *P. sitchensis* woodland Coed y Brenin (plot 1) introduced in Figure 4.15, for example, the standard deviations of the stem diameters of the two main species are 2.8 and 7.3 cm, respectively. The corresponding coefficients of variation are 0.15 and 0.39. Apparently, *P. sitchensis* has a much greater variability than *Betula* spp. which well reflects the impression conveyed by Figures 4.15b and 4.16.

Table 4.4 Simple but useful dispersion measures quantifying size diversity applied to stem diameter d. $\bar{d}$ is the arithmetic mean stem diameter.

Name	Formula	Unit	R code
Standard deviation	$s_d = \sqrt{\frac{\sum_{j=1}^{n}(d_j-\bar{d})^2}{n-1}}$	cm	sd()
Coefficient of variation	$v_d = \frac{s_d}{\bar{d}}$	–	sd() / mean()
Range	$r_d = \max(d) - \min(d)$	cm	range()
Skewness	$s_d^* = \frac{\sum_{j=1}^{n}(d_j-\bar{d})^3}{(n-1)\times s_d^3}$	cm	skewness()[a]

a) This R function requires the R package moments.

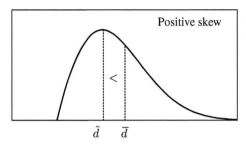
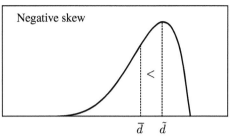

Figure 4.17 Skewness indicated by the relationship between median, $\tilde{d}$, and mean stem diameter, $\overline{d}$.

A stem-diameter distribution is skewed to the right or has a positive skew, if the right tail is longer than the left whilst the mass of the distribution is concentrated on the left of the figure. In that case, $s_d^* > 0$ (cf. Table 4.4) and the tail indicates a legacy of very large, dominant trees. A stem-diameter distribution is skewed to the left or has a negative skew, if the left tail is longer than the right whilst the mass of the distribution is concentrated on the right of the figure. In that case, $s_d^* < 0$ and the tail points towards a few very small trees. A value $s_d^* \approx 0$ indicates a symmetric diameter distribution (Figure 4.17).

An approximate way to discern positive and negative skew is to calculate the median, $\tilde{d}$, and arithmetic mean diameter, $\overline{d}$. A positive skew occurs when the median is smaller than the arithmetic mean. In the opposite case, the skew is negative (Figure 4.17). In the case of symmetric size distributions, mean and median are approximately the same. In the mixed *Betula* spp. – *P. sitchensis* woodland Coed y Brenin (plot 1) introduced in Figure 4.15, for example, the three diameter distributions have a slight skew.

The smallest skew occurred with *P. sitchensis* where the empirical diameter distribution is almost symmetric. Median and mean for this species are $\tilde{d} = 19.1$ cm and $\overline{d} = 18.6$ cm, respectively. For the total population $s_d^* = 1.1$ and the corresponding values of median and mean are $\tilde{d} = 9.1$ cm and $\overline{d} = 12.3$ cm, respectively. Another characteristic of size diversity from general statistics is the *mode*.

The mode is the value of a variable which occurs most frequently.

1 mode	→ unimodal diameter distribution
2 modes	→ bimodal diameter distribution (Figure 4.18)
> 2 modes	→ multimodal diameter distribution

The term mode more formally denotes (local) maxima or peaks of a diameter distribution. The more modes the greater usually is stand diversity. Modes often coincide with what in silviculture we know as storeys, i.e. overstorey, mid-storey and understorey. In Figure 4.14, stem-diameter distributions (a)–(d) are unimodal whilst (e) and (f) are bimodal. Distribution (e) indicates a stand with two storeys, a strongly developed, probably closed overstorey and an under- or mid-storey. Distribution (e) shows what probably are the remnants of a light overstorey and a large population of smaller trees, perhaps an under- or mid-storey.

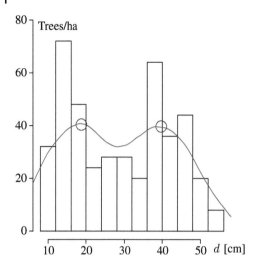

Figure 4.18 A bimodal empirical stem-diameter distribution (*d*) and the corresponding density (red) for *F. sylvatica* L. in the Swiss monitoring site Zofingen AG, Mühlethaler Halde, Baan (41-155) in 1972. The modes are indicated by blue circles at stem diameters $d = 18$ cm and $d = 40$ cm.

Bi- and multimodal stem-diameter distributions can be the result of two or more species with distinctively different size distributions (Figure 4.15) or exist in one and the same species population (Figure 4.18). For example, *F. sylvatica* stands often develop bimodal stem-diameter distributions where the same species contributes to different canopy storeys (Gadow et al., 2021).

To obtain measures of size diversity, some authors have suggested applying the Shannon (Eq. (4.1)) or Simpson indices (Eq. (4.2)) to the proportions of trees in classes of quantitative measures such as stem diameters, tree heights, volume, or biomass as opposed to species. This cannot be recommended because any choice of class or bin introduces an element of arbitrariness and in the author's experience, the results from such applications are inconsistent when comparing the size structure of different forest stands or the size structure of the same stand at different times even if the size classes were kept the same.

Lorenz Curve and Gini Index In the context of economics, Lorenz (1905) and Gini (1912) developed two related characteristics of relative concentration. These were used to analyse the income distribution within the population of a country. The two characteristics have been applied to various problems in forest ecology (de Camino, 1976; Lee et al., 1999; Damgaard and Weiner, 2000; Stöcker, 2002).

For computing these characteristics, all n size observations are sorted from smallest to largest according to $x_1 \leq \cdots \leq x_n$. The sum of all observations is $\sum_{i=1}^{n} x_i > 0$. For each $j = 1, \ldots, n$ and proportion $u_j = j/n$ of individuals there is a corresponding cumulative size proportion

$$v_j = \frac{\sum_{i=1}^{j} x_i}{\sum_{i=1}^{n} x_i}. \tag{4.6}$$

The cumulative size proportion v_j can alternatively be expressed in terms of cumulative tree stem volume (Kramer, 1988; Bachofen, 1999) or by any other size variable of interest. The Lorenz curve connects the points $(0,0), (u_1, v_1), (u_n, v_n)$ and $(1,1)$. The connecting lines are straight, and their slopes are always positive with an increasing tendency.

The Lorenz curve is convex, i.e. it curves downwards towards the abscissa. Applied to trees the Lorenz curve depicts the relative increase in tree size against the proportion of trees. A size variable often used in forestry applications of the Lorenz curve is individual-tree basal area, i.e. $x_i = g_i$, but volume or biomass are also suitable choices. The Lorenz curve results when the basal-area, volume or biomass percentile is plotted over the stem-number percentile (Lee et al., 1999).

The basic idea of the Lorenz curve is to study which cumulative size proportion each individual of a population accumulates. If proportion u_j always corresponds to the cumulative proportion v_j, the resulting Lorenz curve is a straight line coinciding with the diagonal of the graph, i.e. the 45° or 1 : 1 line, and indicating that trees are approximately of the same size (see Figure 4.19). Damgaard and Weiner (2000) referred to this line as the line of equality. The smaller the cumulative proportions of the smallest trees are the more the Lorenz curve deviates from the 45° line and the more diverse is the size structure (Fahrmeir et al., 2016). One Lorenz curve indicates more diversity than another one, if it always runs below the other one throughout its range.

The degree of concentration is expressed by the distance between the Lorenz curve and the graph diagonal. An intuitive summary characteristic of concentration is therefore the area between the 45° line and the Lorenz curve, also referred to as *Lorenz area*. The Lorenz area can be related to the total area between abscissa (u axis) and the graph diagonal. The resulting characteristic is the Gini index G (Fahrmeir et al., 2016). For individual observations sorted from smallest to largest according to $x_1 \leq \cdots \leq x_n$ the Gini index can be calculated as

$$G = \frac{2 \sum_{i=1}^n i \times x_i}{n \sum_{i=1}^n x_i} - \frac{n+1}{n}. \tag{4.7}$$

For homogeneous size distributions $G \approx 0$ indicating a maximum of size equality. The larger G the larger the size inequality in a given forest stand. Extreme values of G include $G_{min} = 0$ and $G_{max} = \frac{n-1}{n}$. Since maximum G apparently depends on the number of observations, a standardisation of the Gini index with values between 0 and 1 is often considered as

$$G^* = \frac{G}{G_{max}} = \frac{n}{n-1} \times G. \tag{4.8}$$

G and G^* can be interpreted as resulting from the transformation $1 - 1/\beta$ of the allometric coefficient β in the equation $v_j = u_j^\beta$ (Lee et al., 1999; Adnan et al., 2021). Adnan et al. (2021) demonstrated that for a uniform size distribution with $G = 0.5$ the allometric coefficient β is 3.

de Camino (1976) proposed the reciprocal of G^*, i.e. $G' = 1/G^*$ as the *homogeneity index* and a more sophisticated measure of size inequality, see Eq. (6.47) in Section 6.5. The homogeneity index is useful when comparing forest stands with only small differences in size diversity. The larger G' is the more homogeneous a population, the smaller G' becomes the greater is the size inequality. For very heterogeneous forest stands, G' is around 1.0, while extremely homogeneous stands can have values of G' up to 10.

For getting started with the Lorenz curve and the Gini index, consider the simple data example in Table 4.5. The five tree diameters were sorted in ascending order, and the rank numbers (used to compute u) assigned accordingly. Using Eqs. (4.7) and (4.8), $G = 0.74$,

Table 4.5 Simple example for the calculation of the Lorenz curve.

Rank	*d* [cm]	*g* [m²]	*u*	*v*
1	5.5	0.00238	0.2	0.00588
2	8.5	0.00567	0.4	0.01993
3	9.2	0.00665	0.6	0.03640
4	16.2	0.02061	0.8	0.08744
5	68.5	0.36853	1.0	1.00000

d – Stem diameter, *g* – basal area, *u* – proportion of individuals, *v* – Eq. (4.6).
Source: Adapted from Katholnig (2012).

$G^* = 0.93$ and $G' = 1.08$ indicating high-size diversity and low homogeneity. Generally, large values of the Gini index indicate a heterogeneous distribution of basal area in a forest stand, whilst small values point to homogeneous distributions of basal area or volume. For $G = 0$, the corresponding Lorenz curve coincides with the 45° line. The code for calculating Eqs. (4.6)–(4.8) is provided in the listing below:

```
1   > myData <- data.frame(dbh = c(5.5, 8.5, 9.2, 16.2, 68.5))
2   > myData$ba <- pi * (myData$dbh / 200)^2
3   > myData$u <- seq(from = 1, to = length(myData$ba),
4   + by = 1) / length(myData$ba)
5   > myData$v <- cumsum(myData$ba) / sum(myData$ba)
6   > Gini <- 2 * sum(myData$ba * seq(from = 1,
7   + to = length(myData$ba), by = 1)) /
8   + (length(myData$ba) * sum(myData$ba)) -
9   + ((length(myData$ba) + 1) / length(myData$ba))
10  > Gini * length(myData$ba) / (length(myData$ba) - 1)
```

The stem-diameter data are read into a data frame in line 1. Basal area is calculated from stem diameter in line 2. The proportions $u_j = j/n$ of individual trees are calculated in lines 3f. Eq. (4.6) is computed in line 5 using the R cumsum() function. Then the Gini index (Eq. (4.7)) is calculated in lines 6–9 followed by the standardised Gini index (Eq. (4.8)) in line 10. An independent, alternative calculation of homogeneity index G' closer to Eq. (1) in de Camino (1976) is offered in the following listing:

```
u <- cumsum(rep(1 / length(myData$ba), length(myData$ba)))
v <- cumsum(myData$ba / sum(myData$ba))
(G.prime <- sum(u[-length(v)]) / sum(u - v))
```

The graph visualising the Lorenz curve can be produced with the following listing:

```
1   > par(mar = c(2, 3, 1, 1))
2   > plot(c(0, myData$u), c(0, myData$v), pch = 16,
3   + xlim = c(0, 1), ylim = c(0, 1), axes = FALSE, lwd = 2,
4   + xaxs = "i", yaxs = "i")
5   > lines(c(0, myData$u), c(0, myData$v), lwd = 2)
```

```
6  > abline(0, 1, lwd = 2)
7  > axis(1, lwd = 2, cex.axis = 1.7)
8  > axis(2, lwd = 1, las = 1, cex.axis = 1.7)
9  > box(lwd = 2)
```

The commands in line 4 adjust the axes to start immediately at zero origin and to finish at 1. Line 6 produces the 45° line. The result of the listing can be seen in Figure 4.19. From the Lorenz curve it is, for example, possible to understand that tree #4 with a stem diameter of 16.2 cm (Table 4.5) represents 80% of the ranked tree list but only 8.7% of the stand's basal area (Katholnig, 2012).

Sterba and Zingg (2006) found that the Gini index is significantly correlated with the mean quadratic difference, a deviation measure for equilibrium models (see Section 6.5). The mean quadratic difference is a measure quantifying whether a stand under consideration is in a steady demographic state and/or how far it is removed from it (Katholnig, 2012). Sterba and Sterba (2018) demonstrated that the Gini index is closely related with the parameters of the q factor model, see Chapter 6. The Lorenz curve and associated Gini indices reliably indicate different types of population size structure (Figure 4.20).

The diverse selection forest of mainly *Pseudotsuga menziesii* at Gwydyr Forest (cf. Figure 2.4) has far greater size inequality than the *P. sitchensis* plantation of Tyfiant Coed plot 1 at Clocaenog Forest. Apparently, the standardisation of Eq. (4.8) does not play a major role when many trees are involved. In that case the Lorenz curve is also much smoother than in Figure 4.19. The stem-diameter coefficient of variation apparently highlights the

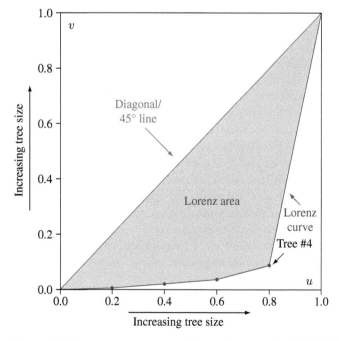

Figure 4.19 Lorenz curve corresponding to the example of Table 4.5. Variables u and v are explained in the text. Source: Adapted from Katholnig (2012).

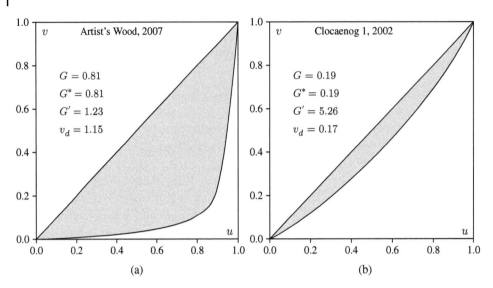

Figure 4.20 Lorenz curve, Gini indices (G, G^*) and stem-diameter coefficient of variation (v_d) in a mixed-species selection forest (a) and in a monospecies plantation (b). Variables u and v are explained in the text. See also the photos of these two stands in Figures 2.4 and 1.11a.

same tendency of size inequality, however, the range of values differs. Pommerening et al. (2016) showed that the Lorenz curve and its variants can also be computed separately for tree species populations that are part of the same forest stand.

Lorenz curve and Gini index can be applied to data from natural forests, plantations as well as forests managed for CCF. A monitoring of the size structure of a forest stand can give valuable insights on the dynamics of a forest ecosystem. The Gini index helps summarise size-structure data for a longer time period, although it is always a good idea to refer to the Lorenz curve when marked changes occur, e.g. after human interventions and natural disturbances. As an example we show here the temporal development of size structure for the Zofingen *F. sylvatica* stand 41-155 that was monitored from 1898 to 2001 (Figure 4.21). The monitoring of this simple-structured woodland started with a Gini index of approximately 0.3 and reached a peak larger than 0.5 before 1920. After that the index has gradually declined (Figure 4.21a). It is interesting to see that Gini index and coefficient of variation deliver very similar results that only differ a bit more in the time between 1981 and 2001. It is also clear that G and G^* give almost identical results throughout the monitoring period. However, starting around 1940 the differences between G and G^* increase. This is when the absolute number of trees falls below 272. In 1981, the number of trees is 64 and decreases towards 31 in 2001. Apparently, these are numbers of observations where the difference between G and G^* matters. These differences are largest when the differences between Gini index and coefficient of variation are largest, too.

Gini index and coefficient of variation are undoubtedly tightly correlated so that the coefficient of variation can act as a substitute of the Gini index which is interesting, since v_d is easier to compute (Figure 4.21b). Also, simple linear functions can be constructed to estimate the Gini index from the coefficient of variation. Again, on first sight, the difference

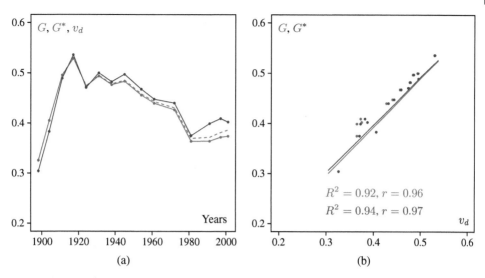

Figure 4.21 Comparison of Gini index (*G*), standardised Gini index (*G**) and coefficient of variation (*v_d*) when applied to the Swiss *F.sylvatica* L. time series Zofingen AG, Mühlethaler Halde, Baan (41-155). R^2 – adjusted coefficient of determination, *r* – Pearson correlation coefficient. The dashed red line in the left graph represents *G**. The stand also included minor proportions of *Quercus robur* L. and *Betula pendula* Rotʜ.

between G and G^* seems academic, however, it slightly influences both the coefficient of determination and the correlation coefficient. Differences between G and G^* are greatest, where v_d is small.

Analysing 85 forest stands in Romania, Duduman (2011) found that $G \leq 0.35$ indicated even-aged size structures, $0.35 < G \leq 0.43$ accounted for bimodal size structure whilst $G > 0.43$ was obtained for uneven-aged size structure. The corresponding thresholds of the diameter coefficient of variation were quite similar, i.e. 0.34 and 0.44. These findings confirmed the trends of earlier results by O'Hara et al. (2007) and later results by Valbuena et al. (2013) for other parts of Europe.

Growth Dominance Based on the Lorenz curve and the Gini index, Binkley et al. (2006) proposed another important statistic, i.e. the *growth dominance* characteristic. Growth dominance characterises the contributions of different tree sizes to total population growth. An individual tree is growth dominant, if its growth contributes more to stand growth than its size contributes to cumulative stand size (Bradford et al., 2010). This implies that the growth dominance statistic requires information on individual absolute growth rates (AGR) in addition to tree size. If past stem diameters were not recorded, it is possible to establish growth rates using increment boring on sample trees and modelling a data imputation function that can account for missing growth-rate observations. To construct the growth dominance characteristic, all tree basal areas are again arranged in ascending order and their cumulative basal-area proportions v^x are computed according to Eq. (4.6). Over a certain time period, Δt between two surveys each tree of a forest stand that has grown in size by a mean AGR y, which is the difference in size of a tree between two surveys divided by

Δt. The cumulative proportions of AGRs v^y are computed in the same way as those of basal area; however, here, it is important that the vector of growth rates, y, follows the order of x (basal area in our application) in Eq. (4.6), i.e. the ranking of trees is determined by their size alone. It is now possible to plot v_j^y as a function of v_j^x, yielding the growth dominance curve (West, 2014). The growth dominance curve would follow the 1 : 1 line, if the proportional contribution of each tree were the same for size as for growth (Figure 4.22). The degree to which the observed curve departs from that line represents the degree to which some trees dominate population growth (West, 2014). If, for example, the growth contribution falls short of the size contribution and large trees accounted for a greater proportion of population growth, the growth dominance curve would fall below the diagonal (concave upwards). The growth dominance would arc above the 1 : 1 line, if the growth contribution were larger than the size contribution, and relatively low growth rates of large trees accounted for a smaller portion of total population growth than of population basal area (concave downwards), indicating *reverse growth dominance* where small trees dominate the growth of the population (Binkley et al., 2006; Pommerening et al., 2016). The asymmetry of the growth dominance curve can be described in relation to the graph diagonal (from top left to bottom right) running perpendicular to the 1 : 1 line. Damgaard and Weiner (2000) referred to this diagonal as the *axis of symmetry* (Figure 4.22). If the point of maximum concavity is not on the axis of symmetry, growth dominance is said to be asymmetric. This concept of asymmetry is in some ways similar to skewness in the traditional distribution-moment analysis (Damgaard and Weiner, 2000). Naturally, the axis of symmetry can also be applied to the Lorenz curve. Valbuena et al. (2013), for example, pointed out that understorey recruitment, i.e. a strong regeneration phase, can often contribute to asymmetric Lorenz curves.

Drawing on these cardinal shapes the growth dominance and on subsequent, extensive analyses of different forests, Binkley (2004) and Binkley et al. (2006) proposed a conceptual model of the progression of forest development phases (see Pommerening and Grabarnik, 2019, Appendix A) and the corresponding shapes of the growth dominance curve in natural forests (Figure 4.22): Early in forest development before canopy closure (phase 1), competition between trees is low and each tree's contribution to total population growth is proportional to its size (1 : 1 line). In phase 2, after canopy closure size differentiation sets in, larger trees have gained substantial dominance and suppressed the growth of smaller trees accounting for the departure of the curve below the diagonal. Later towards stand maturity in phase 3, growth dominance subsides as one or more factors cause a reduction in growth of dominant trees. Late in stand development, when the old growth phase prevails and regeneration takes hold, this trend finally results in reverse growth dominance (phase 4). Here, the contribution of small- and mid-size trees to total population growth is greater than their proportional sizes. Reverse growth dominance may be caused by an acceleration of growth in smaller trees, by a decline in growth of dominant trees or it may be the result of both phenomena (Binkley et al., 2006). Some forests, however, are known not to fit this conceptual sequence of patterns, such as monoculture plantations of *Eucalyptus* (Doi et al., 2010; Pommerening et al., 2016).

Considering the curve labelled with phase 2 in Figure 4.22, approximately 20% of the largest trees contribute about 40% of total growth, whilst with the curve labelled with phase 4 the 20% largest trees contribute only 5% of total population growth. Looking at the reverse

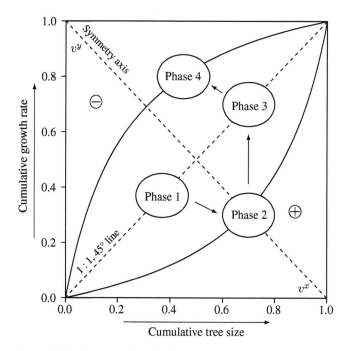

Figure 4.22 Conceptual model of how growth dominance indicates four phases of forest development (see Pommerening and Grabarnik, 2019, Appendix A). The plus and minus signs show the areas of *positive* and *negative* (or *reverse*) growth dominance. Further explanations can be found in the text. Source: Binkley et al. (2006) and Pommerening et al. (2016).

growth dominance curve in Figure 4.22, we can conclude that the 20% smallest trees make a contribution of more than 50% to total population growth.

Binkley (2004) proposed that the patterns indicating phases of forest stand development are the result of changes in resource use between different tree sizes, but the exact underlying physiological processes are still largely unknown. Pommerening et al. (2016) found that reverse growth dominance is not only common in old-growth forest stands but also in any forest stands that develop towards natural forest stands and exhibit more complex forest structure. The more inhomogeneous the structure of a forest stand or the more mixed the species composition, the more likely it is that the corresponding growth dominance curve is reverse (Katholnig, 2012). This property renders growth dominance ideal as an indicator for monitoring the transformation of plantation forests to CCF.

In analogy to the Gini index of a Lorenz curve, the growth dominance curve can be summarised in a single number, the *growth dominance index* G_D. Based on the trapezoidal rule for area calculation, West (2014, Eq. 2) recommended the estimator

$$G_D = 1 - \sum_{i=2}^{n} (v_j^x - v_{j-1}^x) \times (v_j^y + v_{j-1}^y), \tag{4.9}$$

where x commonly is chosen to be basal area in tree applications (but theoretically can also be volume or biomass) and y, as previously explained, represents the corresponding growth rates.

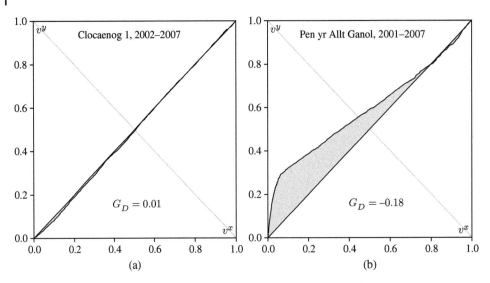

Figure 4.23 Growth dominance curves and growth dominance indices (G_D, Eq. (4.9)) of the woodlands (a) Clocaenog (plot 1) and (b) Pen yr Allt Ganol (both in North Wales, UK). Mean annual absolute basal area growth rates were used for v^y and basal area in 2002 and 2001, respectively, for v^x.

Figure 4.23 offers a comparison of the growth dominance curves and indices of a typical *P. sitchensis* plantation (Clocaenog, Tyfiant Coed plot 1) and a mixed *P. sylvestris* – *P. sitchensis* forest stand. In the latter, the understorey was heavily colonised by mixed broadleaved species (mainly *Betula* spp. and *Sorbus aucuparia* L).

Clocaenog (plot 1) represents a *P. sitchensis* plantation before transformation to CCF (see Section 2.4) and in such conditions, each tree's contribution to total population growth is more or less proportional to its size. Therefore, the corresponding growth dominance curve largely follows the 1 : 1 line, although the age of the plantation was 51 in 2002 (Figure 4.23a), i.e. the stand was not young but quite mature in the context of British forestry.

Pen yr Allt Ganol was planted in 1931 mainly as a *P. sylvestris* plantation with a minor component of *P. sitchensis* in wet depressions. These two species form the main canopy which is rather light so that a dense understorey of mainly *Betula* spp. and *Sorbus aucuparia* successfully colonised the stand. These mixed broadleaves are mainly responsible for the (moderate) reverse growth dominance effect in Figure 4.23b. (G_D values of −0.30 to −0.40 are not uncommon with strong reverse growth dominance.) Towards the end of the abscissa (axis labelled v^x), in the region of large trees, we can see the plantation legacy where the growth dominance curve crosses and straddles the 1 : 1 line. Particularly, for Pen yr Allt Ganol, we can also observe asymmetric growth dominance where the point of maximum concavity deviates markedly from the axis of symmetry. Such asymmetric patterns were frequently observed for Białowieża Forest (Poland) by Pommerening et al. (2016) and can often be attributed to underlying (negative) exponential size distributions. As such asymmetric growth dominance offers an additional clue for understanding ongoing transformation and highlights that the broadleaved regeneration is particularly dynamic.

The following listing gives the R code for calculating growth dominance:

```
1  > myData$ba2002 <- pi * (myData$dbh2002 / 200)^2
2  > myData$ba2007 <- pi * (myData$dbh2007 / 200)^2
3  > myData$AGR <- (myData$ba2007 - myData$ba2002) /
4  + (2007 - 2002)
5  > myData <- myData[!is.na(myData$AGR),]
6  > myData$AGR[myData$AGR < 0] <- 0
7  > myData <- myData[order(myData$ba2002,
8  + decreasing = FALSE), ]
9  > cumG <- cumsum(myData$ba2002) / sum(myData$ba2002)
10 > cumInc <- cumsum(myData$AGR) / sum(myData$AGR)
11 > narea <- 0
12 > for(i in 2 : length(cumG))
13 > narea[i] <- (cumG[i] - cumG[i - 1]) * (cumInc[i] +
14 + cumInc[i - 1])
15 > GD <- 1 - sum(narea)
```

In lines 1–2, basal area is calculated for all trees recorded in 2002 and 2007. In line 3f., mean annual AGR is calculated for all trees. In lines 5–6, implausible or unsuitable AGR results are excluded or modified. The tree data are then arranged in ascending order according to basal area in 2002 (line 7f.). Next, cumulative basal area and basal-area AGR proportions are calculated in lines 9 and 10, respectively, using again the R cumsum() function as in the listing for the Lorenz curve. Finally, the growth dominance index (Eq. (4.9)) is calculated in lines 11–15. In a way similar to the Lorenz curve, the growth dominance curve can now be plotted with command plot(cumG, cumInc, type = "l", ..., xaxs = "i", yaxs = "i"). The symmetry axis can be produced using command abline(1, -1, lwd = 2, col = "grey").

Studying data from 83 even-aged and uneven-aged forest plots in Switzerland, Katholnig (2012) found that Gini and growth dominance indices are related in a nonlinear way; however, for the Swiss data, this relationship they used was not strong and turned out to be noisy. The general trend is that the growth dominance index increasingly drops below zero for increasing values of the Gini index, mostly for $G > 0.5$.

The Hirschlacke Forest stand is a large monitoring and management demonstration site at the Schlägl forest estate in North-West Austria that has been monitored by Prof. Hubert Sterba of BOKU University Vienna since 1977 (Sterba and Zingg, 2001; Sterba and Sterba, 2018). The forest stand has been transformed to a two-storeyed high forest by Reininger (2001), see Sections 3.6 and 5.5. Over the decades, the forest has increasingly adopted a forest structure that is typical of the single-tree selection system. Applying the growth dominance index, we can see an overall trend towards increasing negativity for the whole population, although the index did increase in a few years (Figure 4.24). Such fluctuations are probably to be expected, since the management was not specifically for optimising growth rates or for the growth dominance index but had to take many criteria into consideration. Approximately, the same negativity trend can be seen for the main species *P. abies* and *A. alba*. The latter species reaches a value of $G_D = -0.31$ in 2012. The growth-dominance index curves of *P. abies* and *A. alba* are, however, for the

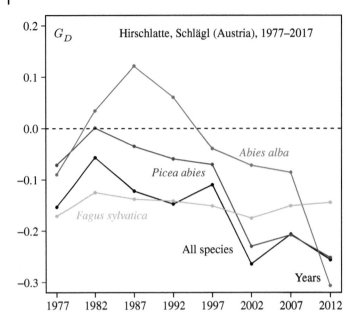

Figure 4.24 Growth dominance index (G_D, Eq. (4.9)) monitored in the Hirschlacke forest stand (Austria) for all species together (total population regardless of species) and separately for the three main species *P. abies* (L.) H. Karst., *Fagus sylcatica* L. and *Abies alba* Mill.

most part located above the population curve, and the index values are therefore less negative. It is interesting to note that the growth-dominance index curve for *F. sylvatica*, the second most abundant species after *P. abies* is almost constant throughout the years with a mean value of $G_D = -0.15$. In this example, the different curves hint that the component species of a forest stand can make quite different contributions to overall growth dominance. Another possible way to carry out growth dominance analyses is to compute Eq. (4.9) separately for different height bands as, for example, defined in Table 4.3.

4.4.2.2 Size Diversity Based on Tree Heights

In forestry, both horizontal and vertical forest structure are important to consider. Horizontal forest structure gives important information on canopy closure or site occupancy. In the special context of CCF, one could, however, argue that vertical forest structure matters more than horizontal forest structure, since it is related to light availability in different canopy strata, to microclimate, habitats, resilience and also to the principles of biological rationalisation, see Section 2.6. Arguing from an ecophysiological point of view, growing in height and raising their photosynthetically active biomass well above other vegetation has been an important evolutionary strategy of trees (Falster and Westoby, 2003). Vertical forest structure also plays a major role in the definition of transformation targets, cf. Section 2.4. Schütz (2001b) stated that in selection forests tree crowns tend to occupy the whole vertical growing space and do not touch each other (see Section 5.4). This tendency of filling the vertical growing space with tree crowns can certainly be generalised for many forms of CCF, particularly when compared to plantations.

Population Height-Diameter Curve

Between stem diameters and heights of a tree population, there is a nonlinear statistical relationship, which is motivated physically and physiologically (Pommerening and Grabarnik, 2019). This relationship is evident when depicting measured tree heights over the corresponding stem diameters in a diagram, see Figure 4.25. With increasing diameters, tree heights increase at a lower rate than the corresponding diameters until the height curve finally levels off (in mathematical terms the height curve is approaching an asymptote). As height measurements are often more difficult than diameter measurements, it is quite common to measure heights on a sample basis only.

A general recommendation is to sample between 30 and 40 tree heights across the range of diameters per species per population (e.g. a forest stand) (Laar and Akça, 2007; Philip, 1994). Especially, for the assessment of interspecific competition, for volume and for research purposes, the height–diameter relationship of a tree population should be analysed separately for each species. However, height/diameter relationships can also be studied for different canopy layers or height bands.

A number of models have been developed for the quantitative description of height–diameter relationships. Some important ones are listed in Table 4.6. More models can be found in Laar and Akça (2007), Weiskittel et al. (2011) and Scaranello et al. (2012). Height curve functions help estimate total tree heights for trees where only diameter information is available. In that situation, they can be used for *imputing* missing height observations (Weiskittel et al., 2011). Height-diameter curves are also the basis for the estimation of forest population heights such as mean stand heights and top/dominant heights. In the context of structural analysis, they can, however, also be used for studying the vertical structure of forest stands, and this is the focus here. The logarithmic function

Table 4.6 Example height curve functions. a_0, a_1 and a_2 are regression coefficients, i.e. model parameters.

Eq. #	Name/author	Main version	Linear version
(4.10)	Logarithmic	$h = a_0 + a_1 \times \log_e d$	$h = a_0 + a_1 \times \log_e d$
(4.11)	Petterson (1995)	$h = 1.3 + \left(\frac{d}{a_0 + a_1 \times d} \right)^3$	$(h - 1.3)^{-3} = a_0 + a_1 \times \frac{1}{d}$
(4.12)	Michailoff (1943)	$h = 1.3 + a_0 \times e^{\frac{-a_1}{d}}$	$\log_e (h - 1.3) = \log_e a_0 + a_1 \times \frac{1}{d}$
(4.13)	Clutter et al. (1983, p. 13); Burkhart (2012, p. 281)	$h = e^{a_0 + a_1} \times \frac{1}{d}$	$\log_e h = a_0 + a_1 \times \frac{1}{d}$
(4.14)	Chapman-Richards (Pienaar, 1973)	$h = 1.3 + a_0 \times \left(1 - e^{-a_1 \times d}\right)^{a_2}$	–
(4.15)	Wykoff et al. (1982)	$h = 1.3 + e^{a0 + \frac{a_1}{1 + d}}$	–

d – diameter at breast height [cm], h – (total) tree height [m], $\log_e$ – natural logarithm, e – base of the natural logarithm.

is one of simplest height functions but quite inflexible and therefore often unsuitable especially in diverse forests. The Clutter function is similar but more flexible than the logarithmic function. Petterson and Michailoff functions have turning points which allow a better adaptation to diverse irregular forests. In Scandinavia, the Petterson function is also referred to as Näslund function (Kangas and Maltamo, 2002). The Chapman-Richards model often produces very good results, however, the function requires three model parameters. The good performance of this model was also confirmed in the study on tropical Atlantic moist forest trees in South-Eastern Brazil by Scaranello et al. (2012). The Wykoff function also produces good results.

Nonlinear regression is applied to identify suitable model parameters a_0, a_1 and a_2. In the listing below, the R code for a non-linear regression using the logarithmic function is provided in lines 1–3. The start parameters in line 2 sometimes need to be adjusted to avoid singular gradients and similar errors. Line 4 gives basic regression results and the commands in lines 5–10 produce the height–diameter scatter plot which is overlaid by the regression line in lines 11–14. In lines 15–21, bias and RMSE (root mean square error) are calculated.

```
1   > nls.log <- nls(h ~ a + b * log(dbh),
2   + data = myData, start = list(a = -18.4, b = 12.1),
3   + trace = T)
4   > summary(nls.log)
5   > par(mar = c(2, 3, 1, 1))
6   > plot(myData$dbh, myData$h, las = 1,
7   + cex.axis = 1.7, lwd = 2, xlim = c(0, 65),
8   + ylim = c(0, 35), pch = 16, xlab = "",
9   + ylab = "", col = "black")
10  + box(lwd = 2)
11  > curve(summary(nls.log)$coefficients[1] +
12  + summary(nls.log)$coefficients[2] * log(x),
13  + from = min(myData$dbh), to = max(myData$dbh), lwd = 1,
14  + lty = 1, col = "black", add = TRUE)
15  > h.est <- 1.3 + (myData$dbh /
16  + summary(nlm.Petterson)$coefficients[1] +
17  + summary(nlm.Petterson)$coefficients[2] *
18  + myData$dbh))^3
19  > (varres <- var(myData$h - myData$h.est, na.rm = TRUE))
20  > (bias <- mean( myData$h.est - myData$h, na.rm = TRUE))
21  > (rmse <- sqrt(varres + bias^2))
```

For the other height–diameter curves of Table 4.6, lines 1 and 11ff. need to be modified accordingly. In some cases, it may be better to use the more robust optim() function instead of nls() as explained in Pommerening and Grabarnik (2019, p. 393f.). For analysing vertical forest structure the height–diameter models presented here are only used as trend curves to better understand vertical forest structure.

The mixed-species forest stand Pen yr Allt Ganol was already introduced in Figure 4.23b. Species-specific height-diameter curves offer information on the vertical structure involving different species populations of a forest stand (Figure 4.25). Both *P. sylvestris* and *P. sitchensis* were planted at the same time in 1931.

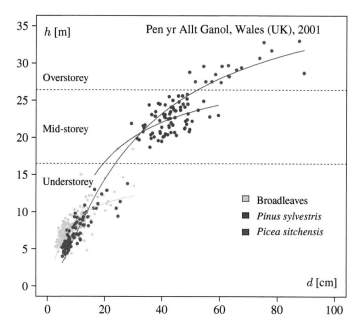

Figure 4.25 Vertical structure of the stand Pen yr Allt Ganol indicated by species-specific height–diameter curves. The Petterson model (Eq. (4.17)) in Table 4.6 was used for the trend curves and the relative height bands defined in Table 4.3 for indicating different storeys. The broadleaves are *Betula* spp. and *Sorbus aucuparia* L. *h* – total tree height, *d* – stem diameter.

Due to different growth dynamics, the height–diameter curves of these species varied greatly. The height bands of Table 4.3 suggest that the two species now form two different canopy layers. This is not so obvious in the field, and the quantitative analysis makes this much clearer. Since *P. sylvestris* is more light-demanding than *P. sitchensis* (cf. Appendix B) and the fact that the latter species has overtopped the former, could be cause for some concern. In this woodland it is not, because *P. sylvestris* is the majority species of the main canopy and both species are spatially segregated. This is a good example of a situation where horizontal forest structure matters, too. Figure 4.25 also highlights that *P. sitchensis* is present in both the over- and the understorey. In fact one might argue that it makes sense to fit two independent height–diameter curves separately for the *P. sitchensis* over- and understorey in this situation. Another important component of the understorey is two broadleaved species that have colonised the site for some time due to the openness of the overstorey and the light transmittance of *P. sylvestris*. Judging by the height structure given in Figure 4.25, Pen yr Allt Ganol is a woodland well on its way towards CCF, and goal-oriented forest management can take this process forward and accelerate it.

It is not a major commitment to measure tree heights on a sample basis across the whole stem–diameter range of a forest stand whilst the simple tool of height–diameter curves gives valuable insights on forest stand structure. After 5–10 years, the sampling can be repeated to see what changes may have occurred to vertical stand structure.

Cumulative Abundance Profile Another informative way to analyse and visualise vertical forest structure is using *cumulative abundance profiles* (CAP) (De Cáceres et al., 2019). Stoyan

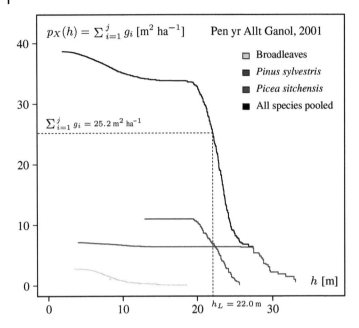

Figure 4.26 Vertical structure of the stand Pen yr Allt Ganol indicated by species-specific cumulative abundance profile, $p_X(h)$ (Stoyan et al., 2018; De Cáceres et al., 2019). The broadleaves are *Betula* spp. and *Sorbus aucuparia* L. h – total tree height, h_L - Lorey's mean height (Eq. (4.18)).

et al. (2018) termed them *coverage functions*. Assume there is a set X of n trees with total tree heights $h_1, h_2, \ldots, h_n$ and corresponding basal areas $g_1, g_2, \ldots, g_n$ per hectare. A useful summary characteristic of such a set X of trees is the function

$$p_X(h) = \sum_{i=1}^{j} g_i, \tag{4.16}$$

which gives the cumulative basal area of trees corresponding to the number j of trees equal to or larger than height h_i. For the presentation of CAP, the trees are ranked according to their heights h_i in descending order and the ranking of trees with equal heights does not matter. Thus, after ranking the height values form a decreasing sequence (Stoyan et al., 2018). In our application, CAP is a function taking a height value, e.g. h_i, as input and returning the cumulative basal area of trees whose height values are equal or larger than h_i (De Cáceres et al., 2019).

CAP does not involve any model, i.e. it is entirely non-parametric, and the height data are not summarised in height classes. The curves of functions $p_X(h)$ show very clearly how the trees ranked by height contribute to overall population basal area and offer information on the ranges of both heights and basal area (Figure 4.26). Commonly, a comparatively gradual, initial decrease of cumulative basal area is followed by a rapid descent towards the end of the height range. It is obvious that *P. sitchensis* contributes the largest trees to the forest stand, as the species-specific curve coincides with the stand curve for trees larger than approximately 26 m. However, the total basal area of *P. sitchensis* is much lower than that of *P. sylvestris* due to the much larger number of trees of the latter species. CAP of the broadleaved species is very shallow throughout its range indicating that the contributions to both stand height and basal area are very small. Applying Lorey's mean height (Eq. (4.18)) to stand CAP yields a cumulative basal area of 25.2 m^2 ha^{-1} that corresponds to this height, i.e. h_L marks the 0.67-height and the 0.65-basal area percentiles and lies in the interval between the 130 and 131 tallest trees. The following listing shows how $p_X(h)$ can be computed:

```
1  > udata <- myData[order(myData$h, decreasing = T), ]
2  > pX <- cumsum(udata$ba)
3  > l <- which(udata$h == max(udata$h[udata$h < hL]))
4  > u <- which(udata$h == min(udata$h[udata$h > hL]))
```

In line 1, the tree data are ranked according to their heights in descending order. $p_X(h)$ is computed in line 2 as cumulative basal area. Indices l and u in lines 3 and 4 give the indices marking the interval in which Lorey's mean height is situated. After applying the principle of right continuity, pX can be plotted over udata$h to give the graphs in Figure 4.26.

Vertical Structure Indices

In a similar way as for stem diameter, tree-height size structure is often characterised by applying the Shannon index (Eq. (4.1)) to the proportions of trees in different arbitrary height classes, but for the same reasons as for stem diameter, this cannot be recommended. Barbeito et al. (2009) proposed an index for characterising the vertical size structure of trees based on the differences of their total height from mean height:

$$H = \sum_{i=1}^{n} p_i \times \sqrt{|h_i - \overline{h}_L|} \tag{4.17}$$

Originally Barbeito et al. (2009) termed this index SQRI. In Eq. (4.17), p_i is the basal-area proportion of tree i, i.e. the ratio between basal area of tree i and the total population basal area. n is the total number of trees in the population and h_i is the total height of tree i. In the original index, the arithmetic mean population height $\overline{h}$ was included, but for consistency of concept, we suggest using *Lorey's mean height*, $\overline{h}_L$, instead, which is a mean height weighted by basal area (Lorey, 1878):

$$\overline{h}_L = \frac{\sum_{i=1}^{n} g_i \times h_i}{\sum_{i=1}^{n} g_i} \tag{4.18}$$

As mentioned before, usually mean population heights are not directly calculated from the sampling data but are indirectly derived from height–diameter curves by inputting population diameters into the height-diameter model equation (Laar and Akça, 2007). This is different for Lorey's mean height. Here, it is recommended to estimate missing height measurements from the corresponding height-diameter curve (cf. Table 4.6) first and then to calculate $\overline{h}_L$. However, the more height estimations compared to field-based height measurements are used the less Eq. (4.18) reflects the real vertical size diversity. H measures the deviation of individual tree height from Lorey's mean height, and an indication of this diversity is the variability of data points around the height–diameter curves in Figure 4.25. For very low height diversity, H values are around 1 and keep increasing with height diversity.

The calculation of H is straightforward. The listing below gives the calculation of individual-tree basal area g_i in line 1 followed by the computation of Eq. (4.18) in line 2. For simplicity, the code assumes that all individual-tree stem diameters and total heights have been measured and are available. Individual-tree basal area proportion p_i is calculated in line 3.

```
1  > myData$ba <- pi * (myData$dbh / 200)^2
2  > hL <- sum(myData$ba * myData$h) / sum(myData$ba)
3  > p <- myData$ba / sum(myData$ba)
4  > H <- sum(p * sqrt(abs(myData$h - hL)))
```

Finally, index H (Eq. (4.17)) is computed in line 4.

Given the tree height information in the mixed-species forest stand Pen yr Allt Ganol in Figure 4.25, H for *P. sylvestris* is 1.11. For *P. sitchensis*, $H = 1.73$, and for the broadleaved species in the understorey, $H = 1.35$. Judging by Figure 4.25, these results are plausible when considering the range of height values and their variation in relation to the height–diameter curves. Apparently, the species with the lowest vertical size diversity is *P. sylvestris*. According to H, the broadleaved species show more height diversity than *P. sylvestris*, although they only occur in the understorey. For the total stand, $H = 1.69$.

An alternative to Eq. (4.17) is the coefficient of variation of $g \times h$, i.e. tree height multiplied by corresponding basal area, denoted by $v_{g \times h}$. In R, this can simply be calculated as sd(myData$ba * myData$h)/hL, i.e. the standard deviation is divided by Lorey's mean height (Eq. (4.18)). However, the standard deviation, $\hat{S}_{g \times h}$ of $g \times h$ is internally estimated as

$$\hat{S}_{g \times h} = \sqrt{\text{Cov}(g^2, h^2) + (\text{Var}(g) + \bar{g}^2) \times (\text{Var}(h) + \bar{h}^2) - (\text{Cov}(g, h) + \bar{g} \times \bar{h})^2}, \quad (4.19)$$

where Cov and Var are covariance and variance, respectively. Luckily R applies Eq. (4.19) without a need for the user to code this equation explicitly when using sd(myData$ba * myData$h). For *P. sylvestris* in Pen yr Allt Ganol, $v_{g \times h}$ is 0.05. For *P. sitchensis*, $v_{g \times h} = 0.15$ and for the broadleaved species in the understorey $v_{g \times h} = 0.01$. This coefficient of variation gives the impression that the broadleaves show the lowest vertical diversity followed by *P. sylvestris*. The results of the two indices differ because index H uses height differences in relation to Lorey's mean height, whilst the coefficient of variation, $v_{g \times h}$, is based on observed height values and therefore possibly better reflects the varied canopy structure of the given example woodland. For the total stand, $v_{g \times h} = 0.09$.

5

Interacting with Forest Structure

Abstract

Forest management is essentially a goal-oriented manipulation of forest structure triggering processes that support biological rationalisation (see Section 2.6). In this chapter, the focus is on deliberate, professional interactions with tree vegetation. This has a long tradition in forestry and forms an important part of CCF, since this type of forestry is largely defined in terms of forest structure. Methods of interacting with forest structure mainly include thinnings and silvicultural systems. Both are broad groups that involve a wide range of specific methods which can be flexibly applied in isolation or in combination to meet varying management objectives. Whilst thinnings are generic methods for manipulating forest structure at almost any point in stand development, most silvicultural systems are larger management programmes or frameworks for achieving natural regeneration designed to enrich forest structure and to bridge the transition between two forest generations. Naturally, silvicultural systems include thinnings, but thinnings do not depend on silvicultural systems. Thinnings and silvicultural systems play a crucial role in the transformation of plantations to CCF (Section 2.4) and in the maintenance of CCF (Section 2.5). This chapter outlines the basic principles of manipulating forest structure in CCF which should be part of the toolbox of every forest manager interested in sustainable forest management.

5.1 Introduction

When reviewing the history of CCF in Section 1.7, we came across the fact that forestry has its origins in agriculture and that the process of emancipation from agriculture took a very long time. In many countries, this emancipation has not yet advanced much or is incomplete. Part of this process was guided by the experience that interacting with forest structure rather than leaving a forest stand completely alone after planting only to come back to it for final harvesting can both accelerate the production process and increase the value of the ecosystem goods and services planned for (Smith et al., 1997; Assmann, 1970). Assmann (1970), for example, was the first author to show that – depending on species and site conditions – an average reduction of stand basal area towards a residual basal area of 80% or 70% of maximum basal area, i.e. the carrying capacity on a given site, would increase stem-volume growth of *Picea abies* in Southern Germany by 15–30% compared to a forest stand of the same species on the same site that remains unthinned. Similar trends were

Continuous Cover Forestry: Theories, Concepts, and Implementation, First Edition. Arne Pommerening.
© 2024 John Wiley & Sons Ltd. Published 2024 by John Wiley & Sons Ltd.

also found for *Fagus sylvatica* and other species. These relationships between stand growth rates and residual basal area depend on species, environmental factors and average tree size (Pretzsch, 2019). Apparently, thinnings have a stimulating effect on growth processes. This fundamental insight gradually gave rise to the design of many different methods of modifying forest stand structure. These were studied and refined in forest growth and yield science, the field of science founded with the explicit objective to optimise thinnings and other tending methods on a quantitative basis (Pretzsch, 2009) whilst silvicultural systems largely remained to be the primary domain of traditional silviculture.

> The most basic rationale of thinnings and silvicultural systems is the steering of interactions between trees and between the trees and their environment to meet the objectives of forest management. These interactions form forest structure and forest structure at the same time contributes to these interactions.

Both thinnings and silvicultural systems can more generally be termed *interventions* or *human disturbances*. As part of the continued emancipation from agriculture, many different motivations and specialised questions were put forward why thinnings should be considered and what objectives they should help address. The most prominent ones of these include

- Removing trees that can cause damage to others, e.g. diseased or damaged trees,
- Maximising production in terms of stem volume or biomass in
  - individual trees,
  - forest stands.
- Modification of forest structure,
- Creating/maintaining habitats for animals and plants,
- Steering the tree species composition,
- Increasing or maintaining biodiversity,
- Maximising carbon sequestration in individual trees and in forest soils,
- Steering the interaction between individual trees by adjusting available growing space,
- Promoting individual-tree resilience to wind and snow,
- Modifying the forest microclimate including light, temperature and humidity in different canopy storeys and at the forest floor.

The more recent, continued emancipation of forestry from agriculture includes non-timber objectives and their sustainability. For example, as part of conservation, specialised thinnings are carried out in North Wales (UK) to revive and maintain rare vascular plants growing at the forest floor. Another alternative objective of thinnings can be the improvement of the light regime at the forest floor to enhance the production of forest berries, which are often considered as non-timber forest products.

In the past and until quite recently, natural disturbances caused by *disturbance agents* such as wind, fire, ice/snow, floods and bark beetles were largely perceived as nuisance and threat by forestry staff and particularly by forest owners, whilst ecologists are typically more focussed on the opportunities that disturbances typically provide (Puettmann et al., 2009). Historically, much effort was directed towards improving collective stand and individual-tree resilience against wind and controlling forest fires. Increasing disturbance events and climate change have forced silviculturists in recent years to change their attitude towards disturbances and to integrate them more positively in their planning. Natural disturbances, for example, often contribute to a diversification of tree species composition and to an increase of forest structure (Binkley, 2021). Natural disturbances and the changes they impose on forests are in fact a great source of inspiration for developing new forest management methods by emulating their effects. With time foresters realised that natural disturbances can complement forest management and that there are similarities between the response of tree vegetation to natural disturbances and to human interventions (Palik et al., 2021; Bose et al., 2014). The most common type of disturbances is localised and infrequent (episodic) to frequent (chronic). Such disturbance types are easy to integrate in forest management. Only few and rare natural disturbances operate at large scale and are *stand-replacing*. Even these calamities have been successfully included in forest management, e.g. in Switzerland where after severe gales at the end of the 1990s, many sites were completely left to natural devices for 20 years and only then traditional forest management resumed (Ammann, 2020).

Franklin et al. (2018) and Palik et al. (2021) defined *biological legacies* as the organisms, organic materials and organically derived patterns in soil and vegetation that persist from the pre-disturbance ecosystem into the post-disturbance environment. Such biological legacies include, for example seeds, spores, rhizomes, live remnant trees as well as dead biomass. The primary function of biological legacies is perpetuating plant and animal species, a process sometimes referred to as *lifeboating*. This function enables successful re-colonisation after disturbance impact. Residual trees, retained habitat/seed trees and deadwood after selective thinnings and harvesting events in CCF assume a role similar to the biological legacies created by natural disturbances (Bose et al., 2014; Gustafsson et al., 2012).

In managed forest ecosystems, modifications of forest structure caused by management usually have a far greater effect on forest development than natural tree growth and disturbances (Gadow, 1996). In CCF, *thinnings*, *silvicultural systems* and *underplanting* (cf. Section 2.4.1) are the main tools of making transformation happen and for maintaining the forest structure required for the specific variant of CCF a forest manager has chosen. Thinnings, however, constitute the most important tool, since they are repeated many times throughout the lifetime of a forest generation and are carried out within a very short amount of time. In CCF, final harvesting of mature trees is in principle also a thinning, since it is selective and many trees stay behind. The importance of thinnings in transformation to CCF can also be gauged from the concept of graduated-density thinning, see Section 2.4.2, which was designed as a transformation method. Silvicultural systems also mainly consist of thinnings in the main canopy of the overstorey. Therefore, thinnings form a very important part of CCF and are also much considered in CCF training (see Chapter 8).

5.2 Thinnings

In the classic Central European literature, thinnings are introduced as part of a larger group of forest management activities that is collectively referred to as *tending* (Assmann, 1970;

Table 5.1 Methods of tending forest stands according to Burschel and Huss (1997) and Nyland (2002).

Tending method	Definition
Thinning	*Young stands*: respacing, pre-commercial thinning
	Stands after canopy closure: thinning and selective harvesting (target diameter harvesting).
Habitat management	Creating and maintaining habitats such as forest margins, streamside buffers (cf. Section 7.3.3) etc.
(Improvement) underplanting	Planting an understorey of shade-tolerant trees supporting the light-demanding overstorey (cf. Sections 2.4.1 and 7.3.1).
Artificial pruning	Accelerating natural pruning and thus improving timber quality through the mechanical removal of branches in the lower part of tree stems.

Bartsch et al., 2020; Burschel and Huss, 1997; Köstler, 1956). Helms (1998) suggested that tending is generally any management intervention that enhances growth, quality, vigour and composition of a forest stand after establishment or regeneration and prior to final harvesting. According to Bartsch et al. (2020) tending includes *respacing, pre-commercial thinning, thinning, artificial pruning* and *fertilisation*. Other tending definitions have been suggested by Burschel and Huss (1997) and Nyland (2002) (cf. Table 5.1).

Respacing and pre-commercial thinnings are essentially early thinnings applied to juvenile forest stands or parts of forest stands (cohorts) before canopy closure with the prime objective of managing growing space through density reduction. The term 'pre-commercial' stems from the fact that these thinnings traditionally did not produce much financial return if any in commercial forestry and were more regarded as an investment in future timber quality or volume production and in resilience to wind and snow (Helms, 1998). Respacing is often seen as a synonym of pre-commercial thinning, although in Britain, it is perceived as a very early treatment, e.g. in dense thickets stemming from abundant natural regeneration. The influence and attitude of foresters practising commercial forestry become very clear in the synonyms 'thinning to waste' and 'unmerchantable thinning' (Hart, 1991). Due to their focus on comparatively small trees, respacings and pre-commercial thinnings often involve techniques different from other thinnings, e.g. pollarding (also termed 'decapitation'), girdling, or chemical injection. Whilst the latter is clearly not compatible with CCF, pollarding and girdling are perceived as cheap and low-impact options in frame-tree management (cf. Figure 2.25). Pollarding involves cutting trees at heights well above ground level (cf. Section 2.6) and girdling is the term for removing a broad band of bark all around the stem including all phloem so as to kill or at least weaken the tree (Helms, 1998). This blocks the flow of carbohydrates from crown to roots and causes the roots and eventually the whole tree to die within a few years (Smith et al., 1997). Pollarding is often faster to achieve in young forest stands, whilst some of the pollarded trees can later after resprouting help nurse frame-tree stems (Ammann, pers. comm.). Within CCF, there is an international trend towards abandoning respacing and pre-commercial thinnings in favour of biological rationalisation (cf. Section 2.6). These early thinnings are still in use in young stands on exposed sites involving species with flat root systems that are susceptible to wind or snow damage, but also as a prevention for avoiding insect calamities.

In CCF, selective harvesting operations are essentially also thinnings and trees removed in earlier thinnings are often put to commercial use in the same way as those obtained from harvesting operations. Therefore, in the context of CCF, the separation of the terms thinning and harvesting is somewhat academic. In RFM, this separation is more appropriate, see Section 1.8.

Management activities aimed at habitat creation and improvement are a typical element of modern forestry and particularly of CCF. Details on what kind of work they entail are provided in Section 7.3.3. Underplanting is another frequent tending activity used as a silvicultural tool for improving timber quality which is discussed in Sections 2.4.1 and 7.3.1, but is also a key method in transformation and conversion, see Section 2.4.

Artificial pruning (cf. Figure 5.1) is the process of removing branches along the stem axis with a view to invest in long-term timber quality. This is a possibility in CCF; however, it is usually very expensive and therefore applied only in rare circumstances where tree species on a given site have the potential to grow comparatively fast and timber (market) value is expected to be high. For example, in the Sellhorn Forest District of Northern Germany, the decision was made to prune the stems of frame trees in *Pseudotsuga menziesii* stands up to a height of 6–9 m, whilst all other trees where not pruned (cf. Figure 5.1). Given the local site conditions, it is anticipated that the *P. menziesii* frame trees can reach their target diameters of 60–80 cm within a period of 100 years, which is considered a comparatively fast production in the area, and accordingly, the pruning is economically viable given the market prices for quality *P. menziesii* timber. A positive side-effect of pruning frame trees is that they do not need to be marked otherwise, since the clean stems help distinguish them from non-frame trees. Fertilisation is often argued not to be compatible with CCF and national certification schemes, but this depends on regions and countries. All these tending operations manipulate forest structure, and this manipulation is carried out to meet defined silvicultural objectives.

Figure 5.1 *Pseudotsuga menziesii* Mirb. Franco woodland in the Sellhorn Forest District (Lower Saxony, Germany). Only the stems of the frame trees (cf. Chapter 3) were pruned up to a height of 6–9 m, and the matrix trees were not pruned. Source: Courtesy of Tim Summers.

Most forest stands start with a comparatively large number of small trees that originated from planting or from natural regeneration or from a combination of both. Initially, allowing for many trees to grow at close proximity is an application of biological rationalisation (cf. Section 2.6) because through vertical, but mostly through lateral competition, the seedlings and saplings are forced to grow primarily in height, form a single stem, and lose branches early, a phenomenon referred to as *natural pruning* in silviculture. All these properties are most desirable in commercial forestry, but to some degree have also their uses in other forms of forestry. As the young trees keep growing, they begin to touch their neighbours. Eventually, they close up into a thicket stage with a closed canopy. Often, this happens faster with conifers than with broadleaves, but this depends on species and sites. By this time, competing ground vegetation is thoroughly suppressed by the trees. Branches intermingle and deprive one another of light. As a result some branches die. In commercial forestry, this is desirable as the mutual pruning leads to the formation of clean stems, which is a pre-requisite for achieving timber quality. Because of differing morphology, self-pruning effects are usually more important for broadleaved than for conifer species. In the thicket-stage, competition for existence among trees becomes keen, the vigorous dominate, and the weak succumb (Hart, 1991). This is the time when size differentiation, the formation of size hierarchies and self-thinning, i.e. increased size-density dependent natural mortality, kick in and begin to transform the size structure of the young forest stand. This process of natural size differentiation is commonly employed in CCF and is part of biological rationalisation (cf. Section 2.6). In the past, forestry much more interfered with the size differentiation process and particularly limited natural self-thinning thus increasing management costs. In particular, it was thought that these natural processes act randomly thus conflicting with the idea of ensuring the survival of trees important to management objectives (Bartsch et al., 2020). This approach of applying intensive early thinnings gave rise to forest stands with uniform structure. Out of necessity in the face of changing timber markets and climate change along with more frequent, severe disturbances, this habit has changed in the last few decades. The requirement for biological rationalisation needs to be balanced with other important concerns such as individual-tree resilience to wind and snow. Leaving forest stands to their natural devices for a long time promotes the development of collective stand resilience (cf. Section 2.2). When then the decision is made to carry out the first selective thinnings or to continue with individual-based forest management (cf. Chapter 3), many trees may be lost to windthrow and to crown breakage particularly in species with flat root systems such as *Picea* spp. and *Betula* spp. Therefore, in exposed woodlands involving the two aforementioned species occasional earlier thinnings may help avoid the loss of trees from such disturbances, since trees are then trained to rely less on collective stand resilience (Cameron, 2002). Sadly selective thinning and harvesting operations can damage residual trees, although the exact damage rates much depend on the skills of the machine operators involved. For example, in the German state of Baden-Württemberg, on average one in four residual trees is damaged in such operations (Pretzsch et al., 2017). Keeping this damage to a minimum is an important management consideration.

Helms (1998) defined thinning as a cultural treatment made to reduce stand density of trees primarily to improve growth, enhance forest health, or to recover potential mortality. As discussed in Section 5.1, this is a very traditional definition with commercial forestry in mind that misses out on many of the other purposes of thinnings. Also, density is only a small part of forest structure, see Chapter 4.

Looking into the history of forestry, silvicultural thinnings had a bit of a late start in Central Europe and many planted forests remained largely unthinned until the beginning of the nineteenth century (Mantel, 1990; Hasel and Schwartz, 2006). After that thinnings, for a long time continued to be carried out with fairly weak intensities. This may be another indicator for the fact that early forestry heavily leaned on agricultural practices. Another reason for the hesitation to introduce thinnings was the conservative behaviour of forest authorities at the time after centuries of forest devastation. The benefits of thinnings were only understood much later and this insight constituted another benchmark on the path of gradual emancipation of forestry from agriculture.

5.2.1 Thinning Regimes

The term 'thinning' regime refers to whether the focus of thinning is on the whole population of trees in a forest stand or on individual trees. Generally, we can distinguish between two fundamental *thinning regimes*, (i) *global* and (ii) *local* or *individual-based* thinnings (Figure 5.2). Global thinning types address the whole forest stand as a population and only marginally consider tree interactions at local scale. The prime objective of global thinnings is the overall reduction of stand density. Global thinnings therefore follow a top-down approach (stand level → tree level), whilst local thinnings pursue a bottom-up strategy (tree

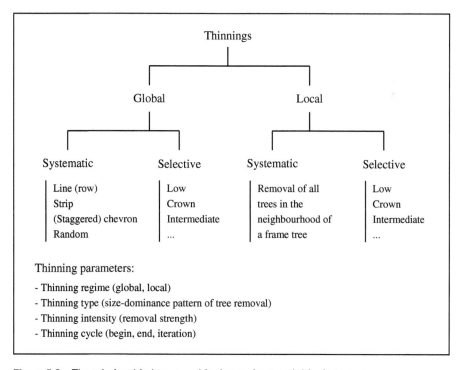

Figure 5.2 The relationship between thinning regimes and thinning types.

level → stand level). Local thinnings aim at modifying tree interactions and at local density reductions in the vicinity of frame trees (cf. Chapter 3).

A classic representative of global thinning regimes is the *systematic, schematic,* or *mechanical thinning,* which are all synonyms of the same thinning type. Smith et al. (1997) referred to this thinning type as *geometric thinning.* As part of line or row thinnings, trees are removed from across the whole stand area without considering their individual characteristics (area-control method, cf. Section 1.8). One application of this approach is the removal of complete rows of (planted) trees to open extraction lines during the first intervention (Pommerening and Grabarnik, 2019). Creating a network of extraction racks is a common way to start stand management. Other patterns include the removal of strips or corridors, the cutting of chevrons, a wedge pattern, that cuts across the planting rows. In *chevron thinnings,* parallel extraction racks are put in place and lines of trees at angles to the main rack are removed forming a herringbone pattern. In standard chevron thinnings, these lines are opposite each other and in staggered chevron thinnings, the lines off the main rack are alternate (Hart, 1991). Random tree removal comes close to the effect of localised disturbances (wind, snow), the results of which often appear like a random tree removal.

Systematic thinnings can be useful in young stands growing at high densities where differences in individual tree properties are not apparent to the naked eye. They are particularly common in plantation management even beyond the first intervention. Global thinnings, however, can also be selective, i.e. individual trees are selected according to certain criteria often related to their size and morphology, but without taking neighbour-hood relationships much into account. Global systematic and selective thinnings can also be carried out in combination, e.g. the complete removal of every seventh row of trees in a plantation combined with a selective removal of individual trees in the stand matrix between these rows. According to Figure 5.2, the graduated-density thinning in Section 2.4.2 is a combination of two variants of global systematic thinning, i.e. line and random or selective thinning. Intermediate thinning is a combination of global crown and low thinning (Rollinson, 1987), as will be explained in Section 5.2.2.

Local thinnings, sometimes also termed 'free' (style) thinning or crop-tree release, are mostly selective, because by definition and according to the size-control method (cf. Section 1.8) only those nearest neighbours are usually removed that are absolutely necessary to take out, see Section 3.5. However, when local thinnings are applied to fairly young forest stands or to the under- and mid-storeys, where size differentiation has not yet well developed, it may be a good strategy for a first intervention to remove all neighbouring trees within a certain radius around the frame trees or within a quadrate with the frame trees marking the centre (Figure 5.2). This method is a systematic (area-control) variant of the local thinning regime and sometimes referred to as halo thinning (Short and Hawe, 2018). Local thinnings generally stimulate the growth of selected frame trees (cf. Chapter 3). These favoured trees grow to a given size in a shorter time, or to a larger size in a given time, than they would with a regime of global low thinnings (Smith et al., 1997). Although seemingly the same thinning types, e.g. crown, low and intermediate thinnings, can be applied as global and local thinnings, and the results are very different because the thinning regimes fundamentally differ. Yield and density characteristics may even be very similar, but size differentiation and structure are usually more developed in local thinnings (Pommerening et al., 2021a). Because they are local and leave parts of the forest stand far away from frame trees untouched, local thinnings can be interpreted as a flexible form of variable-density thinning (cf. Section 2.4.3).

Up to the 1990s, there were some silvicultural schools in Central Europe recommending the combination of local and global thinnings. According to them, the frame tree neighbourhoods should be subjected to local thinning regimes, whilst the matrix between the frame trees should be thinned globally. This mix of regimes, though theoretically possible, led to management inconsistencies where often frame-tree competitors were unduly promoted and stand structure was homogenised (Klädtke, 1993), see Section 3.5.

5.2.2 Thinning Type

The term *thinning type* refers to the pattern of size dominance of trees which are removed in thinnings. Traditionally, size dominance has been defined in terms of crown classes (cf. Section 4.2). As discussed in Section 4.2, different systems of crown classes exist, and although they are susceptible to subjectivity, they still are an important pre-requisite for defining thinning types.

> Originally based on the crown-class system, two thinning archetypes developed, *thinning from below* or *low thinning* and *crown thinning* or *thinning from above* along with a number of variants and combinations.

In thinnings from below, lower crown classes, typically classes 4–5 or SD-S (Section 4.2), are targeted, see Figure 5.3a. The idea of this method is to concentrate growth on larger diameter trees. Canopy closure is usually not interrupted. Thinnings from below aim to maximise the area-related volume, biomass production, or carbon sequestration in trees, generally lead to the development of a uniform, single-storey stand structure and promote the collective resilience of the entire stand. Assmann (1970) termed this the 'thinning method of nature', since in natural self-thinnings the same crown classes contribute to the majority of mortality cases. 'Tidying up the forest' is a way British forest practitioners often refer to low thinnings. This phrase gives the impression of primarily removing badly shaped, sick and dead trees. In a similar way, Smith et al. (1997) referred to low thinnings as the 'wholesale elimination of the weak'. For a long time thinning from below was also known as the German method, since German forestry preferred this thinning type until well into the second half of the twentieth century (Bauer, 1962).

Thinnings from below were originally not designed to affect forest structure, the interaction regime between trees and the production process in commercial forestry, but aimed at utilising trees that negatively differentiated towards lower size-dominance classes before they died naturally and then would be 'wasted' (Schütz, 2003). At higher thinning intensities of low thinnings, some co-dominant trees need to be removed as well.

In crown thinning, also referred to as high thinning, French method and thinning in the dominant (Smith et al., 1997), interventions mainly affect dominant trees and thus reduce crowding within the main crown canopy. This involves a selection of dominant and co-dominant trees (crown classes 1–3, D-SD), see Figure 5.3b. As a consequence, crown

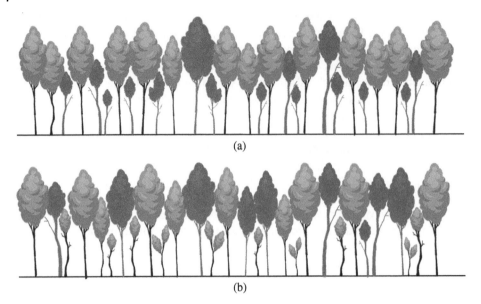

Figure 5.3 The principles of thinning from below (a) and crown thinning (b) targeting trees of different size and dominance for removal (grey). The visualisation also shows the residual stand after thinning (green) and thereby highlights the fundamental difference the two thinning types make to forest structure. Source: Courtesy of Zeliang Han.

closure is interrupted for some time. The rationale of this thinning type involves granting residual dominant space to grow rapidly by gradually removing competing, dominant neighbours. Crown thinning sometimes is colloquially referred to as 'creaming' in the forestry profession (Price and Price, 2006).

Crown thinnings only intervene in trees of the upper crown classes whilst leaving those in the lower untouched, which, if they survive, form a mid- or understorey and thus diversify forest structure (Smith et al., 1997). Unlike thinnings from below, crown thinnings result in measurable growth acceleration of the promoted trees and provide early revenues in commercial forestry. Crown thinnings aim to achieve maximum volume, biomass, or carbon of individual trees rather than high overall yields. They usually lead to more open canopies than low thinnings and to diverse, often multi-storeyed stand structures (Kramer, 1988). Crown thinnings also promote individual-tree resilience (Bartsch et al., 2020; Davies et al., 2008). For all these reasons, crown thinnings are typically preferred in CCF. Crown thinning is a more flexible method than low thinning, and it demands greater skill on the part of the forest manager (Smith et al., 1997). The concept of crown thinning was originally developed in Denmark and France and made its way into Central Europe only towards the 1970s. Most of the evidence indicates that the total yield in terms of standing stem volume is no greater from crown thinnings than from a comparable series of low thinnings. With crown thinnings, however, this stem volume is harvested in fewer trees of greater average stem diameter than with low thinnings (Smith et al., 1997).

Low thinnings are often associated with a *negative selection* of trees, while crown thinnings involve a *positive selection* (Figure 5.4). Negative selection has a focus not only on undesirable, dominated trees of poor form and poor growth but also on wolf trees and

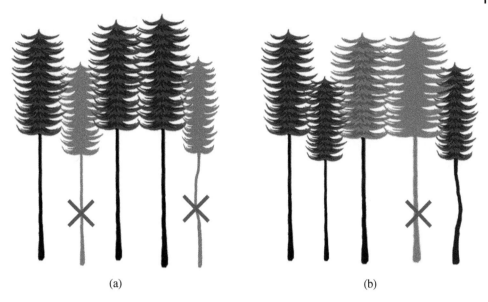

(a) (b)

Figure 5.4 The principles of negative (a) and positive (b) tree selection applied to the same group of trees. Trees selected for removal are shown in grey and have red crosses. The light, olive-coloured tree in the right sketch is a reference tree of sorts, for example, a frame tree (cf. Chapter 3) that is supposed to benefit from the tree eviction. Source: Courtesy of Zeliang Han.

whips (see Section 3.5). This can include forked stems, stems with bents, heavily branched individuals, damaged or diseased trees, or trees of unwanted species. Depending on management objectives, such individuals are perceived as 'negative' and are globally removed throughout the whole stand with a view to benefit the residual trees. The promotion of desired trees is indirectly achieved by evicting these less-desirable trees. Negative selections of trees lead to thinnings that are even more global than thinnings involving positive selection, as they do not require a physical identification of trees they are supposed to benefit, e.g. frame trees (Figure 5.4).

Conversely, positive tree selections follow a more direct approach by removing dominant competitors of the most desirable trees in a forest stand. Desirability criteria can include morphology and growth and depend on management objectives. These can be frame trees (in the context of a local thinning) but do not have to be identified as such (global thinning). Desirable trees eventually form the final forest stand. Positive tree selection usually is more effective than negative tree selection because it allows a concentration of management effort on fewer and larger trees thus reducing costs (cf. Section 2.6). It also promotes a diversity of horizontal and vertical forest structure more efficiently. This again enhances resilience and promotes biological rationalisation. In juvenile forest stands, however, where size differentiation has not yet begun whilst faulty tree properties have already appeared, e.g. in pre-commercial thinnings, negative tree selection can potentially have merits (Schütz, 2003).

Low and crown thinnings have the opposite effect on empirical stem-diameter distributions (Figure 5.5). Because of the preference of selecting trees with low-size dominance, small stem diameters are typically removed from stem-diameter distributions. This is

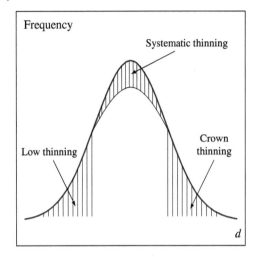

Figure 5.5 The effect of different thinning types on the stem-diameter (*d*) distribution. The hatched areas under the curve are those from which trees are preferably selected when performing the three thinning types indicated. Source: Pommerening and Grabarnik (2019).

contrasted by crown thinnings where trees with high-size dominance are evicted from which a tendency of removing large stem diameters from stem-diameter distributions arises. When applied, both thinning types can theoretically arrive at the same residual basal area; however, the numbers of trees and the mean stem diameter of the trees marked for the two thinning types, are then very different.

Schematic or systematic tree selection (cf. Figure 5.5) has a neutral effect in the sense that there is no particular trend towards the lower or the upper end of the diameter distribution. Schematic thinnings therefore tend to remove trees around the mean stem diameter. Natural disturbances often have a similar effect.

The fact that low and crown thinnings address opposite parts of the stem diameter distribution has been used to quantify thinning types. Quantifying thinning types provides the opportunity to study tree selection behaviour of forestry staff (Pommerening and Grabarnik, 2019, Chapter 7) and is an important tool used in CCF training (Chapter 8). It also helps simulate thinnings at the computer. Murray and Gadow (1991) suggested the proportion of the basal area of selected trees, P_G as an expression of thinning intensity. Naturally, basal area can also be replaced by other measures such as volume or biomass. Kassier (1993) proposed the ratio of P_G and P_N, i.e. the proportion of the number of selected trees, as a measure of thinning type:

$$B = \frac{\text{Proportion of the number of trees selected}}{\text{Proportion of basal area of selected trees}} = \frac{P_N}{P_G} \tag{5.1}$$

This measure quantifies the thinning type by comparing the numbers of trees selected with their cumulative size. If $B < 1$, a smaller proportion of trees has been selected compared to their proportion of cumulative basal area. This typically indicates a crown thinning or a similar approach, and the trees selected show a tendency of being in the upper part of the empirical diameter distribution (Figure 5.5). A larger proportion of trees is selected compared to their proportion of basal area, if $B > 1$. This is consistent with a thinning from below or a similar approach, and trees were preferably selected in the lower part of the empirical diameter distribution (Pommerening et al., 2018; Pommerening and Grabarnik, 2019, Figure 5.5). A result $B \approx 1$ is often a consequence of natural disturbances, schematic and intermediate thinnings.

A similar measure of thinning type is Magin's k factor (Eq. 5.2). Wenk et al. (1990) defined a variant of this measure as the ratio of the mean volume of trees marked for thinning, $\bar{v}_{sel}$, and the mean volume of all trees before marking, $\bar{v}_{init}$.

$$k = \frac{\bar{v}_{sel}}{\bar{v}_{init}} \tag{5.2}$$

According to Wenk et al. (1990), $k < 1$ indicates a thinning from below, $k = 1$ an indifferent thinning and $k > 1$ a crown thinning or a similar approach. Alternatively, it may also be an option to plot the Simpson diversity index (Magurran, 2004, see Section 4.4.1) calculated for tree number proportions against the same index calculated for basal area proportions. Schütz (2003) mentioned a coefficient k' first defined by Hiley (1959):

$$k' = \frac{\bar{d}_{sel}}{\bar{d}_{res}}, \tag{5.3}$$

where $\bar{d}_{sel}$ is the arithmetic mean stem diameter of trees selected for thinning, and $\bar{d}_{res}$ is the mean stem diameter of the residual trees. In the case of a crown thinning or a similar method, $k \approx 1$, whilst for low thinnings k lies between 0.7 and 0.8 or is even less. Smith et al. (1997) reported a similar index where the (quadratic) mean stem diameter of the trees marked or removed is divided by the (quadratic) mean diameter before marking or thinning. All these indices are easy to compute and practical examples are given in Pommerening and Grabarnik (2019, Chapter 7). Whilst they help indicate the thinning archetype, they cannot distinguish between thinning regimes. For the latter purpose, a spatial approach is necessary.

There are a number of combinations and variants of low and crown thinnings. The most important ones are described in the following paragraphs. Graduated density thinning (GDT, cf. Section 2.4.2) and variable density thinning (VDT, cf. Section 2.4.3) were not listed here, as the author of this book interpreted them primarily as transformation methods, but it is certainly possible to argue otherwise.

Combined thinning. Known as *gestaffelte Durchforstung* in German, the combined thinning has originally been designed for forests dominated by *Picea* spp. and *Pinus* spp. The idea is to take both wind risk and growth patterns into account. The concept suggests heavy crown thinnings in juvenile forest stands or juvenile patches of *Picea*. When stand height has reached a critical value, the thinning regime gradually changes to moderate crown thinnings and eventually to thinnings from below. This combination also addresses the growth dynamics of *Picea*, *Pinus*, *Larix*, *Fraxinus*, *Acer* and *Quercus* spp., where growth is comparatively strong in the juvenile phase of the development of these species followed by a rapid decline with increasing maturity. The combination was first proposed by Wiedemann (1951), and later described by Burschel and Huss (1997) and Bartsch et al. (2020). In contrast to intermediate thinnings, this is a combination of low and crown thinnings in time.

Group thinning. This thinning type was already mentioned in Chapter 3 because the idea was developed in the context of frame-tree management and local thinnings (cf. Figure 3.8). Traditionally, particularly in global thinnings, residual trees were more or less uniformly spaced so that each residual tree has similar growing space. Based on the observation of natural clustering of dominant *F. sylvatica* trees in Central

Europe, Busse (1935), Kató (1973) and Mülder (1990), however, suggested a clustered arrangement of two to three frame trees with large distances between the clusters. This method can certainly also be applied to many other species and can even be carried out as a global thinning method. Particularly on sites with heterogeneous environmental conditions, e.g. variable soil conditions, this is an appropriate approach. Therefore, group thinnings have often been recommended in mountain forests where topography and soil depth vary much even at short distances. Natural groups are likely to exist where trees of the same species appear to form a common canopy without crown shyness which otherwise may indicate competition. Within each group, hardly any thinnings take place apart from limited, weak low thinnings as rare exceptions. A combination of global and local crown thinnings is carried out between the groups. They particularly ensure that no trees outside the groups start competing with the group, i.e. the groups as a whole are released as if they were individual trees with ample distances between the groups (Rittershofer, 1999). Some similarities with VDT (cf. Section 2.4.3) exist.

Halo thinning. For newly selected frame trees in comparatively young stands but also for releasing mature, broadleaved trees which are remnants of former broadleaved forests but are now situated in dense conifer plantations, it is recommended to remove all other trees with a certain radius of these trees. Looking from above the main canopy, this gives the impression of halos cut around the frame trees (Short and Hawe, 2018).

Heavy release thinning. In principle, a moderate-to-heavy crown thinning method (Kramer, 1988; Bartsch et al., 2020) is often applied to frame-tree management (cf. Chapter 3) in the second half of a forest generation's lifetime. The idea is to grant frame trees abundant space for maturing and putting on increment before leaving them completely alone until target diameter harvesting starts. As a consequence of this thinning type, the main canopy is permanently interrupted. In German, this method is known as *Lichtwuchs-durchforstung* and can also be part of seed-tree/shelterwood systems for initiating natural regeneration (Mayer, 1984).

Intermediate thinning. This thinning type is a combination or mix of low and crown thinnings. This is a traditional approach of conifer management in Britain (Rollinson, 1987; Hart, 1991). As in low thinnings, most of the suppressed and sub-dominant trees are removed to promote the growth of larger trees. In addition, as in crown thinnings, the main canopy is opened up by breaking up groups of competing dominant and co-dominant trees. The effect of intermediate thinnings on the stem diameter distribution is similar to that of systematic and random thinnings. In contrast to combined thinnings, this is a combination of low and crown thinnings in space.

Selection thinning. Not to be confused with selective thinning, a selection thinning is a specialised method for transforming forest stands to selection stands or for maintaining them (cf. Section 2.4). Known as *Plenterdurchforstung* and *Plenterung* in German, the method is in principle a heavy, extreme crown thinning mainly targeting predominant trees that reach above the main canopy or are in the process of doing so. Smith et al. (1997) referred to this method as thinning of dominants. As such the selection thinning is also a selective thinning and not all predominant trees are removed in one intervention.

Structural thinning. Detailed in Section 3.6, structural thinning was proposed and successfully implemented by Reininger (2001) and Goltz (1991). The method involves two sets of frame trees in two different canopy layers thus aiming at a permanent two-storeyed high forest (cf. Section 5.5) in combination with target diameter harvesting.

Target diameter harvesting. Although seemingly not a thinning but a harvesting method, there is little difference between thinning and harvesting in CCF. Both are selective, i.e. they leave many residual trees behind and trees removed in thinnings can be sold for commercial purposes in the same way as trees in target diameter harvesting. The method was already defined in Section 3.4 and is listed here for completeness. Instead of stem diameter, other size variables such as volume or biomass can theoretically be used; however, stem diameters are easier to check up upon in forest practice. Target diameters are determined by the forest industry or by conservation bodies. There are, however, also regional guidelines (cf. Table 3.1). Each frame tree can be harvested, once it has reached the species-specific target diameter, but it does not necessarily have to be. Due to different microsites and genetics, frame trees usually reach their target diameters at different times, i.e. the tree removal is typically staggered in time. The next forest generation gradually develops under the canopy of the frame trees whilst they wait for their turn to be felled and until that time they simultaneously act as seed trees. Target diameter harvesting follows the size-control method of sustainability, see Section 1.8.

5.2.3 Thinning Intensity

Thinning intensity describes the rate at which trees in terms of certain quantities, e.g. their total number, basal area, volume or biomass, are removed in any one thinning (Hart, 1991). In traditional forestry carried out for commercial purposes, thinning intensity is based on stem volume. However, basal area is also commonly used both in research and in practice. Basal area has the advantage that it is easier to compute and does not involve modelling assumptions concerning taper or additional height measurements.

Thinning intensity not only depends on overall tree density but also on species-specific growth patterns. For example, vigorously growing, intermediate and shade-tolerant species such as *Picea* spp. and *Pseudotsuga* spp. require greater thinning intensities than less vigorously growing and light-demanding species such as *Quercus* spp. and *Pinus* spp. (Bartsch et al., 2020, cf. Appendix B). The idea to consider growth dynamics also implies that early thinnings in young stands should be more intense than those in older trees after their height growth has culminated. Mayer (1984) pointed out that interventions with great intensities, whilst releasing residual trees, can potentially reduce average stand growth for an extended time, particular in middle-aged and mature stands or tree cohorts.

In many parts of Central Europe, thinning intensity was historically defined in terms of so-called 'thinning grades'. Thinning grades were fuzzy qualitative descriptions of thinning intensity based on crown classes (cf. Section 4.2), i.e. forest managers relied on them to roughly define from which crown classes to recruit trees for global thinning methods. The selected classes coupled with some additional descriptions resulted in different grades or thinning intensities. These were largely combinations of thinning types and thinning intensities and received qualitative labels such as weak thinning from below (grade A), moderate thinning from below (grade B), heavy thinning from below (grade C), heavy release thinning (grade L), weak crown thinning (grade D) and heavy crown thinning (grade E) (Wenk et al., 1990; Bartsch et al., 2020). The thinning intensities labelled 'weak',

Table 5.2 Guidelines for residual basal area [m² ha⁻¹] in Central European thinning trials.

Species	Low thinning			Heavy release thinning	Crown thinning	
	Weak	Moderate	Heavy		Moderate	Heavy
Fagus spp.	37–47	26–35	20–26	17–24	20–27	18–24
Picea spp.	53–60	43–55	33–45	25–35	34–43	
Pinus spp.	33–43	28–35	24–29	18–23	24–29	
Quercus spp.	28–37	24–30	17–23	14–18	20–25	17–23

Source: Wiedemann (1951).

'moderate' and 'heavy' were also defined in terms of crown classes but were additionally affected by the intended thinning type. For example, thinning intensity 'moderate' had different implications for the residual basal area, i.e. the sum of basal areas of the remaining trees after thinning, depending on the thinning type, cf. Table 5.2. The lower the residual basal area, the greater usually is the thinning intensity assuming similar basal areas before thinning. The numbers in Table 5.2 may serve as an illustration of how these combinations of thinning types and intensities translated into residual basal area. These numbers are historic and only of local significance, but they can help understand how species, thinning type and thinning grade were thought to interact.

Smith et al. (1997) pointed out that optimum residual basal area much depends on light demand, i.e. the photosynthetic and growth efficiency of species, cf. Table 5.3. The more shade-tolerant a species, the larger the residual basal area theoretically can be. Shade-tolerant deciduous species have a wider range of appropriate basal areas than shade-tolerant evergreens. The lowest basal area levels apply to light-demanding deciduous trees. This suggests that a mix of light-demanding and shade-tolerant tree species is likely to be quite flexible in the response to thinnings (cf. Appendix B). Naturally, residual stand basal area is not constant over time. It typically reaches a first maximum in young, unthinned stands and is then consequently much reduced by thinnings. Later towards maturity, residual basal area is allowed to increase again. Where differences in environmental factors occur, appropriate basal area for stands of a given species would be less on a poor site than on a good one. In arid forest regions (see Figure 3.3) or in the

Table 5.3 Approximate ranges of appropriate residual basal area, after thinning, for middle-aged stands.

Species group	Residual basal area [m² ha⁻¹]
Shade-tolerant evergreens	30–35
Light-demanding evergreens	18–30
Shade-tolerant deciduous	16–37
Light-demanding deciduous	12–18

Source: Adapted from Smith et al. (1997).

boreal forest, where either water or nutrients are limited, full closure of the root systems is possible with very low canopy cover (Smith et al., 1997). Therefore, basal area guidelines known from temperate forests need to be applied in such conditions with great care.

The historical qualitative definitions of thinning intensity discussed earlier in this section left much room for interpretation. Therefore, based on the considerations of the last few paragraphs, residual basal area was often considered as a quantitative criterion for thinning intensity to ensure certain standards (Pommerening and Grabarnik, 2019). Basal area per hectare as a measure of stand density or crowding has the advantage of being negatively correlated with below-canopy light levels whilst being easier to measure (Burschel and Huss, 1997), cf. Figure 5.12. Using this many guide-curve systems and silvicultural prescriptions were developed for both forest practice and research applications in forest simulators. These were often based on functions of residual basal area that depended on tree size (Pretzsch, 2009). A useful measure of thinning intensity is P_G, as introduced in Eq. (5.1). This relative measure of basal area to be removed can also be related to comparable natural stands with the same species composition growing on the same forest sites. In that case, P_G expresses the thinning reduction in terms of the natural carrying-capacity basal area (Döbbeler and Spellmann, 2002). Such a measure is very natural in the context of CCF, since any management input and impact should be limited to the bare minimum and 'correct' natural development only ever so slightly (cf. Section 1.4). Since the definition of thinning intensity by thinning grades does not clearly define the number of trees or the amount of basal area/volume to be removed, *stem-number guide curves* were proposed. These provide information on the recommended density of the residual forest stand initially dependent on age, later they were made dependent on mean tree size, e.g. quadratic mean diameter or stand height, to generalise the relationship for different environmental conditions. Information for producing such guide curves was obtained from thinning and spacing trials (Pretzsch, 2009). The *crown spread ratio*, i.e. crown width divided by tree height, is a good example variable used to produce such guide curves and has often been applied to describe growing space requirements of trees depending on stand height.

Based on these studies, height-based tree-number guide curves have been developed for practical use in forest management (see Figure 5.6). With square spacing, $N = 10\,000/$ crown spread ratio$^2 \times H_o^2$ (where H_o is dominant stand height) and for other tree dispersion patterns, this relationship can be adapted (Pretzsch, 2009; Kramer, 1988).

A similar measure leading to guide curves is *relative spacing*, RS (Kramer, 1988; Gadow et al., 2021):

$$RS = \frac{\sqrt{\frac{10\,000}{N}}}{H_o},$$
(5.4)

where $\sqrt{\frac{10\,000}{N}}$ calculates mean tree distance of regularly spaced trees, H_o is again dominant top height, e.g. the mean height of the 100 largest trees per hectare and N is the number of trees per hectare. The concept of the stand density index (SDI) is a related approach quantifying the number of trees in relation to stand quadratic mean diameter (Gadow et al., 2021). Naturally, it is also possible to derive stem-number guide curves for individual species populations in a mixed-species forest (Pretzsch, 2019). In a similar way, it is an option to establish basal-area guide curves which – particularly in mixed-species woodlands with high

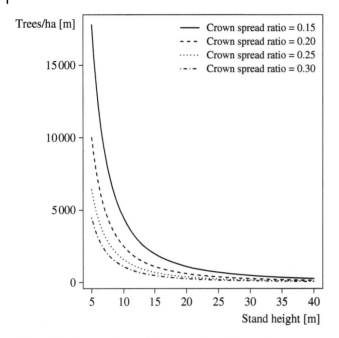

Figure 5.6 Tree number guide curves depending on the crown spread ratio (assuming square spacing). Source: Pommerening and Grabarnik (2019).

structural diversity – would be more appropriate, since they offer a more general approach than stem-number curves.

However, using such general density guide curves somewhat contradicts the principle of local thinnings, where overall density reductions are less important than reducing local densities in the vicinity of frame trees (see Section 3.5) in a form of spot release. In local thinnings, thinning intensity is therefore defined by the number of competitors to be removed in the vicinity of each frame tree, see Figure 3.10. This number can obviously differ from frame tree to frame tree depending on the local release requirements of the tree in question (see Figure 3.6), however, mean numbers can be defined as targets: On average a removal of 0–2 frame-tree neighbours qualifies for a weak thinning, 2–3 neighbours for a moderate thinning and 3+ neighbours for a heavy thinning, cf. Section 3.5. In practice, some forest managers measure the stem diameters of the trees they mark for thinning to estimate the total basal area of the trees selected for removal. Local thinnings typically follow the growth dynamics of forest stands, particularly the tree crown-width development. Therefore, thinning intensity in local thinnings is heavy early in stand development and weak towards mature, old-growth stages when tree growth is much reduced (Bartsch et al., 2020).

5.2.4 Thinning Cycle

Thinnings start at some point during stand development and are typically repeated at certain intervals. Starting point and time between two successive thinnings define the thinning cycle (see Figure 5.7). Thinning cycles usually are a compromise between biological processes (e.g. growth) and economic or logistic restrictions. Few heavy

Figure 5.7 Using a hypothetical 'sawtooth' curve of stand basal over time to explain the concepts of thinning intensity and thinning cycle. Periods with undisturbed stand growth (linear interpolation) are followed by sudden reductions of basal area in thinnings. Source: Pommerening and Grabarnik (2019).

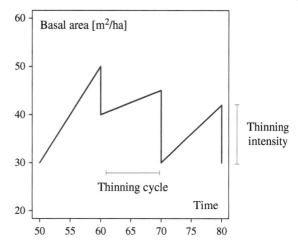

thinnings with long thinning cycles are preferred in economically driven management scenarios, but moderate, more frequent interventions are often better for the forest stand in question because the trees can more gradually adapt. For practical reasons and convenience, silvicultural planning in temperate forests often operates on the basis of five-year intervals; however, with fast-growing species such as *Eucalyptus spp.*, *Pinus sitchensis* and *Pinus radiata* D. Don these can also be shorter. In forest ecosystems with slow growth processes, such as the boreal forest, thinning cycles of ten years, and more are often appropriate. In accordance with species-specific growth patterns, it is also possible to apply shorter cycles of 3–5 years at young ages or early development stages and longer ones (8–10 years) in mature stands (Pommerening and Grabarnik, 2019). In pursuit of biological rationalisation (cf. Section 2.6), there is currently a tendency to extend thinning cycles.

Thinning cycles can also be defined by intervals of stand height. If height intervals (usually stand dominant or top height is referred to) are used instead of time intervals, environmental conditions and changes in growth patterns are automatically taken into account, i.e. this helps generalise silvicultural prescriptions. Height intervals of 2–4 m commonly apply to determine the timing of thinnings during early phases of stand development. When height growth eventually slows down in later stages of stand development, thinning cycles are more likely to be defined by the development of stand basal area, canopy closure, or the competitive situation of frame trees.

Light-demanding species generally require another thinning when the canopy has closed following the previous thinning, whereas shade-bearing species can usually tolerate a certain degree of crown pressure and need interventions only when the crowns start to obstruct each other (cf. Appendix B). Again, on sites where other environmental factors such as water are more limited than light, tree crowns may be less suitable for deriving thinning requirements.

In areas of high wind risk, it is generally preferable to apply lighter thinnings and repeat them more often. Apart from density characteristics, the necessity for thinning can also be gauged from changes in allometric indicators, such as the h/d ratio (Eq. 2.2) or the crown ratio (Eq. 2.3) of frame trees (Pommerening and Grabarnik, 2019).

5.3 Regenerating Forest Stands with Silvicultural Systems

Methods that rely on establishing and releasing propagules of both seed and vegetative origin from sources within or adjacent to the stand being regenerated have been termed *natural regeneration* or *reproduction methods*. A regeneration method is a treatment to a stand and its structure in order to establish or renew it. Each method consists of the removal of all or part of the old stand, the establishment of a new one and any site treatments that are applied to create and maintain conditions favourable to start regeneration (Smith et al., 1997). In these methods, forest structure manipulations are carried out to favour or discourage certain kinds and amounts of seedlings. These manipulations emulate the type and intensity of natural disturbances that achieve similar results. Clearcut and coppice methods reflect severe, stand-replacing disturbance regimes that remove most if not all tree vegetation. All methods discussed in this section are compatible with CCF and emulate small to medium scales of disturbances with moderate intensity, which is the most common type of disturbances (Figure 5.8). In general, a regeneration or reproduction method is a procedure by which a stand is established or renewed (Smith et al., 1997). From a technical point of view, in CCF, the methods of inducing natural regeneration replace clearfelling and replanting used in RFM. In this book, the definition of Troup (1928) and Matthews (1991) is adopted:

> A silvicultural system may be defined as the process by which stands constituting a forest are tended, removed and replaced by new stands resulting in stands with distinctive structure. In this definition, tending refers mainly to thinning operations carried out in immature stands, in so far as these operations affect the state of the stand at the time of regeneration.

Applying silvicultural systems generally results in a smooth transition from one forest generation to the next and makes extensive use of natural processes in the spirit of biological rationalisation (cf. Section 2.6). Since all forests under normal circumstances eventually regenerate naturally, as with thinnings, the idea of silvicultural systems is to accelerate processes and to optimise outcomes by providing favourable microclimatic conditions and neighbourhood relations (see Section 1.5). In a nutshell, following Mayer (1984) the basic silvicultural systems can be described as follows (cf. Figure 5.8):

Clear-cut system. Simultaneous removal of all trees in a forest stand. No protection of regeneration, the whole stand area is affected.

Group system. Removal of irregular groups of trees thus opening gaps of different sizes and shapes in the main canopy that are usually staggered in time. Temporary lateral and sometimes also vertical protection of regeneration in groups. Cuttings tend to be localised.

Strip system. Simultaneous removal of all trees in a comparatively narrow strip. Temporary, mostly lateral protection of regeneration in the strips. Cuttings tend to be localised.

Uniform shelterwood system. Uniform thinning and opening of the main canopy by removing individual trees. Temporary, occasionally longer lasting mostly vertical protection of regeneration. The whole stand area is affected.

Clearcut system

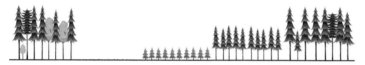

Uniform shelterwood system

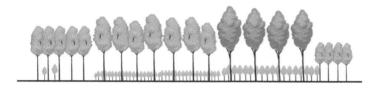

Strip system

Group system

Selection system

Figure 5.8 Visual impression of the structure involved in basic silvicultural systems in progress. The clear-cut system is not part of CCF, but is included here for comparison. Source: Courtesy of Zeliang Han.

Selection system. Irregular individual-tree to group-wise removal of trees on a continuous basis to encourage vertical forest structure. Long-lasting mostly vertical protection of regeneration. Cuttings affect the whole stand area.

Vanselow (1949) distinguished three basic principles or situations of regenerating forest stands naturally, i.e.

- *shelter*,
- *margin*,
- *bare land*.

These basic building blocks occur in all silvicultural systems. Obviously, the shelter principle prevails in all shelterwood systems, regeneration at margins is the dominant principle of the strip system and regeneration on bare land is the typical situation in clear-cut systems and in the centre of large gaps. However, the principles of margins and bare land also occur in group systems.

Silvicultural systems are an important element of continuous cover forestry and often constitute the starting point in the transformation to CCF, see Section 2.4, since they introduce new cohorts of forest structure. Apart from the advantages of working with natural ecosystem processes and limiting the management impact on forest ecosystems, they also help save expensive planting costs. As alluded to at the beginning of this section, silvicultural systems are essentially methods used to mimic small- to medium-scale natural disturbances and to carefully manage light, microclimatic conditions and nutrient supply as part of the process.

In shelterwood systems, the main canopy is heavily thinned in a series of interventions, which can extend over several decades. Depending on the duration allocated to these systems, they can result in the establishment of a comparatively uniform cohort of advance regeneration under the partial shelter of residual main-canopy trees (Smith et al., 1997).

In any silvicultural system, an appropriate part of the biomass of a forest stand is removed so that the productivity of the ecosystem remains undiminished in the long term. At the same time, the main canopy is adjusted and the fellings and extraction of trees are controlled in such a way that the microclimate of the site and condition of the soil are favourable for regeneration with trees of suitable species. In these circumstances, the disruption of the forest ecosystem is minor and its negative effects are short-lived (Matthews, 1991).

In the Anglo-American literature, the impression is often conveyed that silvicultural systems are treatment programmes defining the development and progression of a forest stand indefinitely (Matthews, 1991; Smith et al., 1997; Nyland, 2002). For example, in the IUFRO terminology of forest management, Nieuwenhuis (2000) defined silvicultural systems as 'planned programmes of treatments extending throughout the life of forest stands that include harvesting, regeneration and tending methods or phases. They cover all activities for the entire length of a stand's lifetime'. A very similar definition can be found in Smith et al. (1997). This view is only really true for the selection system and the two-storeyed high forest (see Sects. 5.4 and 5.5). The Central European school of silviculture clearly sees silvicultural systems mainly as broad regeneration methods (Burschel and Huss, 1997; Rittershofer, 1999; Bartsch et al., 2020), and this was indeed the original purpose for which most of them were invented (Mantel, 1990; Hasel and Schwartz, 2006). For an overview and for quick orientation see Figures 5.8–5.10. Particularly for the group and selection systems an increasing 'verticalisation' can be noted, i.e. an increase of the tendency that the vertical growing space is filled with tree crowns (Schütz, 2001b, Figure 5.8), which is often desirable in CCF. Troup (1928) and Matthews (1991) pointed out that silvicultural system include three main aspects: (i) The method of regeneration of individual forest stands constituting the forest, (ii) the structure of the new forest stand and (iii) the spatial arrangement of stands over the whole forest. The last aspect is important for landscape, wind hazard and considerations of thinning/harvesting efficiency. It is also worth noting that all silvicultural systems had their origin in forest practice rather than in research (Hasel and Schwartz, 2006). The selection system, however, was developed by farmers owning agricultural and forest land (Schütz, 2001b). In selection systems, regeneration establishes continuously over the entire stand area, whereas it is confined to a certain time or space in all other systems. For their special nature and because they are strictly speaking not regeneration methods, the selection system and the two-storeyed high forest are discussed in detail in Sections 5.4 and 5.5. We begin with the main types and then refer the reader to important local and historical adaptations and combinations (Burschel and Huss, 1997; Matthews, 1991; Rittershofer, 1999; Bartsch et al., 2020). Silvicultural systems are mainly characterised by the geometry of canopy openings, i.e. by the spatial gap structure applied (see Figures 5.8 and 5.9). They also differ in the spatial pattern of residual trees, which can be analysed using point process methods (Pommerening and Grabarnik, 2019).

There are marked differences in the spatial and temporal scales of regeneration (Figure 5.10). While the clear-cut system leads to a very rapid regeneration (usually

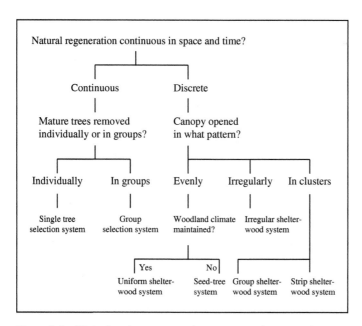

Figure 5.9 Silvicultural or regeneration systems using natural processes. Source: Pommerening and Grabarnik (2019).

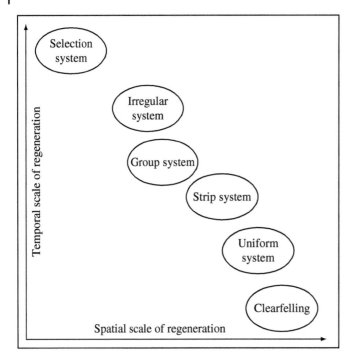

Figure 5.10 Relationship between spatial and temporal scale of natural tree regeneration in different silvicultural systems. The systems denoted as irregular, group and strip also include their shelterwood variants and uniform system is short for uniform shelterwood system. Clearfelling is not part of CCF but is included here for comparison. Source: Adapted from den Ouden et al. (2010).

through planting although natural regeneration is not uncommon, e.g. in stands of *Pinus pinaster* AIT. in Spain and *Picea sitchensis* in Ireland) that affects the whole area, the shelterwood systems retard regeneration and often only affect parts of the stand area at any point in time (den Ouden et al., 2010).

The differences in the shelterwood systems can therefore be also well described by their difference in spatio-temporal regeneration patterns. The irregular shelterwood system, for example, is a variant of the group system and caters for small-scale regeneration that unfolds very slowly. With these characteristics, the irregular shelterwood system comes close to the selection system, where there is uninterrupted regeneration at extremely small scale. The two-storeyed high forest is also close to both the irregular shelterwood and the selection system. On the other hand, the uniform shelterwood is not so far apart from the clearfelling and replanting system with regard to the spatio-temporal scales of regeneration.

There is no single optimal prescription for the regeneration of stands. When applied, each stand must be treated individually and an appropriate regeneration method should be chosen on the basis of a detailed stand analysis. In some cases, the advance natural development of regeneration may suggest a particular silvicultural system. For example, advance regeneration in small canopy gaps might suggest the group shelterwood system, provided the risk of wind damage is not too great. Where the canopy trees are windprone, or at a stand edge exposed to side light by the felling of an adjacent stand, trees sometimes might begin to regenerate in a manner similar to a strip shelterwood system. Also 'local accidents', where unintended and unprecedented interventions triggered the unexpected but promising development of diverse woodland structure, may give vital clues about the silvicultural potential of a forest stand, see Section 2.2.1.

Ideally, the removal of mature trees proceeds in tandem with the establishment of the next tree generation and thinning/harvesting operations are designed to facilitate regeneration at the same time. In this process, it is important that trees with desirable phenotypes have the opportunity to regenerate prior to their removal (Pommerening and Grabarnik, 2019). This strategy also supports tree-size diversification.

Shelterwood systems typically reduce precipitation reaching the forest floor due to the remaining main canopy trees. In group systems, more precipitation reaches the forest floor in the gaps; however, wind turbulences can be strong in large groups and can cause damage on residual main canopy trees at the edges. Large canopy openings can trigger localised waterlogged soils due to a lack of interception by main canopy trees and a lack of water consumption. On the other hand, localised evapotranspiration can be high, particularly at the North-Eastern fringes of gaps (in the Northern Hemisphere), see Figure 5.15. They are also more susceptible to frost events.

In the past, the remaining main canopy trees were more or less simultaneously removed, when regeneration was sufficiently secured. This simple way of removing the remnants of the overstorey has increasingly given way to target-diameter harvesting (see Section 3.4) so that the removal is staggered in time: Once regeneration is established and safe, overstorey trees are individually removed depending on their size (Pommerening and Grabarnik, 2019).

When it comes to silvicultural treatments, a standard set of terminology has developed that applies more or less across the whole range of silvicultural systems. These terms have a long tradition and will be used in the following sections (Matthews, 1991):

Preparatory fellings/cuts. Essentially thinnings with the objective to uniformly reduce the density of the main overstorey so that seed production and regeneration is encouraged.

Regeneration fellings/cuts. Beginning with the seeding and ending when the final felling has been carried out and the young stand is fully established. Tree selection follows the spatial pattern of the silvicultural system.

Seeding (establishment) fellings/cuts. To stimulate and initiate seed production whilst maintaining woodland climate necessary for germination and seedling growth.

Secondary fellings/cuts. To successively remove the overstorey and uncover the regeneration as to provide more light, water and nutrients.

Final or removal fellings/cuts. The last of secondary fellings. All remaining overstorey trees are removed, and the fully established regeneration remains.

Preparatory fellings typically prepare the overstorey trees for seed production and foster individual-tree resilience. They commonly do not follow the spatial pattern of the regeneration system. Uniform shelterwood cuts, for example, are often used as preparatory fellings, whilst the actual silvicultural system may be a group or strip system.

Preparatory fellings made shortly before the seeding felling are not necessary, if the stand in question has had a good thinning history with regular, not too weak interventions in the past so that the crowns of the overstorey trees are well developed and small gaps between adjacent crowns exist. The regeneration fellings typically follow the spatial pattern of the silvicultural system, i.e. uniform removal or gap, strip alignments.

The basic silvicultural systems as visualised and briefly characterised in Figures 5.8–5.10 are hardly applied in isolation. They are only broad frameworks and theoretical concepts. Smith et al. (1997) described them as working hypotheses. In the process of refinement and local adaptation, two or more regeneration methods are usually implemented in combination, e.g. the group-strip shelterwood system (Mayer, 1984). Vanselow (1949) distinguished between *additive* and *substitutive* combinations of basic silvicultural systems.

Broadly speaking, any combinations implemented at the same time are *additive*, whilst in *substitutive* systems two or more basic silvicultural systems combined follow on from each other in time, i.e. the second system starts when the first one is completed. The aforementioned group-strip shelterwood system (see Section 5.3.4.4), for example, is an additive system. Group shelterwood systems are often preceded by a one-off uniform shelterwood cut (preparatory cut) to encourage advance regeneration. After a few years, this is then followed after by group cuttings. This is essentially a substitutive combination of silvicultural systems. The terms *spatial* and *temporal* combinations of silvicultural systems are probably more appropriate.

It is also possible to use both artificial and natural regeneration in silvicultural systems. This is often considered a good compromise because the planted trees are nursed by the overstorey and the spacing between them can therefore be quite wide, i.e. less plants are needed compared to planting on bare land after clearfelling. This renders such underplanting cheaper than planting on bare land. There is a tendency that the use of artificial regeneration prevails in clear-cut systems while natural regeneration dominates group, strip and uniform systems. Artificial and natural regeneration can be even combined.

At the core of their definitions, silvicultural systems are generic. They can be applied to any forest and tree species. Obviously, a fine tuning to the environmental conditions and species involved at a given local site is absolutely necessary when putting them into practice. However, they provide only a general framework and general guidelines that are useful to know, particularly in the context of CCF. As the following discussions will show, silvicultural systems are not rigid rule sets but rather flexible concepts that come with many variants inviting practical experimentation and adaptive management. Smith et al. (1997) stated that a chosen silvicultural system is but a *working hypothesis*, i.e. the best decision and plan developed after examining all available information. Adaptations and modifications are necessary when the results of monitoring the regeneration process suggest them. When reflecting on his book on silvicultural systems and his work for the British Forest Service in India, R. S. Troup in the preface of his book wrote the memorable words:

'Lest it may be held that systems which have evolved in Europe are not applicable to other parts of the world where totally different conditions prevail, the fact may be mentioned that for more than fifty years past the silvicultural systems of Europe, with suitable modifications, have been applied successfully in many parts of India under a variety of conditions and in many types of forest; and it may be truly said that the great progress which forestry has made in that country during the past half-century has been due to a large extent to the fact that the officers of the higher branch of the Forest Service have received their practical training in the forests of Continental Europe. The close study of European systems does not imply that they should be followed slavishly under all conditions; a proper understanding of these systems, however, is an essential preliminary to their adoption in such modified form as may be indicated by local conditions.'

Troup (1928, p. v)

These words express far more than colonial pride. The short text asserts that the training Troup and others have received has been very useful even under very different conditions and that silvicultural systems indeed have a generic, theoretical core that renders them valuable methods far beyond the regions where they were originally invented. In this generic spirit, the silvicultural systems are discussed in the following sections.

5.3.1 Uniform Shelterwood System

5.3.1.1 Method

This silvicultural system involves a uniform, moderate opening of the crown canopy, either on the whole stand area or in parts of the stand, with the aim of establishing relatively even-aged, uniform regeneration underneath. Uniform shelterwood systems can often be achieved by a simple continuation of selective thinnings in the overstorey of mature forest stands, and this may even be how the system was originally discovered rather than invented. The uniform shelterwood system is one of the oldest silvicultural systems going back to the eighteenth century (G. L. Hartig), when it was primarily designed for regenerating *F. sylvatica*, see Figure 5.11.

Assuming a sufficiently good track record of regular, not too weak crown thinnings, an initial seeding felling improves understorey light conditions suitable for regeneration and a favourable seedbed, although satisfactory conditions may already have been created by regular thinnings preceding the seeding felling, particularly where less light-demanding species (cf. Appendix B) are involved. In stands where seeding is infrequent or irregular, or where admitting sufficient light for regeneration is likely to lead to strong growth of competing ground vegetation, the regeneration felling should be timed to coincide with a good seed year. Following this, secondary fellings at suitable intervals admit more light to seedlings on the ground. The number of these secondary fellings depends upon the growth of regeneration, the resilience of the overstorey trees, and, particularly for broadleaves, the time at which overstorey trees reach their target diameter. If initial regeneration is sparse or if the species to be regenerated is frost tender, it may be necessary to prolong the retention of the overstorey.

Figure 5.11 Typical shelterwood system involving *F. sylvatica* L., the species it originally was invented for, at Bleicherode Forest District (Thuringia, Germany). The photo is taken from an extraction rack. Source: Courtesy of author's student, Bangor University CCF field trip to Germany, 2002.

In mixed-species forests, it can happen that the species that establish are more shade-tolerant than those desired. This is probably a sign that the fellings were not heavy enough or that soil scarification is required. The opposite is true, if unwanted grasses or other competing ground vegetation appear (Smith et al., 1997). For a limited time, two storeys are present in the forest, i.e. the mature overstorey and the new cohort of regeneration trees in the understorey (cf. Figure 5.11). Eventually, the remaining overstorey trees are removed in the last secondary felling, i.e. in the final or removal felling (Matthews, 1991; Smith et al., 1997; Nyland, 2002). Following the seeding felling, the overstorey can be gradually removed over 15–30 years in three to six interventions. In each shelterwood cutting, no more than 50–100 m^3 per ha should be removed. Light conditions in uniform shelterwood systems are uniform, and there is hardly any direct sun light. Natural soil scarification caused by tree extraction or the activity of wild boar (*Sus scrofa* L.) can enhance regeneration success, especially for species that require exposed mineral soil for germination.

The uniform shelterwood approach is particularly suitable for intermediate or shade-tolerant species such as *Fagus* spp. or *Picea* spp. More light-demanding species, such as *Fraxinus* spp., *Quercus* spp., *Pinus sylvestris* and *Larix kaempferi* (Lamb.) Carr., require low canopy cover (<40%) and the rapid removal of the remaining overstorey if regeneration is to be successful (cf. Appendix B). Given the same basal-area density, more light is allowed to penetrate the overstorey with main-canopy *Pinus* spp. and *Larix* spp. species compared to *Picea* spp., *Tsuga* spp. and *Fagus* spp. (Rittershofer, 1999).

5.3.1.2 Ecological and Silvicultural Implications

The uniform shelterwood system leads to a gradual and uniform transition from a closed-canopy mature forest to a young forest. The continuous increase of light and soil temperature promotes germination. Species-specific light requirements can be flexibly adjusted by the density of the residual main canopy. With increasing secondary fellings, the protection effect of the residual canopy of mature trees decreases so that problems caused by early/late frost events can return. Due to uniform canopy openings regeneration trees are exposed to similar ecological conditions. Seedling and sapling growth much depends on the density of the residual main canopy, cf. Figure 5.12. Juvenile trees have a tendency to develop stems that are straighter and less branchy than those of trees grown on bare land (Mayer, 1984; Rittershofer, 1999). As the stand canopy is opened, reducing both residual basal area and the percentage of canopy closure, more radiation reaches the forest floor. This increases the availability of light but also air and soil temperatures. This in turn improves the conditions of germination and growth of ground vegetation and tree seeds and seedlings. As briefly mentioned in Section 5.2, residual basal area and

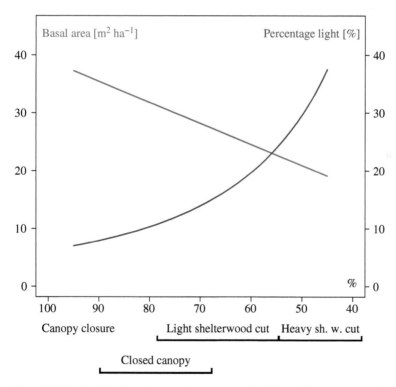

Figure 5.12 Residual basal area and percentage light intensity at the forest floor in mixed-species mountain forests in Bavaria (Germany) on a particular day depending on different percentages of canopy closure as a result of various intensities of shelterwood cuts (abbreviated as sh. w. cut). Source: Adapted from Burschel and Huss (1997) and Rittershofer (1999).

light intensity at the forest floor are negatively correlated (cf. Figure 5.12). Light intensity increases with decreasing stand basal area. Therefore, residual basal area has not only been used as a measure of thinning intensity but also as a surrogate measure of light intensity and likely success of natural regeneration (Hale, 2003). However, the relationship between residual basal area and intensity is not exactly proportional. The example from Bavaria (Germany) shows that even heavy shelterwood cuts involving a basal-area reduction by 50% only lead to an increase of light intensity of up to 40% compared to the light intensity on open land. This implies that even sparse mature canopies still have a marked shading effect on both regeneration trees and ground vegetation and reduce their growth. At the same time, this also means that even light overstorey canopies can maintain woodland microclimates, which is an important tenet of CCF (Burschel and Huss, 1997; Rittershofer, 1999), cf. Section 1.5. It is likely that relationships similar to that between light intensity and basal area exist for water and nutrients (Petriţan et al., 2011; Högberg et al., 2021).

5.3.1.3 Advantages

Under benign conditions, the method is usually very successful and allows the manager to more or less fully rely on natural regeneration. A uniform opening of the canopy is comparatively easy to achieve as a natural consequence of ongoing, regular thinnings and thus very suitable for CCF beginners. Uniform shelterwood systems provide optimal protection against frost and other weather extremes. Due to the uniformity of canopy structure and the reliance on resilient shelter trees this silvicultural system is also quite robust to the adverse influence of wind and snow. In the older British literature, the robustness of the uniform shelterwood system to wind and snow was contested (Hart, 1995; Cameron, 2002); however, according to the author's own experience in Wales, this negative assessment was largely due to an inappropriate preparation of the overstorey trees as a result of the notoriously poor thinning history of British plantations in the past. With the uniform shelterwood system, a maximum of soil conservation and maintenance of woodland climate is ensured. Of all the shelterwood systems, the uniform approach best retains woodland climate. The establishment of dense regeneration promotes timber quality in the successor stand, particularly in broadleaves. In the seeding and secondary fellings, residual main canopy trees are essentially granted a heavy release thinning. The resulting growth response can be used to move promising sub-dominant trees towards maturity (Burschel and Huss, 1997). Such sizeable trees can be interesting for the timber market but can also become excellent habitat trees. Gradually progressing uniform shelterwood systems are also very suitable in forest areas frequently used for recreation, as the forest stands are largely kept intact with little impact visible to laypersons for most of the time.

5.3.1.4 Disadvantages/Challenges

The method partly relies on the coincidence of overstorey thinnings and seed years. However, this requirement has been relaxed in the forest practice of recent years by extending the regeneration process and thus relying on the cumulative regeneration of many years. In the absence of advance regeneration, increased nitrogen depositions can facilitate the rapid growth of competing ground vegetation. Some advance regeneration should therefore ideally be in place before the main canopy is systematically opened (Burschel and Huss, 1997). Regeneration often is abundant, very uniform and even-aged. With *Picea* spp.

and any species with flat root systems this requires expensive respacing thinnings of the regenerated trees for fostering individual-tree resilience. With ongoing secondary fellings, residual overstorey trees may be exposed to sun scorch in some climates, to epicormic growth (*Quercus* spp.) or to windthrow. The residual main canopy is clearly a benefit but also competes with the new cohort of trees for light, water and nutrients. Particularly belowground competition between mature and regeneration trees is potentially high in this silvicultural system. Large elliptic soil patches around the stem base of overstorey trees indicating this competition have often been observed where hardly any regeneration tree can take hold (Bartsch et al., 2020; Schütz, 2003; Burschel and Huss, 1997). This needs to be considered particularly in forests with poor soil nutrient regimes.

5.3.1.5 Variants

Uniform shelterwood systems are often combined with artificial regeneration: Under the light canopy of the overstorey another or the same species can be planted (artificial shelterwood system). Planting of other species should take place in clusters (see Section 4.3) and at an early stage of the system to give them a realistic chance to persist. This possibility is much used in forest restoration and conversion, see Figures 2.17 and 2.18. Another useful variant of the uniform shelterwood system is to introduce more and more overstorey heterogeneity in later secondary fellings, so that some parts of the regeneration receive much light and thus grow more quickly whilst the growth of others is deliberately retarded under a denser overstorey. This is a possibility to break up the uniformity of structure of the new forest generation (Burschel and Huss, 1997).

The *seed-tree system* or *retention system* lies part-way between clearfelling and the uniform shelterwood (Gustafsson et al., 2010, 2012). The system is even older than the uniform shelterwood system and goes back as far as the sixteenth century (Hasel and Schwartz, 2006; Mantel, 1990). It is but a small step from clearfelling and is often used to mitigate the negative environmental and aesthetic impacts of radical harvesting methods (cf. Figure 1.7). There is also a continuous gradient leading from the uniform shelterwood to the seed-tree system depending on how much of the overstorey is retained. Like the uniform shelterwood system, the seed-tree system involves a uniform reduction in canopy cover throughout a stand in order to stimulate regeneration. In the seed-tree approach, the same uniform-shelterwood methods apply; however, the overstorey reduction is far more extreme, leaving no more than 5–60 seed trees (also referred to as *standards* or *reserve trees*) per hectare, and the positive effects of canopy shelter are largely lost. The system is most suitable for the regeneration of hardy light-demanding species, chiefly *Betula* spp., *Larix* spp. and *Pinus* spp. species (Burschel and Huss, 1997, cf. Appendix B). Trees selected as reserve trees should have a high natural life expectancy, since they often need to be around for another forest generation. Reserve trees should have been released early in stand development to gradually develop sufficient individual-tree resilience and an extensive root system over several decades. It is not sufficient just to leave a few more or less randomly selected trees standing whilst clearfelling the rest. Light-demanding species with a deep root system are usually preferred (Rittershofer, 1999). In Europe, suitable species for selecting reserve trees include *Acer pseudoplatanus* L. and *Acer platanoides* L., *Fraxinus excelsior* L., *Larix* spp., *P. sylvestris* as individuals and *Quercus petraea* (MATT.) LIEBL. in groups. Often, light-seeded and wind-dispersed species are favoured that produce abundant seed and regenerate easily in open and unprotected environments (Nyland, 2002). Depending on species and ground vegetation, it may be necessary to time the seeding and secondary fellings to coincide with seed years and ground preparation in the form of scarification may be required. Similar to clearfelling the seed-tree system leaves insufficient residual overstorey trees to mitigate changes of environmental conditions (Nyland, 2002). The residual trees left on site serve as seed source and/or as habitat trees. They do not necessarily need to have regular spacing but can also be arranged in small groups. Specimens of rare species or provenances can be deliberately used as reserve trees to ensure the survival of locally adapted or endangered species/provenances (Mayer, 1984). With time, reserve trees often develop considerable stem-diameter increment gains in response to their release. In commercial forestry applications, they can then act as financial reserves to cover sudden unplanned expenditure (Rittershofer, 1999). In Scandinavia and North America, this system is also known as *green tree retention* and mainly applied for diversifying clearfelling sites but also for conservation (Rose and Muir, 1997; Sullivan and Sullivan, 2018). The sparsity of mature tree cover can potentially encourage competing ground vegetation to take hold for some time (Pommerening and Grabarnik, 2019). In some variants of retention systems, the natural regeneration stemming from the reserve trees is not used and later replaced by artificial regeneration. The sparsity of mature tree cover also implies that the reserve trees can potentially be left on site indefinitely, since they are not likely to have a negative effect on the new forest generation. Burschel and Huss (1997) emphasised that large-scale disturbances caused by wind, fire, drought, or insect calamities often leave a few scattered trees and tree groups behind that survive and subsequently are absorbed by the new, upcoming forest generation. Such natural forest structures are very similar to man-made seed-tree and retention systems. Structures akin to a seed-tree or retention system can also simply be the last overstorey remnants of other silvicultural systems (Bartsch et al., 2020).

5.3.2 Group System

5.3.2.1 Method

An alternative to the uniform opening of the main crown canopy is to concentrate fellings in groups, generally in order to encourage the development of regeneration in gaps. This method was first formalised by Gayer (1886), see Section 1.7, and re-designed by him with the explicit objective to regenerate mixed-species forest stands (Mantel, 1990) at a time when large monocultures were the forestry standard in Central Europe. Ideally, the crown canopy is opened only where advance regeneration is in place. This regeneration can, for example, stem from earlier thinnings or from canopy openings caused by wind or insect-induced small-scale disturbances. Where there is no advance regeneration, an initial heavy thinning or uniform-shelterwood cut is recommended as a preparatory felling. Malcolm et al. (2001) pointed out that the existence of advance regeneration prior to cutting gaps is particularly important on fertile sites, because competing ground vegetation would otherwise dominate freshly cut gaps before tree seedlings have a chance to establish. The gaps create different microsites which allow different species to regenerate, see Figures 5.14 and 5.15. It is an important part of the concept to grant species with a slow juvenile development (often shade-tolerant

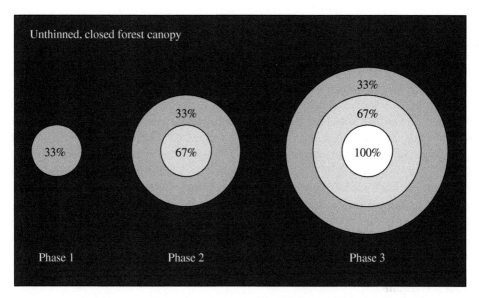

Figure 5.13 Schematic representation of a concentric spatio-temporal gap enlargement in a group shelterwood system designed for regenerating *F. excelsior* L. in Ireland. In each intervention, $\frac{1}{3}$ of the canopy is removed (Joyce et al., 1998). The percentages show how much of the original canopy has been removed in total in each phase. Source: Adapted from Joyce et al. (1998).

species) a head start over those that grow faster (intermediate and light-demanding species) by keeping a light shelter for some time (Burschel and Huss, 1997), see, for example, Figure 5.13. Cutting gaps in the main canopy with no advance regeneration underneath is considered risky and often leads to competing ground vegetation dominating the gaps for many years (see *Variants* of this method).

For regenerating light-demanding species, the group overstorey may be removed entirely in the same seeding felling. For these species, the gaps need to be larger than for regenerating shade-tolerant species. Sometimes this variant is referred to simply as the 'group system', whilst retaining overstorey trees, particularly in larger gaps, leads to the *group shelterwood system*. Canopy gaps can be cut all at the same time or can be staggered in time. The latter results in more diverse forest structure and is therefore more common in CCF. Localised soil scarification in the gaps or at the periphery can encourage tree species requiring exposed mineral soil for germination, e.g. *Pinus* spp. and *Larix* spp. On exposed sites and with species that have flat root systems, e.g. *Picea* spp. and *Betula* spp., it is recommended to start cutting gaps from the opposite side of the prevailing wind direction and in subsequent fellings gradually cut new gaps towards the wind. It has also been suggested to leave a buffer zone for wind protection with a diameter of between 2 and 3 tree lengths free of any gaps at the South-Western edge of the forest stand (Northern Hemisphere) (Mayer, 1984).

With time, the regeneration cones in the gaps expand, i.e. regeneration spreads centrifugally outwards into the unopened main stand in ever-widening circles or ellipses (Matthews, 1991). This natural expansion is supported by felling trees around the edges of each gap. Essentially, this is an application of Vanselow's margin principle of regeneration to groups, i.e. there are elements of the strip system (Section 5.3.3) in this procedure. Such seeding fellings can take the form of a ring or a doughnut at the periphery of the gap when carried out schematically. Initially, these felling zones may be *concentric*, i.e. the new rings have the same width in all directions (cf. Figure 5.13). The rings typically follow the progress of regeneration and release it. It is important and cannot be emphasised enough that felling directions point towards the outside of each group so that the crowns of the falling trees land in stand areas that have not been regenerated yet, see Section 5.3.6. The stems are also extracted through such unregenerated areas so that no damage occurs to regeneration. Joyce et al. (1998), for example, suggested starting with a light removal of 33% of the overstorey in *F. excelsior* woodlands in Ireland to initially create gaps with a diameter of approximately 30 m. When these are enlarged by removing 33% of the overstorey in a concentric ring around the original gap in phase 2, the overstorey in the original gap is also further reduced by another 33%. This pattern is repeated when the next concentric ring is arranged in phase 3. The overstorey in the original gap and in the first ring is then reduced again so that no overstorey trees remain in the original gap (cf. Figure 5.13). More phases and associated cuttings can follow after phase 3 as required.

Often, however, the regeneration progress depends on the environmental conditions in different cardinal directions so that the group extensions take irregular or *eccentric* shapes with time (Vanselow, 1949). For example, an extension of groups in Eastern and North-Eastern directions would promote light-demanding species. Smith et al. (1997) referred to such gap development as *crescentic* spatio-temporal gap enlargement, cf. Figure 5.14. It is

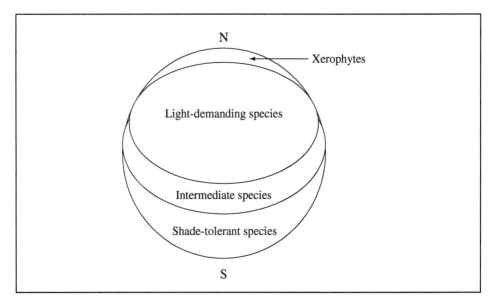

Figure 5.14 Schematic representation of a crescenting spatio-temporal gap enlargement in a group shelterwood system including the differences in microclimatic zones (Smith et al., 1997) in the Northern Hemisphere. The letters give the cardinal directions. See also Figure 5.15.

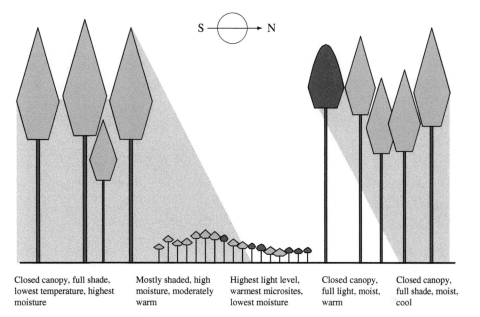

Figure 5.15 Distribution of light and shade in a gap with South-North exposition in the Northern Hemisphere. See also Figure 5.14. Source: Adapted from Rittershofer (1999) and Bartsch et al. (2020).

important to maintain the enlargement of groups on a regular basis to avoid steep edges between regeneration groups originating from different seed years (Bartsch et al., 2020). It is recommended that any stagnation of natural regeneration be complemented with artificial regeneration using wide initial spacings. No management is carried out in the main canopy between the regeneration groups (Rittershofer, 1999).

As we have seen, the successful establishment of regeneration is followed by the successive removal of the group overstorey while the groups are gradually widened. In each intervention, new groups may be created deeper inside the remaining stand matrix. Ultimately, the expanding cones of regeneration coalesce. The regeneration period is likely to be 30–50 years, although it can vary considerably depending on species, site conditions and management objectives. The new cohorts of trees slightly vary in size and typically display a wavy profile due to their varied origin in time. These size differences are usually only of a temporary nature. The longer the regeneration process takes, the more diverse the resulting stand structure becomes. For shade-tolerant species such as *F. sylvatica*, *Carpinus betulus*, *Tilia* spp. or *A. alba* and intermediate species, e.g. *Picea* spp. and *P. menziesii* (cf. Appendix B), initial shelterwood groups should be no more than one tree length in width between the edges of the crowns of the surrounding trees (Figure 5.16). For light-demanding species, groups should be at least one tree length or 30 m wide. The group radius should be increased by 10–20 m in each subsequent intervention. North and North-East patches of groups can be prone to drought effects due to a combination of rain shelter and strong sun radiation in the Northern Hemisphere (Pommerening and Grabarnik, 2019), see Figures 5.14 and 5.15.

Figure 5.16 An emerging *F. sylvatica* L. regeneration cone in a mixed *F. sylvatica* and *Abies alba* MILL. woodland (Wohlen Forest District, Switzerland). Such regeneration cones are typical of group systems. Source: Arne Pommerening (Author).

5.3.2.2 Ecological and Silvicultural Implications

The group shelterwood approach creates a range of microclimates and light regimes suitable for species with different light and climate requirements (cf. Figure 5.15). Therefore, it has often been considered for regenerating mixed-species forests. The environmental conditions in the gaps much depend on topography, height and density of the surrounding stand and gap size. Gaps with a ratio of diameter/stand height <1 are considered small and a diameter/stand height ratio >1 classifies them as large (Bartsch et al., 2020). Malcolm et al. (2001) suggested that under British conditions gap diameter/stand height ratios of at least 2 are suitable for light-demanding species, whilst ratios 1.0–2.0+ are appropriate for intermediate tree species. The authors found that shade-tolerant species grow well when regenerated given a gap diameter/stand height ratio smaller than 1.0.

Precipitation is highest in the centre of the gap. Initially small canopy openings promote the regeneration of shade-tolerant and intermediate tree species whilst successive enlargement of these gaps provide opportunities for light-demanding species (Mayer, 1984). Typical woodland climate is sustained in the gaps up to a diameter of one tree length. In gaps with a diameter larger than 15 m, in Northern and North-Eastern directions, water and nutrients are almost completely used up by the mature trees of the canopy edge whilst the strong exposure to sunlight increases evapotranspiration (cf. Figures 5.14 and 5.15, in the Northern Hemisphere). This combined effect is the stronger, the poorer the soil, and the lower the soil moisture regime. Adverse consequences can to some extent be mitigated by accelerating the felling process (Bartsch et al., 2020). It is also possible to enlarge the gaps successively on the Southern side only (Matthews, 1991). Group systems are close to regeneration processes in natural European forests involving small disturbance patches. As a consequence of these ecological zones, shade-tolerant tree species usually colonise in a crescent along and under the southern edge. More light-demanding and exposure-resistant species find their optima for establishment and height growth in crescents arranged successively northward. At the very Northern edge, there is the aforementioned arid zone, where only specialist species can prevail (Figure 5.14, Smith et al., 1997).

5.3.2.3 Advantages

Microclimate and soils are well conserved through a promotion of natural regeneration that mimics natural processes very closely. Different species with different light, soil-moisture and nutrient requirements can be accommodated and with the necessary skills and experience, the desired species composition can be well balanced (Rittershofer, 1999). Final growth gains on mature trees are encouraged by the system. Through its structural complexity, the new forest generation is resilient and well adapted to disturbances. The silvicultural system is flexible, both a simple and a complex structure of the successor stand can be achieved. It is comparatively straightforward to adapt to unexpected developments.

5.3.2.4 Disadvantages/Challenges

Gaps can give rise to turbulent airflow over the canopy surface and also allow wind entry into the stand potentially leading to increased wind hazard. Therefore, the group system should not be applied on exposed sites and with species that are susceptible to windthrow. Although damage to regeneration can easily be avoided in the early stages by felling away from the centre of gaps, this can become increasingly challenging as the last overstorey trees are felled and extracted. A dense system of extraction racks may be required to extract these trees. There is, of course, the possibility to leave trees that are difficult to extract safely as permanent habitat trees. With species that have infrequent seed years, regeneration groups may not coalesce within reasonable time and limited planting in between them may be necessary. A successful application of this system requires more knowledge and experience than the uniform shelterwood system.

5.3.2.5 Variants

Groups are sometimes expanded elliptically towards the South or South-West to provide more suitable microclimatic conditions for the target species. In mountain forests, ellipses are generally favoured to groups. They are a compromise between light and cover requirements. For example for regenerating *Picea abies* in the Swiss Alps 50–70 m long and 15–20 m wide elliptic openings were recommended for north-facing sites, whilst smaller gaps were proposed to avoid drought effects on south-facing slopes where light is widely available. The main criteria for the design of elliptic groups in mountain forests are to grant seedlings sufficient light, to reduce snow cover and not to trigger avalanches in winter. The gap shape is also believed to avoid the formation of microclimates akin to clearfelling sites (Bayerische Staatsforsten, 2018).

The best orientation of elliptic groups in the Swiss Alps was found to be North-East to South-West (Streit et al., 2009; Pommerening and Grabarnik, 2019). For the Bavarian Alps, the Bavarian Forest Service recommended a gap system where not only elliptic but also nearly rectangular small-sized gaps are cut on either sides of cable crane lines in an alternating way (Figure 5.17). All mature main-canopy trees are removed in these gaps apart from resilient individual trees near the gap boundaries. The elongated gaps are arranged at an angle smaller than 90° towards the cable crane line. With increasing altitude, attention should be paid to the arrangement of the gaps so that they are exposed to as much sun radiation as possible and day temperatures are acceptable. Distances between gaps can vary between 20 and 50 m. In any subsequent intervention, new gaps are cut, and existing ones are widened in an irregular, non-concentric way. Between the gaps, individual trees are harvested according to target diameter harvesting. A similar design can also be applied to lowland forests where cable crane lines are replaced by extraction racks (Bayerische Staatsforsten, 2018). Assuming again the Northern Hemisphere, expanding groups in Eastern and North-Eastern directions by cutting boundary trees favours light-demanding species and those that can sustain themselves in dryer soils. Expanding groups in the other direction promotes intermediate and shade-tolerant species and those that require more soil moisture (Rittershofer, 1999).

In forest practice, although a good starting point, the schematic removal of trees in a 'doughnut' at the edges of gaps (Figure 5.13) has often proved to be a disadvantage. This would include the removal of trees that have not yet reached maturity (target diameter) and also the removal of windfirm trees with low h/d ratios. The newly created forest edge at the new periphery of the edge can then lack resilience to wind with potentially catastrophic consequences. This can threaten the success of the whole silvicultural system. Therefore, it is recommended to cut such 'doughnuts' around gaps selectively with a view to retain individual, resilient overstorey trees with low h/d ratios as long as

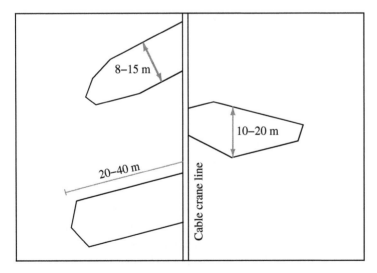

Figure 5.17 Schematic overview of elliptic, 'slit-shaped' canopy openings arranged along a cable crane line in the Bavarian Alps. Source: Adapted from Bayerische Staatsforsten (2018).

possible. This strategy usually does not create problems for the upcoming regeneration that gradually absorbs these few trees. A simplified variant of the group system includes cutting canopy holes in the absence of advance regenerations (see, for example, Figure 5.18) and without enlarging them. Though theoretically possible, this method is often not successful. Ideally such gaps should not be placed in a systematic or regular pattern (Mayer, 1984), which is largely ignorant of differences in microsites and advance regeneration. A more successful variant of this concept is the *artificial group system*, where also canopy holes are cut with a regular or irregular dispersion, but the gap areas are immediately planted up, usually with another species not occurring in the forest stand. Such canopy-hole cuttings were, for example, proposed and successfully tested by Mortzfeldt in Germany (Rittershofer, 1999; Bartsch et al., 2020). It is even possible to temporarily fence gap areas, if there is the danger that large deer populations may feed on the newly introduced species used for planting the gaps. Artificial group systems are used to combine natural and artificial regeneration, e.g. newly planted *Quercus* spp. or *Fraxinus* spp. within parent stands of *F. sylvatica* or *P. sylvestris*. This is also an excellent way to give less competitive or slowly growing species a head start over a majority species that has not started to

Figure 5.18 In a mixed *Picea abies* (L.) H. Karst. and *P. sylvestris* L. forest owned by Umeå Community near Nydala Lake (Northern Sweden) large gaps were cut systematically following a chequerboard pattern. The two colleagues from the Swedish University of Agricultural Sciences (SLU) hold up a poster showing the design. Some of the gaps will be planted, others will be left to natural regeneration. A forerunner of this method (but applied at much larger scale) was described by Franklin and Forman (1987), but it also shares similarities with the Anderson group selection system, see Section 2.3.2. Source: Courtesy of Helena Dehlin.

regenerate yet so that the (re-)introduced species can persist and/or develop (economic) maturity at the same time as the majority species, see Section 4.3. The smaller the gaps in the artificial group system, the larger the head start of the introduced species needs to be, because interspecific competition will be more pressing compared to the situation in large gaps. Artificial group systems are often but not exclusively applied in the context of forest restoration and conversion. For broadleaves used in planting up gaps 1×1 m and for conifers 2×2 m or 2.5×2.5 m spacing are often recommended in forest practice.

Another variant of the group system is the *irregular shelterwood system*. It is often said to be a combination of the group (shelterwood) and the selection system, but clear breakdowns of the method are hard to elicit from the literature. The silvicultural freedom that comes with the system is often emphasised (Rittershofer, 1999; Schütz, 2003). The system is of great regional importance, particularly in the Black Forest (Germany, termed *Badischer Femelschlag*) and in large parts of Switzerland (termed *Schweizerischer Femelschlag*). The transition from regular thinnings to the silvicultural system is very gradual, and there are no preparatory fellings. As part of the thinning programme, individual-tree resilience to wind is promoted by releasing tree crowns well in advance of regenerating the forest. Gaps can vary considerably in size, some of them can be quite large, but the seeding fellings also include individual-tree removals. Therefore, the irregular shelterwood system also includes aspects of the uniform shelterwood system. Individual clusters of regeneration are released by removing or thinning the overstorey, but these gaps are not widened after initial creation. In the long run, forest structure resulting from this system approaches that known from the single-tree selection system. The regeneration process is intentionally long, i.e. between 40 and 80 years. The irregular shelterwood system shares the same advantages with the group system, it is said to be even more flexible. In the long term, the system produces the vertical and horizontal structure akin to that of selection systems and due to the long regeneration process there is the tendency of promoting shade-tolerant at the expense of light-demanding species such as *Pinus* spp. (Mayer, 1984).

5.3.3 Strip System

5.3.3.1 Method

Originally devised as a variant of the clearfell system the central idea of the strip system was to clearfell and regenerate forest stands in small, comparatively narrow portions staggered in time. This is an application of Vanselow's margin principle of regeneration. The strip system is a relatively recent development dating back to the beginning of the twentieth century and important protagonists include Eberhard, Philip and Wagner (Vanselow, 1949). Particularly when mainly concerned with species susceptible to windthrow, progressive strips are cut perpendicular to the main wind direction and strip cutting progresses from the direction opposite of the main wind direction towards the wind. Alternatively, the cuttings progress towards the direction of midday sun so that the upcoming new stand is protected from high temperatures (Smith et al., 1997). The exact orientation of the strips depends on species, elevation and topography and also have to take water availability into account (Figure 5.19). On slopes, strips typically progress downhill. At North and East exposed slopes, the strips are cut with a slight offset added to the slope direction. Strip width usually is 1–2 (max. 3) tree lengths, i.e. approximately 30 m. Depending on growth

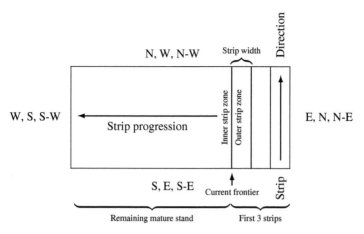

Figure 5.19 Schematic overview of a basic strip system with the first three strips already cut. The capital letters give common alignments of strips and strip progression in the Northern Hemisphere. Linear shapes have been chosen for clarity only. Source: Adapted from Smith et al. (1997).

and regeneration dynamics, subsequent strips are cut within 5–20 years after the most recent strip was cut until the whole area is regenerated (Rittershofer, 1999). The successor stands are even-aged in strip direction (i.e. parallel to the frontiers) and uneven-aged in the direction of strip progression (Figure 5.19). Occasional problems with competing ground vegetation and soil erosion on slopes soon motivated another variant, the strip shelterwood system. The remaining overstorey trees not only in the strips can be dispersed uniformly but also in clusters thus creating regeneration gaps (Matthews, 1991). These standards can be considered for (selective) removal when the next strip is cut but also be retained for a longer period. The shelterwood component is, however, often low, i.e. it is limited to a small number of particularly resilient standards. Leaving standards behind increases the available seed source and also prevents seedling mortality due to droughts on exposed strips. The standards usually respond with increased growth which can be useful for valuable timber species in commercial scenarios (Bartsch et al., 2020). The strip system is a comparatively rapid regeneration method (Matthews, 1991).

5.3.3.2 Ecological and Silvicultural Implications

The lateral protection provided by the adjacent trees of the remaining forest stand reduces the adverse climatic effects acting on bare land. This lateral protection effect depends on the height of the adjacent mature stand, altitude, exposition and topography (Rittershofer, 1999). The canopy removal or reduction on the strips allows the regeneration of light-demanding and tree species with intermediate light requirements (cf. Appendix B). The cuttings involved in the strip system create two distinctive microsites with some degree of woodland climate (Matthews, 1991, Figure 5.19):

Inner strip. Between the new tree frontier and one tree length into the matrix of the remaining stand. A large portion of the overstorey remains in this zone, but the canopy is opened thus allowing light to penetrate the stand vertically. Suitable for moderately shade-tolerant tree species.

Outer strip. Between new tree frontier and the noon shade in summer (approximately one tree length). All or most of the overstorey has been removed. Suitable for intermediate and light-demanding tree species and for species that can tolerate frost.

In addition to or in preparation of the strip cutting, it is possible to open the canopy of the inner strip zone through a light overstorey thinning following Vanselow's shelter principle (in Germany, this variant is part of Wagner's *Blendersaumschlag*, a combination of silvicultural systems). The heavier this thinning is carried out the more light-demanding species can be accommodated whilst the risks of windthrow and snow breakage increase. If the overstorey canopy is already light, it is not necessary to create such an inner strip zone (Bartsch et al., 2020). The residual forest stand beyond the inner strip zone remains untreated.

5.3.3.3 Advantages
The strip (shelterwood) system is a particularly secure system in areas with high wind hazard. Strips are opened and cut against the prevailing wind direction or the direction of midday sun. Damage to upcoming regeneration due to thinning and harvesting operations is kept to a minimum, since it is very straightforward to extract stems through the untreated mature stand. The system is easy to apply even with little experience (Pommerening and Grabarnik, 2019). Varying ecological conditions, especially between the inner and the outer strip zone, allow the regeneration of mixed-species woodlands with different light demands.

5.3.3.4 Disadvantages/Challenges
Strip systems promote the regeneration of species with frequent seed years. It can be challenging to achieve high regeneration densities with species that regenerate infrequently. *Picea* spp. usually respond well to strip cuttings whilst *Pinus* spp. respond a little less favourably. This can sometimes be improved by appropriate ground preparation to create a favourable seedbed (Bartsch et al., 2020). The strip system requires adequate herbivore control, since fencing is not an option.

5.3.3.5 Variants
Many variants of the strip system have been created by diversifying the shape of the tree frontier, i.e. the shape of the strip boundary. Any shape is theoretically possible and nonlinear, undulating frontiers are preferred. Strips can, for example, follow the bends of forest roads that form the stand boundary. Bendy strip shapes but also wedges (see Figure 5.20) are particularly common and typically increase regeneration diversity due to varying ecological conditions. Other variants include short strips arranged in a step-wise pattern, but simple patterns involving long strips have proved to be more efficient (Matthews, 1991). Similar to gaps, the strips also offer the opportunity of introducing artificial regeneration. Greater structural diversity can be achieved by reducing strip width and strip-cutting progress. A slow rate of progress, however, can potentially produce a steeper slope in the profile of the young stand than does a rapid rate of progress (Matthews, 1991). A special variant of the strip system developed by practitioners Eberhard and Philipp in the Black Forest (Germany) is the *wedge system*. In this system, the regeneration frontier is much enlarged compared to the simple strip system thus providing more ecological opportunities. The system is said to be even more windfirm than the standard strip system whilst

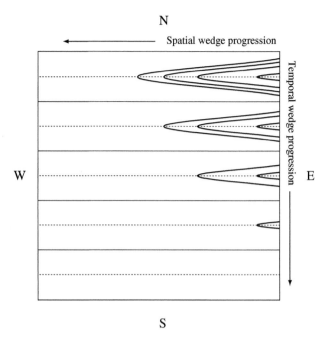

Figure 5.20 Schematic example of a wedge system with a spatial orientation towards West and a temporal progression in Southern direction. The nested wedge lines indicate successive tree-cutting frontiers in time from East to West. Dashed straight lines give the direction of wedge cutting progress and continuous straight lines indicate extraction racks. Inspired by Vanselow (1949), Matthews (1991) and Rittershofer (1999).

potentially increasing tree species diversity. Similar to the standard strip system, extraction of mature trees is easy, but the layout of wedges requires more planning. In this system, strips are initially cut to a width of 2–6 m with large distances (e.g. 80–100 m) between one another and they proceed outwards in both directions. The cutting process is comparatively rapid at intervals between 2 and 3 years. On level ground or gentle slopes, the wedge fellings start by laying out parallel lines running from East to West or from North-East to South-West in the direction of the prevailing wind. Subsequent fellings enlarge the initial strips but at a slight angle to them so as to form gradually widening wedges with their apexes pointing towards the prevailing wind direction. At the same time the apexes are pushed forward towards the edge of the regeneration area (Matthews, 1991). Midway between the initial strips and parallel to them, extraction racks are laid out (Figure 5.20). The wedge edges (frontiers) can, of course, be irregular and do not need to be perfectly delineated according to geometric principles. The example in Figure 5.20 shows another variant of the wedge system, where the wedges are not all cut at the same time but progress at different paces in a Southern direction, which may be suggested by environmental conditions. In this example, wedges are cut right from the start rather than starting with a strip along the dashed lines in Figure 5.20 first. The shape of the wedges and the differentiation in time contributes to a higher species and size diversity in the upcoming regeneration compared to a standard strip system. Over the wedges most if not all overstorey trees are removed whilst seed trees remain for some time in between the wedges. Prior to starting the system, regular substantial overstorey thinnings prepare the forest stand so that advance regeneration is available at least in some places. These modifications and opportunities lead to the *wedge shelterwood system*, which often is considered a combined system in the literature (Rittershofer, 1999; Mayer, 1984).

5.3.4 Combined Silvicultural Systems

The three basic principles of regenerating forest stands naturally, i.e. shelter, margin and bare land, as described by Vanselow (1949) as well as the basic silvicultural systems uniform, group and strip can be freely combined in almost any way imaginable, and the reader is clearly encouraged to think of new ways of combining these principle regeneration methods. When doing this, it is important to consider the properties and requirements of the tree species concerned (cf. Section 5.3.5) as well as local environmental conditions. In practice, combinations of silvicultural systems are far more common than the basic variants, since all of them have certain disadvantages when applied in isolation and combinations increase silvicultural flexibility (Rittershofer, 1999). In some situations, one of the basic silvicultural systems is not sufficient to regenerate a whole forest stand. In the past, many forest practitioners as well as researchers have tried and proposed various combinations and a few of them have proved useful across a range of environmental conditions and over a long period of time. Simple combinations such as adding shelterwood components to group and strip systems have already been mentioned in Sections 5.3.1, 5.3.2, 5.3.3 and some variants of the basic forms of silvicultural systems may also count as combined systems. Most of these combinations are additive or spatial combinations (see Section 5.3) and important examples are discussed in this section.

5.3.4.1 Combining Natural and Artificial Regeneration

Throughout this book, it has been emphasised that CCF can also accommodate the use of artificial regeneration or combinations of natural and artificial regeneration. This is certainly also true in the context of silvicultural systems, although their main objective is to achieve natural regeneration. Even if no natural regeneration would establish, artificial regeneration would involve low densities (wide initial spacings) and thus save planting costs compared to a traditional replanting on clearfelling sites. If natural and artificial regeneration are combined, further savings can be made. In Section 5.3.2, the artificial group system is described, which is such a combination. However, any silvicultural system can be combined with (partial) artificial regeneration. Reasons for involving artificial regeneration include

- Raw humus accumulation, compacted soil, strong competition by ground vegetation, harvesting and extraction damage can sometimes prevent or partially destroy natural regeneration,
- Maintenance of tree species diversity when mainly one or two species regenerate,
- Natural regeneration may not be suitable because it is of an unsuitable (non-native, not-adapted) species,
- Progressive storm or bark-beetle damage requires a fast removal of the main overstorey before the process of natural regeneration is complete,
- A late initiation of silvicultural systems requires a complementation by partial planting.

5.3.4.2 Progressive Silvicultural Systems

It is possible to carry out silvicultural systems progressively in small areas rather than simultaneously across the whole forest stand. A uniform shelterwood system can, for example, be limited to strips of 1–2 tree lengths which progress every 3–7 years. Alternatively, canopy gaps can be arranged in strips which gradually progress towards the wind. Instead of using strips, a progressive uniform shelterwood system can also be exercised in irregular zones delineated according to microsite conditions. Progressive silvicultural systems usually operate at a faster pace than their basic counterparts and also provide more opportunities for light-demanding species (Burschel and Huss, 1997).

5.3.4.3 Group-Uniform Shelterwood System

Following a good thinning history, regeneration is initiated by gap cuttings up to 30 m in diameter to release advance regeneration in groups and to encourage light-demanding species (cf. Appendix B). It is also possible to plant in these gaps, for example, with another species that occurs in the forest stand. *Quercus* spp. groups in a stand matrix of *Pinus* spp. or *Faxinus/Acer* in a *F. sylvatica* stand matrix are typical European examples. An important part of the concept is to grant the regeneration in the gaps a growth head start over that in the surrounding matrix. Outside the gaps, the forest stand is treated according to the principles of the uniform shelterwood system favouring intermediate and shade-tolerant species (Burschel and Huss, 1997; Bartsch et al., 2020), but this is initiated several years after the gap cuttings. Unlike the basic group system, the gaps are usually not expanded by felling trees around the edges.

5.3.4.4 Group-Strip Shelterwood System

A combined silvicultural system that has been successful in a wide range of contexts is the group-strip shelterwood system (see Figure 5.21). In this system, shelterwood groups are opened up to 100–150 m into the stand from the leeward edge. A strip shelterwood is then initiated from the same sheltered edge. As the strips advance towards the windward edge of the stand, existing groups are expanded, and new groups are cut up to 100–150 m ahead of the strips. The advance regeneration in the groups is gradually absorbed by the progressing strips, thus introducing more diverse vertical and horizontal structure. Although canopy gaps are opened and enlarged, potentially risking turbulence and wind damage, they are kept towards the relatively well-protected leeward side of the stand, where in any case the advancing strip shelterwood fellings remove all but the most resilient individuals. This combination of methods provides a wide range of growing conditions for tree species with differing demands and can give rise to heterogeneous stands (Pommerening and Grabarnik, 2019). As with the group and strip systems, there are opportunities for using shelterwood components on the strips and in the groups. Group size can vary and soil scarification may again support the regeneration effort. The groups can be of natural origin, for example, as the result of a small-scaled disturbance, or deliberately cut. Usually, the groups are given a higher priority than the strips, but in situations where much advance regeneration is already established and a rapid removal of the overstorey is for some reason required, a greater emphasis can also be assigned to the strips (Rittershofer, 1999). Artificial regeneration with low densities can be applied where natural regeneration does not develop well or is not desired, e.g. as part of conversion (see Section 2.4). Strip cutting commences after the

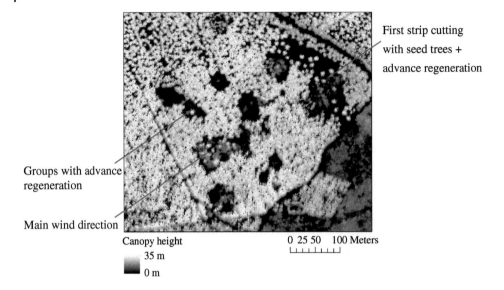

First strip cutting with seed trees + advance regeneration

Groups with advance regeneration

Main wind direction

Canopy height

35 m

0 m

0 25 50 100 Meters

Figure 5.21 Example of a group-strip shelterwood system with Sitka spruce (*P. sitchensis* (Bong.) Carr.) in progress. LiDAR-derived canopy height model (1 m spatial resolution) for block 3, Cefn Du, Clocaenog Forest, North Wales, UK in 2006. Source: Courtesy of Rachel Gaulton.

first set of groups are in place and regeneration in groups is underway. Whilst new strips are cut, pre-existing groups are enlarged, and new canopy gaps are cut deeper inside the stand matrix. Compared to the basic systems, the regeneration frontier is much enlarged in the group-strip shelterwood system thus providing more ecological and silvicultural opportunities (Mayer, 1984).

5.3.5 Regeneration Methods by Species

In some situations, it is useful to know how best to accommodate individual species in possible silvicultural systems. For example, although part of a mixed forest a given stand may have a dominant species component of *Pinus* spp. with a few specimens of *Picea* spp. In that situation, it may be most important to pay attention to the regeneration of the light-demanding *Pinus* spp. and Table 5.4 can be used as a quick reference. Along similar lines, it may be possible to identify the best compromise or combination of systems for a mix of species to be regenerated. Particularly for the beginner Table 5.4 may offer quick advice.

5.3.6 Operational Aspects of Silvicultural Systems

Making use of natural processes including natural regeneration is one of the main tenets of CCF, see Section 1.5. When managing forests using methods of CCF, trees to be removed for whatever reason are always closely located to small- or medium-sized trees that are still far from maturity. Regeneration and mid-storey trees can be damaged, when residual large canopy trees, e.g. seed trees or reserves, are felled (Matthews, 1991). It is therefore

Table 5.4   Regeneration methods traditionally used with tree genera and species.

Genus/species	Silvicultural system(s)	Suitable sites
Abies alba	Uniform, group, irregular, long regeneration process (20–40 years)	High SNR and SMR, wide range of sites, deep soils, shaded and cool locations
Acer, Fraxinus, Prunus, Ulmus	Group, since light-demanding and fast growing when juvenile, no shelter or very light shelter	Good SNR, particularly on calcareous soils, good SMR, no competing ground vegetation, deep soils
Alnus	Large gaps, germinates on exposed mineral soil only	Good SNR, high SMR
Betula	Large gaps, strip (pioneer species)	Wide range of sites
Carpinus	Uniform, small gaps, few seed trees required	Good-to-moderate SNR, moderate SMR
Fagus	Uniform, group with a very gradual removal of overstorey (10–20 years)	Good-to-moderate SNR with little ground vegetation, otherwise soil scarification required
Larix	Group, strip, light uniform, rapid overstorey removal, exposed mineral soil	Good to moderate SNR, moderate SMR, little raw humus and ground vegetation
Picea	Group, uniform, strip	Little competing ground vegetation
Pinus	Group, strip, uniform, fast overstorey removal (5–10 years)	Low SNR, little raw humus accumulation, little competing ground vegetation
Populus tremula	Large gaps, strip, no shelter (pioneer species)	Average to poor SNR
Pseudotsuga menziesii	Uniform, strip, fast overstorey removal	Moderate SMR, little competing ground vegetation, wide range of sites
Quercus	Uniform, group	Depending on species high to moderate SNR and SMR
Tilia	Uniform	Good to average SNR and SMR

SMR – soil moisture regime, SNR – soil nutrient regime.
Source: Adapted from Burschel and Huss (1997).

imperative to keep the damage from fellings of large trees to other trees to a minimum. This can be achieved by observing a few general rules. An important pre-requisite is a good network of extraction racks and forest roads. The former is usually put in place in early thinnings at distances between 20 and 60 m depending on the felling and extraction methods.

Extraction racks are unfortified more or less linear lines within a forest stand free of or with sparse vegetation that specialised forest machines such as harvesters and forwarders can drive on (Helms, 1998). The width of extraction racks corresponds to the width of the forest machines used plus 1–1.5 m (max. 4 m). In CCF, extraction racks

(Continued)

(Continued)

are often permanent, i.e. they are in use for several generations of forest stands so that soil compaction caused by the weight of machinery is limited to these racks. Temporary brash mats put onto the extraction racks created from branch and foliage residue can further reduce soil compaction.

Extraction racks must be clearly marked (see Figure 1.9) and machine operators should keep to the rack system. Keeping natural regeneration safe is possible by defining *felling directions* of individual trees, i.e. thus achieving *directional fellings*. Felling directions are determined in such a way that the crown of a tree to be felled does not land in areas of established regeneration. Trees should be felled towards extraction racks at a small angle so that felled stems and extraction rack form a herringbone pattern (Figure 5.22). Similar principles apply to skyline racks or trails used in uplands for moving tree logs with a skyline cable instead of ground-based extraction racks where slope inclination exceeds 25%. A concentration of fellings in certain areas (felling zones) as well as a temporary concentration of logs on the forest floor decrease the damage on advance regeneration. The trees to be felled in each intervention should be clearly marked ideally including the envisaged direction of felling. In underplanting, some areas of the stand should be spared to create felling zones where falling trees do not damage planted saplings.

Modern felling and extraction technology based on short stem sections rather than tree-length sections can further reduce damage (Davies et al., 2008). In group (shelterwood) systems, trees are felled with their crowns pointing towards the outside of the group so that their crowns come to fall into unregenerated areas of the stand. Before felled

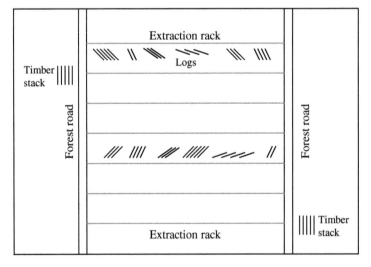

Figure 5.22 Schematic overview of a forest stand situated between two forest roads with extraction racks at regular distances in between. The herringbone pattern formed by extraction racks and logs (cut stems without branches and crown tips) is indicated by two examples. The logs are eventually extracted to the forest roads and temporarily arranged in roadside timber stacks. Lorries sent by the timber processing industry then transport the timber to sawmills and/or factories.

trees are moved, major branches and crown tips are removed. Some silvicultural systems such as the strip system and variants of the group system make the extraction of felled trees easy, but with others, detailed plans need to be made about what felling directions are used, and how the timber is extracted so that minimum damage on live trees occurs. In uniform shelterwood systems, first thinnings can concentrate on trees that are furthest away from extraction racks so that these areas remain unaffected by later interventions. The last remaining trees should be located near the racks to facilitate felling and extraction. In complex forest structures, it may be useful to supplement harvester operations with motor-manual felling, because

- manual fellers can bring down trees that are too big for harvesters or too far away from the harvester's location on the extraction rack (though the harvester may still be able to process these trees),
- difficult trees that require precise directional felling may be more easily brought down by chainsaw operators,
- in cases where dense natural regeneration obstructs the machine operator's view of the tree base, manual fellers can either remove the obstacle or fell the tree.

Such combined operations require good coordination and communication between chainsaw and machine operators (Davies et al., 2008).

Whenever possible extraction is timed to coincide with periods when the bearing capacity of the soil is high, e.g. in winter when parts of the top soil is frozen, and the ground is covered with snow. On soils with low bearing capacity, the aforementioned brash mats are employed so that harvester and forwarders move on a cushion of branches and foliage. Machines should be used that exert low ground pressure and can be manoeuvred in restricted space. In sensitive forests and high profile conservation areas, traditional extraction by horse and other animal may still be a good option (Matthews, 1991).

5.3.7 Wind, Fire and Herbivores

Environmental conditions influence and interact with thinnings and silvicultural systems, specifically with the residual mature trees and with regeneration. Considering these conditions is an important part of adapting the theory of silvicultural systems to local applications. Among the multitude of different environmental factors affecting the outcome of a silvicultural system, there are three that are more important than others: Wind, fire and herbivores. These also act as disturbance agents. Natural disturbances are discrete events providing opportunities and small chronic disturbances that result in the death of individual trees or small groups of trees are overwhelmingly the most common type of disturbance in forests (Palik et al., 2021). Legacies of wind and fire disturbances include high stumps, snags, downed boles and exposed mineral soil and are important elements of forest ecosystems.

5.3.7.1 Wind
Trees grow tall and the destructive power of wind increases exponentially with increasing distance above the friction effect of the ground surface. The force exerted by wind also increases exponentially with wind speed (Smith et al., 1997). Wind is an environmental and disturbance factor that any forest ecosystem around the world is exposed to. Wind results

from the differential heating of the atmosphere, which in turn results from differential heating of the earth's surface. As the earth rotates, areas are sequentially warmed and cooled, giving rise to global variations in air pressure which drive the global air circulation systems (Kimmins, 2004). There are also many regional systems, e.g. valley and land-sea circulation systems. Wind disseminates pollen, spores and seeds. It also influences tree physiology and plant morphology. The distribution of wind speeds through the year and across years shows a typical pattern of mostly light-to-moderate winds with certain periods of high winds, e.g. in the winter months (Binkley, 2021).

Wind has always been a major factor in the management of forests and has played an important role in the development of forestry. Wind damage has wrecked early attempts to provide secure future timber supplies in Europe, and this environmental factor can sometimes limit the range of suitable silvicultural methods or the rotation length of plantations in RFM, for example in Britain and Ireland. Wind was also a major stimulant behind various harvesting and silvicultural systems (Kimmins, 2004). The studies by Dvorak and Bachmann (2001) and Hanewinkel et al. (2014) have shown that complex CCF woodlands generally tend to be more windfirm than even-aged plantations. The tree crown bears the wind load and the stem transfers the resulting forces to the root system. The roots are connected to the soil and anchor the tree in the ground. The strength and flexibility of all the tree's parts enable it to withstand great wind forces (Pommerening and Grabarnik, 2019). The swaying of trees in heavy gusts causes adverse stress in rapidly alternating directions, which damage the fine root system and eventually lead to the gradual loosening of the roots from the surrounding soil. Exceedingly violent winds can result in *windbreak*, (*windsnap*) and *windthrow*. With windbreak, the stem is snapped off at some height above the ground, while the lower part of the stem remains upright, still anchored by its roots, cf. Figure 2.13. Windthrow describes wind damage where the roots are pulled off the ground and the intact trees are toppled over (Kimmins, 2004). Both windbreak and windthrow are common where trees previously sheltered in dense stands are suddenly exposed to wind by thinnings or clear-cut harvesting in adjacent stands. This is why forest stands need to be prepared for transformation to CCF (cf. Section 2.4) by regular thinnings fostering individual-tree resilience to wind. Windthrow is more common in shallow-rooted tree species such as *Picea* spp. and *Betula* spp. Deciduous tree species tend to be less susceptible to wind damage than non-deciduous species, because the seasonal occurrence of severe storms coincides with a time when deciduous species are usually bare of foliage and therefore subjected to smaller wind loads (Davies et al., 2008). In addition, most deciduous tree species tend to have deep root systems.

Windthrow also occurs in trees with damaged root system and trees growing in shallow soils over bedrock. Rooting can also be limited in waterlogged soils. Forest management and wind interact and it is possible to reduce wind damage on trees through appropriate silviculture. Interventions can prepare a forest stand to face average and slightly more than average wind speeds, however, it is not possible to prepare for catastrophic events (Figure 5.23). Indicator values for assessing individual-tree resilience to wind have already been introduced in Section 2.2.

A good way to build up individual-tree resilience to wind are regular, not too heavy thinnings. Thinning operations temporarily weaken the collective resilience of a forest stand. The heavier these interventions are and the later in a stand's lifetime they occur, the longer

Figure 5.23 Severely bent *P. sylvestris* L. trees, windbreak and windthrow casualties in the Ostrów Mazowiecka Forest District (Poland) as the result of an unexpected catastrophic summer storm in 2011. Source: Arne Pommerening (Author).

it takes for the stand to re-establish the pre-intervention resilience level (Figure 5.24). Wind risk typically increases as stands grow taller and the risk is highest immediately after the interventions. After the initial three years of adaptation, further resilience gains, particularly for taller stands, are only small. Therefore, it is good advice to thin early in the growing season immediately after the storm season has passed so the trees have a full vegetation period to recover a little before facing fresh gales. The combined thinning approach (cf. Section 5.2.2) was developed as a strategy to maintain resilience in *Picea* spp. forests.

Forest margins can help reduce the wind load at the boundaries of forest stands with the open landscape, cf. Section 7.3.3. On exposed sites where the forest stands mainly involve susceptible tree species, fellings should be organised in such a way that they progress towards the main wind direction, see Section 5.3.3. Silvicultural systems are usually more susceptible to wind damage than ordinary thinnings, since the stand canopy is markedly opened up. Group and seed-tree systems can be considered risky on such sites, whilst the uniform shelterwood system and the strip system are safe options under these conditions.

5.3.7.2 Fire

Wild fire is an important environmental factor in many forest ecosystems throughout the world. Fires occur naturally and can then be considered natural disturbances. Increasingly, fires are accidentally or deliberately started by humans. With ongoing climate change and associated rising temperatures and increasing frequency and duration of droughts, arson is clearly a great threat to forest ecosystems.

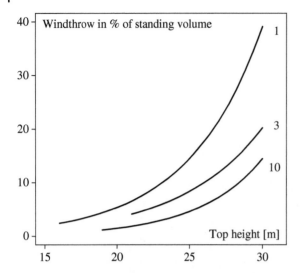

Figure 5.24 Wind damage in stands of *Picea abies* in relation to stand height. The numbers next to the curves give the years elapsed since the last thinning. Source: Modified after Kramer (1988).

However, fire often is a natural disturbance agent with high-frequency or short return intervals (O'Hara, 2014). In that case the idea is to implement silvicultural systems in such a way that they work around the disturbance. This is fairly straightforward by incorporating gaps that owe their existence to a small fire with localised effects. Larger and more devastating (stand-replacing) fire disturbances can, however, destroy large forest areas and then have an effect similar to clearfelling. Such a situation is less straightforward and then the options available for CCF on bare land would apply, cf. Section 2.3, since even advance regeneration may be destroyed to a large extent (O'Hara, 2014). Fire disturbances have for a long time been suppressed because they were considered a threat to forest resources. The suppression of natural fires in fire-adapted forest ecosystems often led to a build-up of fuel loads and consequently to catastrophic fires. Also, the competitive status of tree species depending on fires was often worsened by fire control (Cogos et al., 2020; York et al., 2021). Fires also influence the abundance of tree species in mixed forests and favour diversity (O'Hara, 2014). Some tree species with *serotinous cones* such as *Pinus banksiana* Lambl., *Pinus contorta* Dougl. ex Loud. and *Pinus mariana* Mill. rely on fires to melt resin that keeps the cone shut otherwise so that fires are essential to seed dispersal. In addition, fires can have marked effects on soil biogeochemistry and their absence can potentially change the whole woodland community (Kimmins, 2004; Palik et al., 2021). Thinnings (cf. Section 5.2) can reduce the risk of fires through increasing crown spacing and the promotion of fire-tolerant tree species. However, thinnings result in different residual spatial patterns than those caused by fire and the bio-chemical effects cannot be mimicked by thinnings alone (Crecente-Campo et al., 2009; Franklin et al., 2018).

The appreciation and more thorough understanding of fire ecosystems of recent decades has granted us the opportunity to simulate fire disturbances in a controlled way and thus to use fire in forest conservation and timber production. *Prescribed* or *controlled burning* involves low-intensity surface fires that burn rapidly and sweep quickly over an area that is limited using fuel breaks (e.g. ditches). As such foresters mainly rely on prescribed burning to reduce logging residue and other downed woody materials as a fire prevention measure, to facilitate subsequent activities (e.g. planting), as a site preparation tool to kill back interfering understorey vegetation, to prepare seedbeds, or to control pests and diseases that proliferate rapidly in woody debris (Nyland, 2002; Smith et al., 1997). Prescribed burning has also been used for conditioning a stand for the subsequent regeneration method but also in ecological restoration. Surface fires generally burn off just the litter layer and the aboveground parts of herbs and shrubs (Kimmins, 2004). Prescribed burning, however, requires knowledge and skills (Smith et al., 1997). In addition, prescribed burning is only possible within a rigid set of parameters and conditions that permit its containment to a predetermined area, which limits the flexibility in using fire as a silvicultural tool (Nyland, 2002). Despite the general ability to use fire proactively, there are only few studies attempting to make fire a part of the design of silvicultural systems. Apart from broad concepts and ideas, clear implementation details for the combination with silvicultural systems are lacking. This is surprising, since Smith et al. (1997) stated that fire is the most common of the regenerative disturbances of natural forests.

York et al. (2021) suggested and defined the term *pyrosilviculture* as the direct use of fire to meet management objectives. In their study involving mixed-conifer forests of the Sierra Nevada in California, the authors promoted the combination of gap systems and prescribed burning so that the gaps are spared from the fire by using surface fuel breaks.

In the Northern parts of Sweden, particularly in the Malå district of Västerbotten County, Joel Wretlind and others advocated prescribed burning in seed-tree systems for regenerating the boreal forest, particularly in areas with much competing ground vegetation and raw humus accumulation. Prescribed burning in these stands reduces fuels and the chances of unplanned fires (O'Hara, 2014). The development of this method happened in the first half of the twentieth century and was mainly used as a soil preparation tool to expose the mineral soil and to make nutrients available to tree seedlings (Cogos et al., 2020). Post-fire environments can be good for seedling establishment, since the soils are warm because of high light input and dark surfaces. Absence of plants and presence of charcoal lead to an accumulation of nutrients and water in the soil. Full sunlight can support high rates of photosynthesis for the colonising plant (Binkley, 2021). Prescribed burning transforms the top soil into a mull humus that is more suitable to seedling growth. Prescribed burning in Northern Sweden promotes the regeneration of *P. sylvestris* but also improves the early growth of *Picea abies*. Wretlind often emphasised that his method of prescribed burning essentially mimicked natural processes. He was particularly interested in maintaining mixed-species forests including *Betula* spp. (Cogos et al., 2020). Similar examples of prescribed burning also exist in North America (O'Hara, 2014).

These examples show that there is considerable scope for developing combined silvicultural systems involving prescribed burning in forest ecosystems. As the range of fire ecosystems may expand with ongoing climate change, there may be more demand for pyrosilviculture in the future.

5.3.7.3 Herbivores

Trees, particularly when they are young and small, but also the foliage and bark of older trees are an important food source of herbivores. Ungulate populations increased drastically in the past decades in many parts of the world (Kuijper, 2011; Binkley, 2021). High densities of deer and/or other herbivores can seriously diminish the success of regeneration. Since natural and to some degree also artificial regeneration are important processes that CCF relies on, the control of herbivores is not unimportant. High deer densities can also endanger tree diversity through heavy selective browsing (Ramirez et al., 2018) and modify tree species composition (Brzeziecki et al., 2016). Introduced or re-introduced tree species that are not common in a given area, are often particularly targeted by deer populations. Regeneration growing in canopy gaps, for example, is more often browsed by ungulates than regeneration outside gaps (Kuijper, 2011). Downed large deadwood logs irregularly dispersed in large quantities can sometimes act as natural fences and protect pockets of regeneration trees. A regular monitoring of browsing damage on regeneration can help establish whether there is any need for action. Options for dealing with a situation of excessive browsing of regeneration or bark stripping include the introduction of predators, culling, tree guards and temporary fencing. The application of sheep wool is another alternative, see Figure 4.5. Needless to say that large-scale livestock grazing and browsing in forests can cause serious problems to CCF attempts unless regeneration is prolific. Particularly sheep and goats should be avoided on forest land. On the other hand, a diversification of forest land through the adoption of CCF will increase the availability of specialist ungulate habitats and may relieve some of the browsing pressure in the long term (Kuijper, 2011). Large ungulates do not always need to be the main culprits. In the United Kingdom and Ireland, for example, bark-stripping grey squirrels are a major hindrance to the CCF management of broadleaves (Nichols et al., 2016).

5.4 Selection System

The discussion of the selection system is deliberately placed outside Section 5.3 to emphasise that it is not a system designed specifically for achieving regeneration but is rather close to the Anglo-American understanding of silvicultural systems, where the system constitutes a programme of treatments that covers an entire forest generation and beyond. The selection system has also often been described as the ultimate or even classic variant of CCF (Pukkala and Gadow, 2012), although this perception is not true, cf. Section 1.6. More than any other form of CCF the selection system ensures the self-sustainability of size structure at stand level, i.e. if appropriately managed, the size structure includes a wide range of tree sizes and over time remains more or less unchanged (see Figure 5.25). Selection forests have no distinct forest generations as typically exist when using any of the silvicultural systems or their combinations discussed in Section 5.3. Many aspects and structures of selection systems are also relevant to or can be observed in other forms of CCF. Therefore, the study of this special silvicultural system is instructive, even if the reader may not be interested in applying it. In the German literature (Bartsch et al., 2020; Burschel and Huss, 1997; Rittershofer, 1999), the selection system is known as *Plenterwald* and the term 'plenter' is often used also in the English literature.

5.4.1 Method

Originally invented by farmers with small upland forest ownerships, the single-tree selection system is characterised by the fact that all tree sizes and development stages are present and intimately mixed in a single forest stand and that all forest operations simultaneously affect all of these stages (Pommerening and Grabarnik, 2019; Rittershofer, 1999, cf. Figure 5.25), although interventions are always local and never global (see Section 3.5 and Section 5.2). It is likely that the selection system gradually evolved from coppice with standards (see Section 1.3) (Schütz, 2001b).

> 'A forest managed as selection system includes trees whose crowns do not touch, but tend to occupy the whole vertical growing space.'
>
> Schütz (2001b)

The above statement by Jean-Philippe Schütz is a very appropriate characterisation of the visual, structural appearance of selection forests (cf. Figures 2.4 and 5.26) and helps identify them in the field. The most common intervention is the felling of few individual, mostly large trees at fairly short but irregular thinning cycles. In these interventions, no interruption in the main canopy should be much larger than the extent of one dominant tree crown. Subsequently, regeneration spreads throughout the stand and occurs simultaneously with other development stages (Burschel and Huss, 1997; Schütz, 2001b).

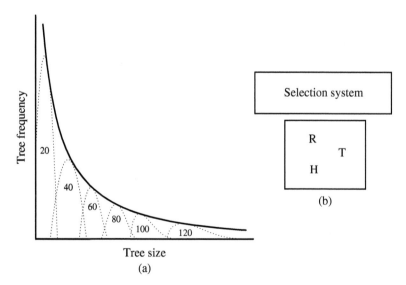

Figure 5.25   (a): When pulling even-aged stands representing different age classes (indicated by the numbers) from a forest estate together in a single graph, a negative exponential size distribution is achieved akin to that often observed in selection forest stands. However, in selection systems such distributions exist in single forest stands. (b): Given the (negative) exponential size distribution, all management activities such as regeneration (R), thinning (T) and harvesting (H) are integrated in space and time (cf. Section 1.8). Inspired by Smith et al. (1997) and Assmann (1970).

Figure 5.26 Mostly monospecies *Picea abies* (L.) H. Karst. selection forest in compartment 326 of the Harz National Park near the Oderteich Lake (St. Andreasberg, Lower Saxony, Germany) at 750 m asl (Petersen and Guericke, 2004), cf. Table 5.5. The site has been monitored since 1935. Source: Courtesy of author's student, Bangor University CCF field trip to Germany, 2003.

Large trees removed are rapidly replaced by mid-storey trees which in turn are supplemented by regeneration trees in the understorey. The system can be thought of as a kind of *process conservation*, where a forest stand is permanently kept in a disturbance and regeneration phase. It is common to temporarily see structures typical of selection forests in the old-growth and regeneration phase of many mixed-species forests. In selection forests, it is possible to keep regeneration and to some degree even mid-storey trees in a stand-by reserve position for a long time until needed (Rittershofer, 1999). Regeneration rates are moderate and just sufficient to replace trees that die. Thinnings in selection systems are mainly selective crown thinnings aimed at pre-dominant trees; however, they are carried out to maintain the particular demographic equilibrium of tree sizes that retain a forest stand in the aforementioned permanent disturbance and regeneration phase.

Since this thinning approach is quite special, thinnings in selection systems are termed *selection thinnings* to distinguish them from the more general selective thinnings (Schütz, 2001b; Nyland, 2002). Occasionally, other dominant and sometimes even mid-storey trees also need to be removed as the need arises. Ideally, individual trees are harvested when they reach or have surpassed their target dimensions and regeneration occurs in the small gaps created by their removal leading to a completely uneven-aged stand in which many tree sizes and cohorts are intimately mixed. Target diameter harvesting alone, however, may not preserve the desired stand structure or ensure sustained size distributions. The selection-thinning principle is also applied to other CCF forms such as the irregular shelterwood (referred to as *Badischer* and *Schweizerischer Femelschlag* in German) and the two-storeyed high forest (Mayer, 1984).

Interventions are essential for maintaining the size structure of selection forests: Too few or too weak interventions lead to a dense overstorey with closed canopy (similar in appearance to a vaulted hall densely packed with large stone columns) which ultimately will destroy the under- and mid-storey through suppression. Too many or too heavy interventions reduce the overstorey of dominant trees to a minimum, give rise to too much regeneration ultimately resulting in an even-aged successor stand. Typical selection forests in Central Europe have densities between 500 and 600 trees per hectare, 30–50 m^2 per hectare basal area and standing volumes between 300 and 500 m^3 per hectare. These numbers depend on species composition and environmental conditions. The better the environmental conditions and the more shade-tolerant the species involved are the higher the density numbers can be. The transformation of even-aged stands to selection forest stands takes at least 50–100 years (Mayer, 1984; Pommerening and Grabarnik, 2019; Rittershofer, 1999).

The classic species *Picea abies*, *A. alba*, *F. sylvatica* of Central Europe are confined to the mountain regions of the Vosges, Jura, Alps and Carpathians, where permanent forest cover on steep valley sides is vital to guard against avalanche and rock fall (Matthews, 1991), but are theoretically also applicable in lowlands elsewhere. This combination of species seems ideally suited to management by single tree selection (Anderson, 1960); however, the selection system is certainly not limited to them (see, for example, Schütz and Pommerening, 2013). *P. menziesii* can be mixed with *Tsuga heterophylla* and *Thuja plicata*. Other species that have been successfully grown in selection forests include *A. pseudoplatanus* and *F. excelsior* (Rittershofer, 1999). In the elevated continental valleys of the Alps, mixed *Larix decidua – Pinus cembra* L. selection forests can be found (Schütz, 2001b). Selection forests mostly include a range of tree species, but there are also well-known examples of monospecies selection forests based on *Picea abies* (see Figure 5.26) and *F. sylvatica* (e.g. in Western Thuringia, Germany). In the traditional areas of selection forests, shade tolerant *A. alba* plays an important stabilising role tolerating periods of suppression between 40 and 60 years (in rare cases up to 200 years, and more) (Mayer, 1984).

National proportions of selection forests are small (3% in Austria, 1% in Germany and France, 13% in Greece, 12% in Slovenia, 8.4% in Switzerland), i.e. selection forests are comparatively rare (Schütz, 2001b; Mayer, 1984). In the areas where they occur, they are very significant and form a part of the regional culture. Half of all the selection forests surveyed by Schütz (2001b) are located above 1400 m asl. Only 13% of these were deliberately established as selection systems. In lowland selection forests, there can occasionally be problems with competing ground vegetation, bark beetles and xylophagous fungi. A frequent reason for losing selection-forest structure is the application of too few and too weak interventions (Schütz, 2001b).

More than any other form of CCF, the size distribution in selection forests tends to take the shape of a (negative) exponential distribution (see Figure 5.25). The fascination with the single-tree selection system and the requirement to keep this size structure in balance has tempted many researchers to develop demographic guideline models describing a supposedly ideal exponential size distribution. This fascination was fuelled by the conviction that the concept of self-sustainable size distributions can be generalised across a large part of the wide continuum of CCF, see Section 1.5. The original idea of these demographic models was to assist forest managers in deriving thinning intensity and thinning cycles so that the self-sustainability of size structure is ensured. These types of models are discussed in detail in Chapter 6.

Most selection forest stands typically wander in and out of what can be considered an equilibrium or steady state of structure and yield data. This even happens when experts or researchers are involved in their management. In the *Picea abies* selection forest in compartment 326 of the Harz National Park (Lower Saxony, Germany, see Figure 5.26 and Table 5.5), for example, the number of trees along with volume and basal area was allowed to steadily increase over the decades.

Table 5.5 Stand monitoring data relating to the *Picea abies* (L.) H. Karst. selection forest in compartment 326 of the Harz National Park (Lower Saxony, Germany), cf. Figure 5.26.

| Survey | Age range | Residual stand | | | Removed | | |
		N	G	V	N	G	V
1935	1–205	316	18.0	222			
1952	1–222	516	25.7	311	23	0.7	8
1961	1–231	693	39.2	310	7	0.1	
1990	1–260	386	34.2	348	188	9.3	86
2000	1–270	361	36.5	386	18	2.7	30

N – number of trees per hectare, G – basal area in m^2 per hectare and V – standing volume in m^3 per hectare, cf. Figure 5.27.
Source: Adapted from Petersen and Guericke (2004).

Selection systems have no tradition in the area. The accumulation of volume in the upper and mid-storey reduced and by the year 1990 stopped the regeneration process so that heavy thinnings in the mid-storey along with a light thinning of dominant trees (amounting to 86 m^3) were necessary in 1990 in order to restore the structure of the selection forest. The steady decline of the size structure typical of selection forests can be clearly seen in Figure 5.27, where after 1956 the tree numbers steadily decreased in small stem-diameter classes and the curves flattened. The intervention in 1990 helped re-start the regeneration process, but judging by the distribution of the year 2000, there is still a long way to go. The history of this and other selection forests emphasises that thinnings in selection forests are often too weak, but also that mistakes can be corrected with the necessary dedication (Petersen and Guericke, 2004). The reason for undercutting selection forests is often related to a conservative behaviour of the managers involved who are worried about timber sustainability but are also due to logistic difficulties associated with the frequently required interventions in selection forests.

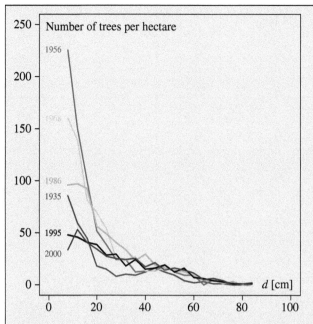

Figure 5.27 Temporal development of the empirical stem-diameter distributions relating to the *Picea abies* (L.) H. KARST. selection forest in compartment 326 of the Harz National Park (Lower Saxony, Germany), cf. Figure 5.26 and Table 5.5. *d* – stem diameter. The survey years are given in the colour of the corresponding diameter distribution curve. Source: Modified from Roloff (2004).

5.4.2   Ecological and Silvicultural Implications

The selection system represents the understorey reinitialisation and old-growth phases in the qualitative development stages defined by Oliver and Larson (1996). With regard to the stand as a whole there is no significant change in ecophysiological conditions; within the stand there is large small-scale variability (Schütz, 2001b). The system requires moderate soil nutrients but good soil moisture regimes to be able to sustain lower canopy strata (Rittershofer, 1999). The juvenile phase is very slow but mid-storey development can be rapid (Burschel and Huss, 1997). Tree age is even more undefined than in any other forms of CCF. Not only large trees but also small understorey or mid-storey trees can sometimes be several hundred years of age so that individual tree size is only poorly correlated with age. Single tree selection is particularly suitable for relatively shade-tolerant species, as regeneration is often obliged to germinate and grow in comparatively small gaps created by the removal of individual trees, where light is limited. However, if the main canopy is kept sufficiently open, intermediate and light-demanding species can also be accommodated (cf. Appendix B). If harvesting of large trees is carried out frequently, it is possible to take advantage of many seed years. In this system, much of self-pruning and density regulation is taken care of by biological automation, i.e. natural processes, cf. Section 2.6.

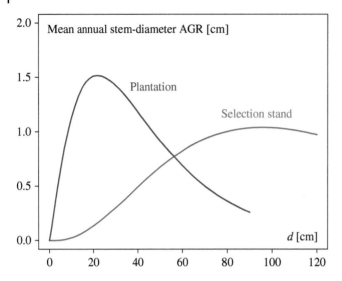

Figure 5.28 Differences in mean annual stem-diameter absolute growth rate (AGR) between a comparable plantation and a selection forest stand in Switzerland. *d* denotes stem diameter. Source: Adapted from Schütz (2001b).

The maximum size of trees is often larger in selection forests than in any other form of CCF (Rittershofer, 1999). Up to a stem diameter of 50 cm, a selection forest typically has 30% fewer trees than occur in comparable plantation forests. The selection forest has 60% more trees with diameters of 50 cm and larger than even-aged forest stands (Schütz, 2001b). A selection forest depends much on regular, frequent human interventions (every five to eight years) in order to retain its characteristics. Natural processes are ideally linked with management activities, but a rather high human input is required to maintain the system.

'Natural selection forests' only occur where moderate but frequent natural disturbances constantly modify forest structure. In selection forests, competition between trees is more vertical than in plantations, where it is typically stronger in the lateral domain. In plantations, absolute stem-diameter growth rates tend to increase rapidly, reach a maximum early and then decrease markedly (cf. Figure 5.28). In selection forests, absolute stem-diameter growth rates are low for an extended period and then gradually increase. Following this, stem-diameter growth rates tend to be more or less constant across a broad range of tree sizes. Trees of the same size can have differences in age up to more than hundred years. Formerly suppressed trees are capable of quickly becoming larger trees after release. The minimum area required for a balanced selection forest is 1–2 ha (Schütz, 2001b).

5.4.3 Advantages

The presence of many canopy strata helps suppress branching on the lower parts of stems, although heavy branching in well-developed crowns leads to large knots in upper stem parts. Selection forests can be extremely aesthetically pleasing, are very suitable to community and recreational forests (see Section 7.3.5), because they conform to many common conceptions of the appearance of natural forest with trees of all sizes standing side by side. In general, selection forests are ideal for meeting the requirements of a management providing many different ecosystem goods and services (Schütz, 2001b). Stand structure and ecological conditions are largely constant even after human interventions and disturbances so that visitors hardly perceive any change. Whilst potentially being very productive, selection forests are usually much less susceptible to natural disturbances due to a consistent promotion of individual-tree resilience (Dvorak and Bachmann, 2001; Hanewinkel et al., 2014). This is partly related to favourable c/h and h/d ratios (see Section 2.2) of large trees (Schütz, 2001b), which are a result of the selection forest structure. Timber sustainability can be ensured for comparatively small units of land (1–4 ha).

5.4.4 Disadvantages/Challenges

Considerable skill and experience are needed for the marking of thinnings and the execution of felling and extraction operations. Continuity of goal-oriented, comparatively frequent management input is particularly important. Potential yields are less easy to forecast than in even-aged stands. A good network of extraction racks is required. Inappropriate management or neglect can quickly destroy the size structure required for a selection forest, whilst the transformation of plantations to selection forests takes a very long time (Burschel and Huss, 1997). Herbivore densities should be appropriately low to ensure that the regeneration can survive.

5.4.5 Variants

The *group selection system* is a variant of the single-tree selection and makes allowances for the regeneration of light-demanding species (Rittershofer, 1999; Nyland, 2002). Here the unit of harvesting is a group of trees rather than an individual. The size of groups depends on the shade tolerance of the species to be regenerated and also on the predicted crown size of trees of target dimensions (Matthews, 1991). Wind risk may be increased by the cutting of large gaps, but surrounding groups are likely to have been exposed to the wind at some stage during their development and, if well managed, should be relatively windfirm. In practice, individual-tree and group cuttings of varying sizes often occur side by side in one and the same woodland. It can be argued that there is hardly any difference between the group selection and the irregular shelterwood system. An extreme area-control form of the group selection system is the Anderson group selection system described in Section 2.3.2.

Lack of regeneration or unsuitable regeneration can be compensated for by planting. Other variants of the selection system include the *coppice selection system* (cf. Section 1.3) and the short-rotation selection system. The management principles are very similar to those applied in the single tree selection system, but the target diameter is much lower and often applies to stems that are surrounded by other stems originating in the same stools. Management focuses on larger stems which are harvested selectively at frequent times. Regeneration is achieved from resprouting and/or limited planting (Matthews, 1991; Coppini and Hermanin, 2007; Sjölund and Jump, 2013). Selection management gives coppice woodlands and short-rotation stands a more natural appearance and increases diversity.

5.5 Continuous Two-Storeyed High Forest

Similar to the selection system (Section 5.4), the two-storeyed high forest is not a regeneration method, but rather a silvicultural system in the Anglo-American sense of the term. Sterba and Zingg (2001) demonstrated that this system applied to mixed-species stands dominated by *Picea abies* with time tends to approach the size structure of selection forests.

5.5.1 Method

A two-storeyed high forest is composed of an upper and a lower storey of trees of seedling origin, growing in intimate mixture on the same site (Matthews, 1991). In contrast to shelterwood systems, where two or more canopy storeys *temporarily* co-exist with the mature overstorey until it is finally removed, in continuous two-storeyed high forests two distinct canopy strata are maintained on a permanent basis. Each storey usually is comparatively homogeneous in terms of species composition, tree size and age, but these parameters differ markedly between the two storeys. The understorey typically includes more intermediate and shade-tolerant species such as *Fagus* spp. or *Picea* spp. while the overstorey can involve light-demanding species such as *Pinus* spp. and *Larix* spp. (cf. Appendix B). Other possible combinations can, for example, involve *P. menziesii* (overstorey) and *Tsuga heterophylla* (understorey) (Matthews, 1991). The understorey can have also the purposes of protecting the soil, preventing the formation of epicormic growth on the stems of overstorey trees, or maintaining soil fertility, and many have originated from underplanting (Matthews, 1991). However, it is also possible that both storeys are largely made up of the same species like, for example, in the case of Reininger's management at the Schlägl estate in Austria, cf. Section 3.6. The more light- and/or nutrient-demanding the understorey species are, the more open the canopy of the overstorey needs to be. The understorey is sometimes initialised by natural regeneration processes or by disturbances. As mentioned by Matthews (1991), it is also possible to plant the understorey after opening up the overstorey or after the overstorey was severely disrupted by natural disturbances. In the long term, the overstorey is replaced by the understorey and a new understorey needs to be initiated. This is often achieved naturally by keeping the overstorey sufficiently open so that regeneration and the survival of regeneration trees is encouraged. It is important to keep the understorey trees in a stand-by position so that they can just survive but do not emerge into the mid- and overstorey too early. This is achieved by finding an appropriate compromise of overstorey density (Burschel and Huss, 1997).

5.5.2 Ecological and Silvicultural Implications

Two-storeyed high forests can resemble shelterwood systems and are often initialised in a similar way. Due to the permanence of the two storeys, an explicit regeneration phase (and associated silvicultural system) is not required. Therefore, this management approach is quite ideal for the concept of CCF. Trees in the light overstorey often benefit from increased growth rates due to heavy release. The lower the soil moisture and/or soil nutrient regimes, the more open the canopy of the overstorey needs to be. The same principle applies to light-demanding species (cf. Appendix B). Two-storeyed high forests are comparatively rare

(Burschel and Huss, 1997), but some of the early CCF management demonstration sites in Germany (see Section 1.7) as well as the transformation carried out by Reininger at the Schlägl estate in Austria (see Section 3.6) aimed at this system.

5.5.3 Advantages

The timber value of trees grown in two-storeyed high forests can be high. Two-storeyed high forests have a high aesthetic value and can also serve as a transitional phase during the transformation of a plantation to a selection forest.

5.5.4 Disadvantages/Challenges

Depending on overstorey density, understorey trees may have reduced stem-diameter but not height growth so that resilience to wind and snow may be reduced. This can be improved by adjusting the overstorey density. A good network of extraction racks is required for the removal of large overstorey trees to minimise the damage on the understorey (Burschel and Huss, 1997). The application of the system requires experience and silvicultural skills, though it is easier to manage than the selection system whilst sharing some of its advantages (Matthews, 1991).

6

Demographic Equilibrium and Guidance Modelling

Abstract

In CCF forest stands, sustainability can be achieved at stand level. In such woodlands involving diverse species and size structures, determining and monitoring demographic sustainability is a challenge. Early approaches were suggested by Gurnaud, Biolley and de Liocourt in the nineteenth century but were eventually replaced by more mechanistic demographic modelling. Systematic work in demographic modelling really took off in the 1930s. Recent approaches to demographic modelling explicitly take growth, mortality and harvesting information into account. The modelling aims at achieving an equilibrium size structure that can help forest managers make informed decisions about timber sustainability and necessary interventions in CCF woodlands. As such, equilibrium size distributions are often interpreted as management guidelines or used for didactic purposes. In addition, demographic models can – within reason – predict growth and other processes in highly diverse forests. Equilibrium and guidance models play an important role in silvicultural monitoring and are often used by forest practitioners and in forest planning, particularly in Anglo-American countries. Appreciating the processes involved in this type of modelling usually leads to a deeper understanding of CCF.

6.1 Introduction

The principle of timber sustainability ensures that not more trees are harvested or die naturally than re-grow in any time period. After centuries of unrestricted forest exploitation and devastation, cf. Section 1.7, this principle has naturally been the prime concern of early forestry. Simple methods heavily leaning on agriculture have been employed in plantation management and one successful method has been the normal forest model, see Section 1.8. Early pioneers of continuous cover forestry (CCF), however, intuitively understood that such a borrowed approach would not work well in forest stands with great size diversity where timber sustainability should be achieved at stand level rather than at forest district or forest enterprise level. As these early pioneers were much criticised by traditional plantation foresters, they felt under pressure to engage in what we now know as demographic modelling in order to replace the normal forest doctrine and similar concepts.

Continuous Cover Forestry: Theories, Concepts, and Implementation, First Edition. Arne Pommerening.
© 2024 John Wiley & Sons Ltd. Published 2024 by John Wiley & Sons Ltd.

The main application of demographic modelling are selection and natural forests, cf. Section 5.4, although they can in principle be applied to any CCF woodland (Lorimer and Frelich, 1984; Salk et al., 2011; Kohyama et al., 2015). Historically, the immediate concern triggering these efforts was to ensure timber sustainability at a time when illegal logging was still widespread (Schütz, 2001b). Compared to plantations where each tree has a geometrically defined location, diverse CCF woodlands look 'irregular', as trees can be all over the place and even hugely differ in size at close proximity. Many trees are obstructed by others and therefore hard to see. The existence of more than one canopy storey limits visibility, and this made many traditional foresters uneasy. For them such forests were difficult to comprehend and it was hard to tell, if a tree or two were missing. Therefore, there was substantial pressure on CCF pioneers to come up with methods that would allow a firm 'control' of CCF forest stands. Later other important considerations and research questions came into play:

1. Is there any ideal size structure, an equilibrium, that would ensure a certain, desired forest structure and sustained harvests in perpetuity?
2. Can demographic models of CCF woodlands be used as guidance for forest managers?
3. Do demographic models even offer direct, hands-on guidance for deciding which trees to mark for removal and which to keep?
4. Can demographic models be used for predicting future growth and yield?
5. Are differences between the observed empirical size distribution and the corresponding demographic guidance model a suitable basis for silvicultural monitoring?

Not all of these questions have been fully answered yet, i.e. there is plenty of opportunity for more research. Those who concerned themselves with these questions and the modelling quickly realised that demographic models are helpful for developing a deeper understanding of how CCF and particularly selection forests work, i.e. what the processes driving the size structure are. The models therefore have a high didactic value. The results simulated with these models provide guidance but – as with any model results – should not necessarily be followed blindly. Personal experience, a critical mind, literature and other more sophisticated models should be consulted as well before making final decisions. With time silviculturists also understood that such models are invaluable in silvicultural training.

6.2   History

Forest practitioners Adolphe Gurnaud (France) and Henry Biolley (Switzerland) were the first to consider a *check method*, which they termed *Méthode du Controle* in French, for sustainable yield regulation in selection forests (Schütz, 2001b). The basic idea was to

Table 6.1   Normative definition of the proportions of standing volume of a selection forest according to Gurnaud's and Biolley's Méthode du Controle.

Diameter class [cm]	Volume proportions
20–30	0.20
35–50	0.30
≥55	0.50

Source: Adapted from Schütz (2001b).

measure all trees of a forest stand or a sample and to check the proportions of standing stem volume against a simple norm that was based on experience (Table 6.1). For this purpose, the trees of a forest stand were classified according to three broad classes of stem diameter. When deviations from the norm occurred, appropriate interventions had to be made to increase or decrease the proportions in the classes where the deviations were observed. If, for example, the proportions were too low in the first two classes more trees with a stem diameter larger 55 cm had to be cut to allow for more light, nutrients and water to be available to seedling growth. If proportions were low in the last class, harvesting of big trees had to be put on hold for a while. Although very much in use and often referred to in the literature, the check method is only a very rough concept. It does not distinguish between different species compositions and environmental conditions.

In Switzerland, equilibrium volume proportions were later refined as shown in Table 6.2 and some versions of the check method also involved volume growth (Knuchel, 1953). Mitscherlich (1952) proposed a similar system to include both tree-number and volume proportions for comparison arranged in three classes that he referred to as *small timber* (7–25 cm), *medium timber* (25–50 cm) and *high* or *large timber* (>50 cm). Whilst the tree number proportions should decrease from small to high timber, the volume proportions should increase (Figure 6.1). A major disadvantage of volume regulation, however, is the inability to monitor the development of small-diameter trees that matter much to the overall demographic sustainability of a forest stand (Guldin, 1991).

Table 6.2   Revised Méthode du Controle with ideal volume proportions in different stem diameter classes.

Diameter class [cm]	Volume proportions
16–24	0.08
26–38	0.16
40–50	0.28
≥52	0.48

Source: Adapted from (Schütz, 2001b).

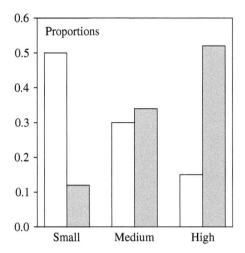

Figure 6.1 Tree-number (white) and stem-volume (grey) proportions in balanced selection forests after Mitscherlich (1952). Small – small timber, Medium – medium timber, High – high timber. For the stem diameter definitions of these three classes see the text.

In Chapter 4, we have already concluded that empirical tree-size distributions are valuable indicators of forest structure and therefore useful for monitoring forest development. It was the French forester François de Liocourt (de Liocourt, 1898) who first noticed that the shape of the stem-diameter distribution curve in mature CCF woodlands can often be described by a decreasing or (negative) exponential function with most trees in the smaller diameter classes (Figure 6.2). This phenomenon is now often known as the *law of de Liocourt*. More generally, the number of trees with increasing size follows a *decay process* where size replaces time with proportionality $\Delta n \propto n \times \Delta d$. For such processes, Eq. (6.1)

$$\frac{\Delta n}{\Delta d} = \lambda \times n \text{ with } \lambda < 0 \tag{6.1}$$

holds. Similar conditions apply, for example, to radioactive decay and to the exponential extinction of species with time. The number of trees is denoted by n and λ is a constant model parameter that in the context of decay processes is also known as *decay* or *disintegration constant*. Tree size represented by stem diameter is denoted by d in Eq. (6.1) and Δd is the change in stem diameter. Where this condition holds, an exponential function best expresses the relationship between the two variables (Leak, 1965):

$$n = n_0 \times e^{-\lambda \times d} \tag{6.2}$$

Model parameter n_0 is the initial number of trees for $d = 0$ or for infinitesimally small d, whilst λ is the aforementioned model parameter and e is the base of the natural logarithm. n_0 is the intercept, i.e. the point where the function defined in Eq. (6.2) crosses the ordinate (see Figure 6.2a). As we will see in Section 6.3.1, λ is a relative growth rate relating to the change in tree numbers with increasing size (Cancino and Gadow, 2002). Shifting n_0

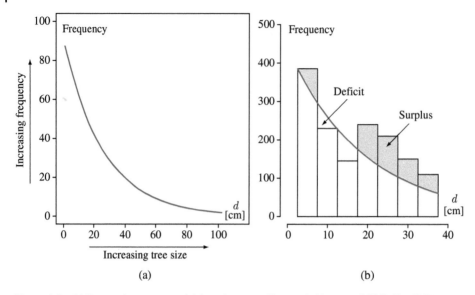

Figure 6.2 (a) Decreasing exponential function according to de Liocourt (1898) (Eq. (6.2)) as a result of applying the R code shown in the following R listing. (b) An equilibrium curve based on Eq. (6.2) overlaid on an empirical diameter distribution. The parts of this distribution to be cut in thinnings are highlighted in grey. *d* - stem diameter at 1.3 m above ground level.

up and down shifts the model curve up and down, which, in turn, shifts stand basal area up and down. Colloquially, the shape of such a size distribution is also labelled as 'reverse J' and 'inversed J-shaped'. By contrast, the classic exponential distribution in statistics has only one parameter λ, i.e. $f(x) = \lambda \times e^{-\lambda x}$. The use of Eq. (6.2) was later promoted by Meyer (1943), himself a Swiss immigrant to the United States of America, with the aim of deriving an ideal or equilibrium diameter distribution model as a silvicultural guideline.

Eq. (6.2) defines a continuous function, which can be interpreted as an empirical diameter distribution with a very small, constant class width. Like empirical diameter distributions, demographic equilibrium or guideline models typically consider discrete size classes, in our case classes of stem diameter *d*.

The basic assumption of any equilibrium model is that they represent a size structure affording demographically sustainable thinning levels. Such a distribution is said to be in equilibrium. In practice, the equilibrium curve is compared to the current empirical diameter distribution of a forest under study, and an allowable cut is generated by the difference between the two (Guldin, 1991), i.e. in one or more interventions, those numbers of trees in their respective size classes are removed that exceed the model curve (Figure 6.2b). The allowable cut for each diameter class is determined by subtracting the number of trees according to the equilibrium curve from the corresponding number in the observed empirical diameter distribution. If this difference is positive, this *surplus* represents the allowable cut for that class. Ignoring natural mortality, the total allowable cut is simply the sum of these differences.

A more refined way of describing the allowable cut that also includes deficits (cf. Figure 6.2b) is discussed in Section 6.6.4. Other terms describing the equilibrium property include *balanced*, *stationary*, *steady-state* and *stable* (Brzeziecki et al., 2016). In theory, the de Liocourt model, if adhered to, produces a demographically balanced stand where tree mortality and removal equal tree growth and each size class occupies equal growing space (O'Hara and Gersonde, 2004). The function described in Eq. (6.2) can easily be visualised using the following R listing (cf. Figure 6.2a).

In line 1 and 2 of the following code, parameters n_0 and λ are initialised. The values do not matter much here and were arbitrarily chosen, since this is just an example to get started. Line 3 gives technical settings for the graph margins.

```
1  > n0 <- 90.735
2  > lambda <- -0.038
3  > par(mar = c(2, 3.5, 0.5, 0.5))
4  > curve(n0 * exp(lambda * x), from = 1, to = 102,
5  + col = "red", lty = 1, lwd = 2, axes = FALSE)
6  > axis(1, lwd = 2, cex.axis = 1.7)
7  > axis(2, lwd = 1, las = 1, cex.axis = 1.7)
8  > box(lwd = 2)
```

The curve() function in line 4 tells R to draw the function defined in Eq. (6.2) using the model parameters assigned to variables n0 and lambda in lines 1f. The code in lines 5–8 essentially includes technical commands for formatting the graphical output. For all subsequent discussions, it is important to understand and remember the frequency and size trends indicated in Figure 6.2a.

Demography is a scientific field concerned with demographics where the structure and development of populations are studied statistically. This study involves the size, age structures and economics of different populations and can be used for a variety of purposes (Bijak et al., 2018).

Key processes in demography are *birth rates*, *mortality* and *migration rates* (Lorimer and Frelich, 1984; Salk et al., 2011; Kohyama et al., 2015). Whilst demography usually studies human populations, in forest ecology and management, the size structure and development of trees is analysed and modelled. Birth and mortality rates are also of interest in the context of tree populations and migration rates can be interpreted as rates of trees moving in and out stem-diameter classes, cf. Figure 6.3. Birth processes are sometimes also referred to as *recruitment* (Schütz and Pommerening, 2013), as even the first possible size class often only considers trees beyond a certain threshold size long after their germination. After a time period of t years, there are n_i trees in stem-diameter class i. This number n_i of trees is the result of the number of trees in class i at the beginning of t, minus the number of trees that died naturally and/or were cut, minus the number of trees that moved on towards the next larger stem-diameter class n_{i+1} (outgrowth) plus the ingrowth from the next smaller diameter class n_{i-1} (cf. Figure 6.3). For the intents and purposes of equilibrium guideline

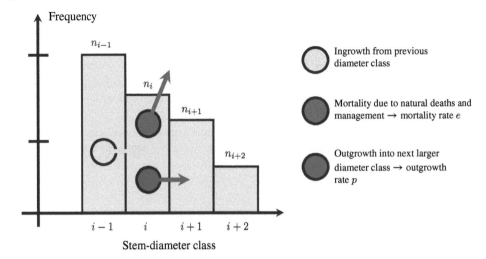

Figure 6.3 Main processes of a dynamic stem-diameter distribution and the corresponding notations used in this chapter. Modified from Schütz and Pommerening (2013), see also Bachofen (1999).

modelling, mortality due to natural processes and mortality as part of human interventions are often expressed in a single rate, as the cause of death is often not recorded in available data. When considering the equilibrium in natural forests, the mortality rate in Figure 6.3 only includes natural mortality. In well-managed forests on the other hand, natural mortality is usually low so that the mortality rate in Figure 6.3 in that case mainly reflects mortality due to human interventions. However, if the cause of death was recorded in a given data set, these two rates can be modelled separately.

Unlike other methods, equilibrium or guideline models were developed by the CCF research community, and there are two main modelling approaches:

- *Static approach*: q factor model family including BDq extensions,
- *Dynamic approach*: Prodan, Schütz and similar models.

The fundamental difference in the two approaches is that the static approach does not explicitly consider in/outgrowth and mortality rates, whilst the dynamic approach includes them as auxiliary functions of stem diameter. Due to a more accurate input, dynamic models are therefore likely to deliver a more accurate output, however, this comes at the expense of a greater demand on data compared to the static models. The static models are simpler than the dynamic models and make certain assumptions that do not hold in all CCF situations. Dynamic models are based on less assumptions and are therefore more flexible.

As already stated, the theories of both modelling approaches are based on discrete stem-diameter classes (cf. Figure 6.3). Therefore, class or bin width is clearly a modelling parameter, i.e. the model results additionally depend on the chosen diameter class width. In selecting an appropriate class width, w, one should identify a number that allows important general trends to be seen whilst not smoothing away too many details. In our experience, bin sizes of 4 and 5 cm have been appropriate in many situations, whilst 1–3 cm classes are often too detailed with very few trees if any in each class and are therefore not recommended (Nyland, 2002). Possible equations from general statistics for estimating an appropriate number of classes, c or bin size include $c \approx \log_{10} n$ and $w \approx \frac{y_{max} - y_{min}}{c}$ (Fahrmeir et al., 2016), where n is the number of observations and y_{max} and y_{min} are maximum and minimum observation. These equations offer only guideline information which should be improved by the modeller based on their experience.

An ecological modelling theory related to the concept of demographic equilibrium is the *metabolic scaling theory* (Enquist et al., 2009).

6.3 Static Equilibrium Models

Static equilibrium models can be estimated from current one-off data surveys without knowing and considering growth and mortality rates. The low demand on data and simplicity has contributed to their popularity, but also to some uncertainty.

6.3.1 Model Theory

Both the q factor model and the underlying de Liocourt model come with the assumption that the rate of frequency reduction in successive stem-diameter classes is constant, e.g.

$$\frac{n_0 \times e^{\lambda \times 32}}{n_0 \times e^{\lambda \times 36}} = \frac{n_0 \times e^{\lambda \times 36}}{n_0 \times e^{\lambda \times 40}} = \cdots = e^{\lambda \times w} = q \text{ with } q > 1,$$

where 32, 36 and 40 are arbitrarily selected 4-cm stem-diameter classes and w is the class width in cm, i.e. 4 cm in this case. Later in this chapter, we will see that this assumption is rather crude and somewhat limiting. Following this assumption, the model type was named after q, which is also known as the *diminution quotient*, because it controls the diminution (reduction) rate of trees in successive stem-diameter classes. As such q controls the steepness of the equilibrium curve. The q factor also determines how the stand basal area is distributed across the range of diameter classes (Powell, 2018). Nyland (2002) stated that foresters in North America generally refer to (negative) exponential diameter distributions as q *structures*. In more general terms, we can write

$$q = \frac{n_i}{n_{i+1}} = \frac{n_0 \times e^{-\lambda \times d_i}}{n_0 \times e^{-\lambda \times (d_i + w)}} = \frac{e^{-\lambda \times d_i}}{e^{-\lambda \times d_i} \times e^{-\lambda \times w}} = \frac{1}{e^{-\lambda \times w}} = e^{\lambda \times w}. \tag{6.3}$$

Here, n_i and $n_i + 1$ theoretically refer to model estimations and should therefore be denoted more accurately as $\hat{n}_i$ and $\hat{n}_i + 1$, since the hat symbol indicates estimation. Without hats, these symbols in principle refer to observed numbers of trees. However, we have decided to relax the notation in this chapter for ease of reading and omitted the hat symbol in most cases.

A q factor of 1.5 means that a given class i has 1.5 times more trees than the next larger stem-diameter class $i + 1$. According to the assumptions of the model described in Eq. (6.2), successive diameter classes of equal width form a decreasing geometric series with ratio $\frac{1}{q} = e^{-\lambda \times w}$.

Using Eq. (6.3), q is usually calculated from models (and theoretically should then more formally be denoted as $\hat{q}$) but can also be computed from empirical stem-diameter distributions. However, for observed stem-diameter distributions that are not in the state of equilibrium, q typically varies from diameter class to diameter class and then the notation q_i is appropriate, since each class has a different q factor.

In principle, q is a summary of complex information involving the aforementioned but commonly unknown growth, mortality and resulting migration rates. Equilibrium curves typically differ from forest stand to forest stand based on different numbers of trees in the smallest n_{min} or largest n_{max} stem-diameter class and on different q factors.

From Eq. (6.3) follows that q varies for the same data, if w changes. For example, assuming $\lambda = 0.0524$, $q = 1.29$ for $w = 5$, $q = 1.69$ for $w = 10$ and $q = 2.19$ for $w = 15$, i.e. all these values lead to equivalent demographic curves (Cancino and Gadow, 2002).

Eq. (6.3) implies that the number of trees in one stem-diameter class can be calculated from the previous (smaller) one by simple division:

$$n_i = \frac{n_{i-1}}{q} \tag{6.4}$$

By contrast, the number of trees in the next smaller diameter class can be calculated from the previous (larger) class by multiplication:

$$n_{i-1} = n_i \times q \tag{6.5}$$

This way of calculation suggests that the q factor can be interpreted as tree-number *growth multiplier* in the context of relative growth rates (cf. Meyer, 1933; Wenk, 1994; Pommerening and Grabarnik, 2019, Chapter 6). Based on Eq. (6.9) in Pommerening and Grabarnik (2019), a *relative growth rate* (RGR) corresponding to q can be calculated as

$$RGR = \frac{\log_e n_{i-1} - \log_e n_i}{\Delta d} = \frac{\log_e n_{i-1} - \log_e n_i}{w}. \tag{6.6}$$

In this specific context, the term 'growth rate' should not be misunderstood in the sense that RGR describes the result of growth processes. RGR in Eq. (6.6) rather describes the way how – starting from the smallest or the largest size class – the number of trees progresses from class to class towards the other end of the demographic distribution. In this application to empirical and model size distributions, RGR is not calculated as a mean in terms of time, but in terms of size difference, which in the majority of applications is the constant width, w, of the empirical stem-diameter distribution. Practically speaking, RGR is the proportion of trees by which the number of trees in the next smaller stem-diameter class is increased. In the q factor model, RGR is constant in the same way as q. Following Eqs. (6.3) and (6.17) in Pommerening and Grabarnik (2019), RGR can be transformed to q:

$$q = e^{RGR \times w} \tag{6.7}$$

Consequently, constant RGR is identical with the second parameter of the de Liocourt model (Eq. (6.2)), λ, since

$$RGR = \frac{\log_e q}{w} = \lambda. \tag{6.8}$$

Using RGR, the number of trees in any stem-diameter class i can be calculated from the number of trees in the first class, n_{min}, with the largest number of trees:

$$n_i = n_{min} \times \left(1 - 1 - \frac{1}{e^{RGR \times w}}\right)^i = n_{min} \times \left(\frac{1}{q}\right)^i \tag{6.9}$$

The term $1 - \frac{1}{e^{RGR \times w}}$ is another type of RGR, referred to as RGR$^-$ in Pommerening and Grabarnik (2019) and is in this context defined as $(n_{i-1} - n_i)/(w \times n_{i-1}) = 1 - 1/q$. Incidentally, Chevrou (1990) stated that the number of trees in the smallest diameter class n_{min} can be computed from

$$n_{min} = N \times \frac{q-1}{q}, \tag{6.10}$$

where N is the total number of trees per hectare recorded, i.e. the stand density characteristic.

Alternatively to Eq. (6.9), the number of trees in all other classes can be calculated from the largest stem-diameter class (with the lowest number of trees), n_{max}, by using Eq. (6.11):

$$n_i = q^{c-i} \times n_{max}, \tag{6.11}$$

where c is the total number of stem-diameter classes. Eqs. (6.9) and (6.11) clearly show that the number of trees either in the lowest or in the largest size class, n_{min} or n_{max}, and the q factor or its reciprocal form a decreasing geometric sequence in which the next n_i is found by multiplying the proceeding terms. Chevrou (1990) pointed out that parameter n_0 of the de Liocourt model (Eq. (6.2)) depends on q in the following way:

$$n_0 = N \times (q-1) \times q^{min(i)-1} \tag{6.12}$$

Here min(i) is the index of the lowest stem-diameter class and N is again the total number of trees per hectare as before. Incidentally, throughout this chapter for $i = 1$, $n_i = n_{min}$ and for $i = c$, $n_i = n_{max}$. Another way to determine n_0 follows from the previous equations in this section:

$$n_0 = \frac{n_{max}}{e^{-\frac{\log_e q}{w} \times d_{max}}} \tag{6.13}$$

In Eq. (6.13), e is again the base of the natural logarithm denoted as $\log_e$. The equations and the example show that the mathematics involved in the basic q factor model is rather simple and straightforward. This has undoubtedly contributed much to its appeal.

6.3.1.1 BDq Approach

In practical applications in North America, it soon emerged that it would be desirable to additionally make the q factor model dependent on an equilibrium basal area (denoted as B) and on a maximum diameter class (notated as D). The notation used for these two variables and that of the q factor coined the name of the approach, i.e. BDq. A synonym is the term *guiding diameter limit* (GDL) (Nyland, 2002). The concept appears to have its origin in Leak (1964) and Leak (1965), although there are at least two different methods. One, presumably the older method, is based on *basal-area weights* or *expansion factors* (Guldin, 1991) and explained in Section 6.3.2 and in Table 6.5 where the Susmel model is described. The other, more recent and more elegant method is elaborated on in this section. Both methods lead to similar but marginally different result.

In the BDq approach, equilibrium curves are defined by (i) the maximum stem-diameter d_{max}, (ii) the equilibrium or residual basal area B, (iii) the q factor and (iv) the number of trees in the stem-diameter class with the largest stem diameter, n_{max}. Three of these four characteristics need to be defined, the missing one resulting from them by simple mathematics. By multiplying n_{max} and the resulting n_i successively by q (Eq. (6.5)), the equilibrium curve is obtained as previously explained (Nyland, 2002; Sterba, 2004). Equilibrium basal area and maximum diameter have a substantial effect on the resulting model and to some degree diminish the effect of q (O'Hara and Gersonde, 2004). The acronym BDq represents not only alphabetical order but also the relative importance of each parameter in the modelling process (Guldin, 1991).

In commercial applications, maximum stem diameter can be thought of as target diameter (cf. Section 3.4) and in natural forests, the maximum diameter is the largest possible diameter of a certain tree species on a given site. Defining equilibrium basal area is usually a matter of silvicultural experience, but is also related to the ecological *carrying capacity* which depends on environmental conditions and tree species. Good advice includes selecting equilibrium basal area B in such a way that regeneration can just survive and develop (Guldin, 1991). Such values for stand basal area are commonly known from previous forest research (e.g. Hale, 2003).

Meeting the requirement of a certain equilibrium basal area per hectare and a maximum stem diameter or final stem-diameter class is achieved by introducing an additional auxiliary parameter k. Cancino and Gadow (2002) defined k as

$$k = \frac{\pi}{40\,000} \sum_{i=1}^{c} q^{c-i} \times d_i^2. \tag{6.14}$$

Auxiliary parameter k is derived from the formula for calculating the equilibrium basal area, i.e. $B = \frac{\pi}{40\,000} \sum_{i=1}^{c} n_i \times d_i^2$ and from Eq. (6.11). The diameter class values are denoted d_i with $d_i = w \times i$ and usually supplied in cm. An alternative to Eq. (6.14) has been suggested by Clarke (1995):

$$\tilde{k} = \frac{\pi}{40\,000} \sum_{i=1}^{c} q^{\frac{d_{max}-d_i}{w}} \times d_i^2 \tag{6.15}$$

Here, w is again the stem-diameter class width. Eqs. (6.14) and (6.15) are equivalent, because $q^{\frac{d_{max}-d_i}{w}} = q^{c-i}$. Auxiliary model parameter k is a key parameter for calculating the entire equilibrium distribution according to the BDq approach (Cancino and Gadow, 2002). The next step in the BDq approach involves the calculation of the number of trees in the largest diameter class:

$$n_{max} = \frac{B}{k}, \tag{6.16}$$

where B again is the anticipated equilibrium basal area. Changing equilibrium basal area does not cause a change in shape or slope of a curve, but modifies the intercept

(n_0 in Eq. (6.2)) and therefore the location of the curve. Consequently, the guideline curve is raised or lowered depending on B. Equilibrium basal area is the most influential factor of the BDq approach (Powell, 2018).

Following the determination of n_{max}, the number of trees in all other stem-diameter classes is calculated using Eq. (6.11). Now, it is possible to calculate the rate-of-change parameter, i.e. RGR, of the de Liocourt model (Eq. (6.2)) using Eq. (6.8). Finally, the intercept parameter of the de Liocourt model (Eq. (6.2)) is computed as

$$n_0 = \frac{n_i}{e^{-\lambda \times d_i}},$$ (6.17)

where i is an arbitrary integer number in the interval $[1, c]$. A quick check can involve inserting the newly calculated parameter λ in $q = e^{\lambda \times w}$ (Eq. (6.3)). Factor q resulting from Eq. (6.3) should then match the q value used in the previous calculation steps.

Modellers can evaluate and decide whether the basic q factor approach is sufficient for their application or whether there is scope for improving the results by applying the BDq approach as well.

6.3.2 Determining q

When applying the q factor model to observed data the question arises how to determine q. Not much field research nor computer-based experimentation have demonstrated how to select a q value that has biological or economic relevance (Nyland, 2002).

Experience with other modelling work may suggest employing nonlinear regression routines for optimising parameters n_0 and λ of Eq. (6.2). This course of action has even been suggested by authors in the past, see, for example, Meyer (1952). In the context of equilibrium and guidance modelling, such a strategy would lead to not very meaningful results. Most observed data used for this type of modelling represent forest stands and structures that are actually not in a steady state. They may be close, but often they are not. Without knowledge about how close or how far the forest stand from which the data were collected actually is to or from an equilibrium, a model fitted to observed data would only describe the current observation but not an equilibrium (Schütz, 2001b). Therefore, this strategy cannot be recommended and other techniques need to be employed for identifying the parameters of the equilibrium model.

It is possible to take values of q from the literature where models for woodlands and environmental conditions similar to the application were published. Another strategy of determining q has been (i) to calculate the q_i values from successive stem-diameter classes of the observed diameter distribution and (ii) to determine q as the arithmetic mean or better as the geometric mean of all empirical q values, since the q factor is a multiplicative concept. However, this procedure is strictly speaking not better than non linear regression, since any mean again only reflects the current size structure rather than the desired equilibrium. When deciding which q factor to use, there are some important general points to consider (Guldin, 1991; Buongiorno et al., 2000; Bruchwald et al., 2016; Powell, 2018):

- Low q factors emphasize large-diameter trees and discriminate against smaller size classes,
- High q factors emphasize small-diameter trees and smaller numbers of trees in large diameter classes are estimated,
- Low q factors result in smaller differences in number of trees between diameter classes; the opposite is true for high q factors,
- Larger q values steepen the equilibrium diameter distribution, smaller q values flatten the equilibrium diameter distribution,
- On productive sites with benign environmental conditions, q factors are low, and they increase with decreasing site fertility.

The most common, objective way to determine q is through using forest stand characteristics, which partly represent the current state of the stand under study and partly the anticipated equilibrium or steady-state structure, as explanatory variables. This approach of estimating q and the parameters of the de Liocourt model is reminiscent of the *parameter-prediction method* described in tree growth modelling textbooks (Clutter et al., 1983; Weiskittel et al., 2011; Burkhart, 2012). There are some examples of such approaches in the literature, and we consider three of them here in detail.

To illustrate the strategy of determining q by the three methods, we will use a data example published in Cancino and Gadow (2002), see Table 6.3. In this example, an equilibrium stem-diameter structure involving an equilibrium basal area of $B = 30$ m^2/ha, a target diameter of $d_{max} = 35$ cm, and a class width of the empirical stem-diameter distribution of $w = 5$ cm are assumed. To achieve an equilibrium stem-diameter distribution, $q = 1.3$ has been determined. The observed stem-diameter distribution includes diameters up to a class with a mid-diameter of 60 cm.

6.3.2.1 Pretzsch q Factor Model

Pretzsch (2009, p. 165) suggested an approach where the parameters n_0 and λ of Eq. (6.2) are estimated from the number of trees in the smallest (n_{min}) and in the largest stem-diameter class (n_{max}) along with the corresponding class diameters (d_{min}, d_{max}):

$$\hat{\lambda} = \frac{\log_e n_{max} - \log_e n_{min}}{d_{max} - d_{min}} \tag{6.18}$$

$$\hat{n}_0 = n_{min} \times e^{\lambda_1 \times d_{min}} \tag{6.19}$$

Stand characteristics n_{min}, n_{max}, d_{min} and d_{max} can be determined using the observed data of a forest stand under study, but more meaningful would be applying desired values that are anticipated to exist when an equilibrium is eventually established.

In the case of our example, we know the equilibrium structure and all associated characteristics from the publication by Cancino and Gadow (2002). Our objective here therefore is to see how well other published approaches of determining n_0, λ and q can reconstruct the published equilibrium. This gives us important insights on how the different approaches

Table 6.3 Tree-number and basal-area characteristics of an observed hypothetical forest stand along with the simulated characteristics according to the Pretzsch (2009, P) and Poznański and Rutkowska (1997, PR) approaches.

Diameter class (cm)	Trees per hectare			Basal area per hectare		
	Obs. n_i	Sim. $\hat{n}_i$ (P)	Sim. $\hat{n}_i$ (PR)	Obs. g_i	Sim. $\hat{g}_i$ (P)	Sim. $\hat{g}_i$ (PR)
5	370	338	392	0.73	0.66	0.77
10	230	260	294	1.81	2.04	2.31
15	145	200	221	2.56	3.53	3.90
20	240	154	166	7.54	4.83	5.21
25	210	118	124	10.31	5.81	6.10
30	150	91	93	10.60	6.43	6.59
35	30	70	70	2.89	6.73	6.73
40	36			4.52		
45	24			3.82		
50	32			6.28		
55	20			4.75		
60	3			0.85		
Total	1490	1230	1361	57	30	32

With both model approaches, $B = 30$ m^2/ha, $d_{max} = 35$ cm and $w = 5$ cm. For the Pretzsch approach, $q = 1.3$, whilst for the Poznański-Rutkowska approach, $q = 1.65$. Values of n_i, g_i, $\hat{n}_i$, $\hat{g}_i$ and totals were rounded. Obs. – observed, Sim. – simulated. Source: Adapted from Cancino and Gadow (2002).

work. The input values for the Pretzsch model are therefore taken from Cancino and Gadow (2002, cf. Table 6.3) and coded as follows:

```
> d.max <- 35
> d.min <- 5
> n.max <- 70
> n.min <- 338
> w <- 5
```

Based on Eqs. (6.18) and (6.19), the following listing computes the results for the Pretzsch model.

```
1  > (lambda <- (log(n.max) - log(n.min)) / (d.max - d.min))
2  [1] -0.052485
3  > (n0 <- n.min * exp(lambda * d.min))
4  [1] 259.984
5  > q <- exp(abs(lambda) * w)
6  [1] 1.30008
7  > (c <- length(seq(d.min, d.max, w)))
8  [1] 7
9  > (ni <- q^(c - seq(1, c, 1)) * n.max)
10 [1] 338.0000 259.9842 199.9757 153.8181 118.3144
```

```
11   [6] 91.0055 70.0000
12   > (B <- pi / 40000 * sum(ni * classes[1: c]^2))
13   [1] 30.0471
```

In lines 1 and 3, model parameters λ and n_0 are calculated according to Eqs. (6.18) and (6.19).[1] Using Eq. (6.3), q is computed in line 5. Function abs() is used here to ensure model parameter λ is processed with a positive sign. Following this, the number of stem-diameter classes is determined in line 7. In line 9, based on n_{max}, the numbers of trees in all other stem-diameter classes are calculated, cf. lines 10f. and Table 6.3. Now that these numbers are established, the equilibrium basal area B is calculated in line 12. We see in line 13, that calculated B is very close to the anticipated target, so we do not need to go any further with our calculations. Apparently, the Pretzsch model is able to reconstruct the equilibrium as published in Cancino and Gadow (2002) quite well.

6.3.2.2 Poznański-Rutkowska q Factor Model

Poznański and Rutkowska (1997) published an alternative approach, which does not rely on any numbers of trees as explanatory variables, but considers the arithmetic mean stem diameter, the lowest diameter class, and the target diameter instead:

$$\tilde{\lambda} = \frac{1}{\bar{d} - d_{min}} \tag{6.20}$$

$$\tilde{n}_0 = \frac{\tilde{\lambda}}{e^{-\tilde{\lambda} \times d_{min}} - e^{-\tilde{\lambda} \times d_{max}}} \tag{6.21}$$

Eq. (6.21) is a rather general equation which has its origin in work carried out by Meyer (1933) and both equations combined estimate probability density functions (Jaworski and Kołodziej, 2004), i.e. the resulting values of the function defined in Eq. (6.2) need to be multiplied by the total number of trees per hectare and by class width w. We found that the Poznański-Rutkowska approach leads to good results without a need for transformation when combined with an equation suggested by Chevrou (1990) based on the mean of the exponential probability density function:

$$q = 1 + \frac{w}{\bar{d}} \tag{6.22}$$

Calculating q with Eq. (6.22), parameter λ can then be estimated from Eq. (6.8) and n_0 from Eq. (6.17). Otherwise using the same settings as before, the application of this modified Poznański-Rutkowska approach leads to the results shown in Table 6.3 (Cancino and Gadow, 2002).

In line 1 of the following listing, $\bar{d}$ previously calculated from the published equilibrium stem-diameter distribution is assigned to variable meand.ide. Factor q is calculated according to Eq. (6.22) in line 2. Using q, model parameters λ and n_0 are calculated according to Eqs. (6.8) and (6.17) in lines 4 and 11.

---

1 Unfortunately, it turned out that Eq. (6.19) in this case produced an unrealistic estimate of parameter n_0. Instead, we therefore calculated the intercept by extrapolating n_i for $d_i = 0$ using Eq. (6.10). The first n_i value (439.4) is then the intercept, i.e. n_0.

```
1   > meand.ide <- 15.0284
2   > (q <- 1 + w / meand.ide)
3   [1] 1.3327
4   > (lambda <- log(q) / w)
5   [1] 0.0574418
6   > (ni <- q^(c - seq(1, c, 1)) * n.max)
7   [1] 392.1911 294.2825 220.8163 165.6906 124.3268
8   [6]  93.2892 70.0000
9   > (B <- pi / 40000 * sum(ni * classes[1: c]^2))
10  [1] 31.6207
11  > (n0 <- ni[3] / exp(-lambda * classes[3]))
12  [1] 522.674
```

In line 6, based on n_{max}, the numbers of trees in all other stem-diameter classes are calculated (Eq. (6.11)). Building on these numbers, the equilibrium basal area B is computed in line 9. We see in line 10, that calculated B is reasonably close to the published target equilibrium basal area. It seems not necessary to additionally apply the BDq approach (cf. Section 6.3.1.1) in order to ensure an equilibrium basal area of 30 m^2. However, theoretically a 'correction', if desired, is possible using the following code:

```
1   > (k <- pi /40000 * sum(q^(c - seq(1, c, 1)) *
2   + classes[1: c]^2)) # Cancino & Gadow (2002)
3   [1] 0.451724
4   > (n.max <- 30 / k)
5   [1] 66.4122
6   > (ni <- q^(c - seq(1, c, 1)) * n.max)
7   [1] 372.0894 279.1991 209.4984 157.1982 117.9544
8   [6]  88.5077 66.4122
9   > (lambda <- log(q) / w)
10  [1] 0.0574418
11  > (n0 <- ni[3] / exp(-lambda * classes[3]))
12  [1] 495.885
13  > (q <- (n0 * exp(-lambda * 10)) / (n0 *
14  + exp(-lambda * 15)))
15  [1] 1.3327
16  > exp(abs(lambda) * w)
17  [1] 1.3327
18  > (B <- pi / 40000 * sum(ni * classes[1: c]^2))
19  [1] 30
```

First, auxiliary parameter k is calculated according to Eq. (6.14) in lines 1 and 2. Next, a new value of n_{max} is computed in line 4 using the desired equilibrium basal area of $B = 30$ m^2.

Clearly the result in line 5 differs from the anticipated value of 70 trees per hectare (cf. Table 6.3). Based on the new value of n_{max}, the numbers of trees in all other stem-diameter classes are calculated in line 6, cf. lines 7f., and Table 6.3. New model parameters λ and n_0 are calculated following Eqs. (6.8) and (6.17), respectively. In line 11, index value 3 was arbitrarily chosen. An index value of 5 would have given the same results. Comparing the results in lines 10 and 12 with lines 5 and 12 of the previous listing yields that λ has

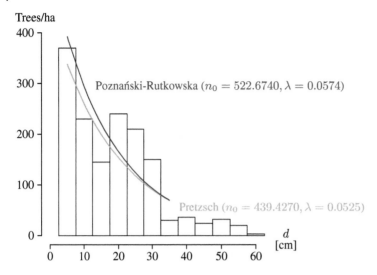

Figure 6.4 Observed stem-diameter (*d*) distribution of a hypothetical forest stand (histogram) along with the simulated stem-diameter equilibrium guide curves according to the Pretzsch (2009, green) and the modified Poznański and Rutkowska (1997, blue) approaches. With both model approaches, $B \approx 30$ m^2/ha, $d_{max} = 35$ cm and $w = 5$ cm. For the Pretzsch approach, $q = 1.30$, whilst for the modified Poznański-Rutkowska approach, $q = 1.33$.

remained unchanged, whilst n_0 has markedly changed after applying the BDq approach. Since the value of model parameter λ has not changed, the q factor has also remained the same. The new equilibrium basal area is calculated in line 18 and the result now corresponds exactly with the published target. However, as previously discussed, the difference between the guideline curve resulting from the original, modified Poznański-Rutkowska approach and the one obtained from the BDq approach (see previous listing) is rather marginal.

The differences between the two q-factor model approaches become clearer when visualised in a graph (Figure 6.4). The Pretzsch variant appears to recommend slightly less regeneration trees than the modified Poznański-Rutkowska approach, but otherwise the differences are rather small and decrease with increasing tree size. The modified Poznański-Rutkowska approach adheres more closely to the observed stem-diameter distribution, which – as discussed before – is not necessarily an advantage.

The guide curves in Figure 6.4 suggest that all trees larger than 35 cm are successively cut. This will provide space, i.e. light, water and nutrients, for regeneration to appear and ensures the long-term self-sustainability of this forest stand. It will also be necessary to cut some trees between 20 and 30 cm that currently form the mid-storey.

Model estimates are usually more accurate when the empirical stem-diameter distribution is not truncated at the lower range of stem diameters, i.e. when small trees are recorded as much as possible. Chevrou (1990) gives estimators for truncated stem-diameter distributions.

6.3.2.3 Susmel's *q* Factor Model

Susmel (1956) intriguingly suggested an approach to estimating q from potential stand dominant or top height, h_{pot}:

$$q = \frac{a}{\sqrt[3]{h_{pot}}} \qquad (6.23)$$

In Eq. (6.23), h_{pot} is the maximum stand height a forest can reach on a given site. Top height usually is calculated as the height corresponding to the quadratic mean diameter of the 100 largest trees per hectare. Potential top height can be gathered from old-growth stands and can be interpreted as a measure of site productivity (Virgilietti and Buongiorno, 1997). Parameter a depends on the woodland community studied and on environmental factors.

Example model parameters are given in Table 6.4 (Susmel, 1980). From Eq. (6.23) with $a = 4.3$, for example, it follows that for $q = 1.3$, $h_{pot} = \left(\frac{4.3}{1.3}\right)^3 = 36.2$ m. All other stand characteristics are expressed as a function of potential dominant or top stand height, see Table 6.4. This is an allometry-based approach for predicting forest structure and tree size distribution (Anfodillo et al., 2013). Obviously, the parameter values depend on the species and on the environmental conditions involved and they cannot be generalised.

In practical applications, first h_{pot} is estimated. Then q is calculated using Eq. (6.23). Next, as explained, d_{max} is calculated from h_{pot} (Table 6.4) and in the corresponding largest diameter class, n_{max} is set to include only one tree, i.e. $n_{max} = 1$ tree per hectare, see the red number in Table 6.5.

Now n_0 can be calculated from Eq. (6.13) with $n_{max} = 1$ and the diameter class corresponding to the d_{max} (Virgilietti and Buongiorno, 1997). The authors proposed calculating all remaining n_i' from q and n_0 through

$$n_i' = n_0 \times \left(\frac{1}{q}\right)^i. \qquad (6.24)$$

Table 6.4 Model parameters of Susmel's q factor model relating to different Italian woodland communities (Susmel, 1980). a – Parameter for calculating q from potential top height, h_{pot}, in Eq. (6.23).

Woodland community	a	$N^{a)}$	G	V	d_{max}
(Mixed) *Picea abies*, *Abies alba*	4.3	~300	$0.96 \times h_{pot}$	$0.33 \times h_{pot}^2$	$2.64 \times h_{pot}$
(Mixed) *Fagus sylvatica*	4.5	210–240	$0.73 \times h_{pot}$	$0.24 \times h_{pot}^2$	$2.30 \times h_{pot}$
Quercus ilex	3.5	~400	$2.35 \times h_{pot}$	$1.50 \times h_{pot}^2$	$5.60 \times h_{pot}$
Mesophilic mixed *Quercus* spec.	4.1	160–200	$0.50 \times h_{pot}$	$0.25 \times h_{pot}^2$	$2.00 \times h_{pot}$

N – trees per hectare, G – Basal area in m² per hectare, V stem volume in m³ per hectare, d_{max} – maximum stem diameter in cm, e.g. target diameter. The authors of scientific tree names were omitted here for clarity.
a)   The number of trees per hectare only includes trees with $d > 17.5$ cm.

Table 6.5 Demographic equilibrium provisional and final numbers of trees per hectare, n'_i and n_i, respectively, and the corresponding basal area, g'_i and g_i.

i	Class	n'_i	g'_i	n_i	g_i
1	5	99.52	0.195	254.46	0.500
2	10	62.82	0.493	160.63	1.262
3	15	39.66	0.701	101.40	1.792
4	20	25.03	0.786	64.01	2.011
5	25	15.80	0.776	40.41	1.983
6	30	9.98	0.705	25.51	1.803
7	35	6.30	0.606	16.10	1.549
8	40	3.98	0.500	10.16	1.277
9	45	2.51	0.399	6.42	1.020
10	50	1.58	0.311	4.05	0.795
11	55	1.00	0.238	2.56	0.607
Sum		268.18	5.710	685.70	14.600
B		14.60			
d_{max}		52.8			
n_0		403.096			
λ		0.092			
q		1.58			

The values were calculated according to Susmel (1956) from Eqs. (6.23), (6.13) and (6.24) for a potential stand top height of 20 m. i – index variable as used in Eq. (6.24), Class – stem-diameter class in cm, B – equilibrium basal area per hectare calculated as $B = 0.73 \times h_{pot}$ in m² ha^{-1}, d_{max} – maximum diameter obtained from $d_{max} = 2.64 \times h_{pot}$ in cm, n_0 – first parameter of Eq. (6.2) calculated from Eq. (6.13), λ – second parameter of Eq. (6.2) calculated from Eq. (6.3), q - q factor calculated from Eq. (6.23) with $a = 4.3$, cf. Figure 6.5.

Eq. (6.24) is in principle only an application of Eq. (6.9), since n_0 is a theoretical form of expressing n_{min} for infinitesimally small d. Here, n'_i is a *provisional* or *unitary* number of trees. Alternatively, n'_i can be propagated from $n_{max} = 1$ by simply using Eq. (6.5).

In a similar way, basal area is first provisionally calculated for each diameter class as $g'_i = n'_i \times \pi \times \left(\frac{d_i}{2}\right)^2$. The g'_i calculated for each class i are then summed over all classes. In a second step, the correct basal area per class is obtained by dividing equilibrium basal area B by the sum of class basal areas of the provisional distribution resulting in a weight or expansion factor as mentioned in Section 6.3.1.1. The basal area g'_i in each class i is now multiplied by this weight, i.e. $g_i = g'_i \times \frac{B}{\sum g'_i}$. The same weighting is now applied to n'_i resulting in the final numbers of trees per hectare, $n_i = n'_i \times \frac{B}{\sum g'_i}$. This procedure of prorating a

Figure 6.5 Equilibrium curves for five different values of h_{pot} produced using the Susmel q factor model (Susmel, 1956) with $a = 4.3$ for mixed *Picea abies* (L.) H. Karst. and *Abies alba* Mill. using Eqs. (6.23), (6.13) and (6.24), cf. Table 6.5. Stem diameter is denoted as d.

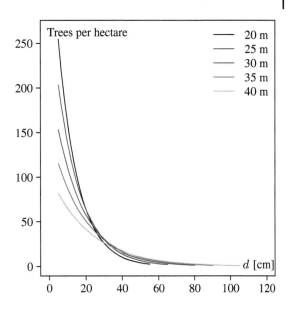

desired basal area over stem-diameter classes was also described in detail by Guldin (1991), and the results are approximately equivalent to those obtained from applying Eqs. (6.14) and (6.16) introduced in Section 6.3.1.1.

Table 6.5 gives the results for $h_{pot} = 20$ m and can easily be re-calculated for didactic purposes using the equations provided in this section. Susmel's model approach is clearly an interesting way of making q dependent on environmental conditions and deserves more attention. Assuming that the Susmel approach is not far off reality, the synthesis in Figure 6.5 suggests that with decreasing potential stand height (and thus with decreasing site fertility) the equilibrium guide curve for small stem diameters moves upwards thus creating a layered structure. For a potential height of 20 m, the corresponding equilibrium curve intersects all other curves at values of d between 20 and 30 cm.

The layered structure based on h_{pot} can also be gauged from the values of model parameter n_0 which are 403.096, 298.881, 211.834, 151.185 and 102.760 for potential top heights 20, 25, 30, 35 and 40 m, respectively. Apparently Susmel (1956) proposed that the theoretical equilibrium strongly depends on environmental factors. For *Abies alba* forests in the Świętokrzyskie Mountains (Poland), Bruchwald et al. (2016) found that the q factor is related to the *site index*, an indicator of fertility calculated (usually in plantations) as the stand top height at a certain base stand age. The q factor decreased with increasing site index, i.e. with increasing fertility.

The R code for calculating the Susmel model is provided below. In lines 1–2, parameter a, and the four potential top height values are assigned. This is followed by the calculation of the four q factors (Eq. (6.23)) in line 3.

```
1  > a <- 4.3
2  > h100 <- seq(20, 40, 5)
3  > q <- a / h100^(1 / 3)
4  > w <- 5
5  > dmax <- 2.64 * h100
```

```
 6  > dsequence <- seq(2.5, 107.5, w)
 7  > c <- cut(dmax, breaks = dsequence, include.lowest = T,
 8  + right = F)
 9  > c <- as.numeric(c)
10  > dclass <- dsequence + w / 2
11  > n0 <- c()
12  > for (i in 1: length(h100)) # provisional n0
13  +   n0[i] <- 1.0 / exp(-(log(q[i]) / w)*dclass[c[i]])
14  > ni.u <- list()
15  > for (i in 1: length(h100)) # provisional ni
16  +   ni.u[[i]] <- n0[i] * (1 / q[i])^seq(1, c[i], 1)
17  > B <- 0.73 * h100 # stand basal area
18  > gi.u <- gi <- list()
19  > for (i in 1: length(h100)) # provisional gi
20  +   gi.u[[i]] <- ni.u[[i]] *(pi *(seq(5, dclass[c[i]], w)/
21  +   200)^2)
22  > sum.giu <- c()
23  > for (i in 1: length(h100)) # sum of provisional gi
24  +   sum.giu[i] <- sum(ni.u[[i]]*(pi *(seq(5,dclass[c[i]],
25  +   w) / 200)^2))
26  > for (i in 1: length(h100)) # final gi
27  +   gi[[i]] <- B[i] * gi.u[[i]] / sum.giu[i]
28  > ni <- list()
29  > for (i in 1: length(h100)) { # final ni
30  +   ni[[i]] <- B[i] * ni.u[[i]] / sum.giu[i]
31  +   n.max[i] <- ni[[i]][c[i]]
32  + }
33  > lambda <- log(q) / w
34  > for (i in 1: length(h100)) # final n0
35  +   n0[i] <- n.max[i] / exp(-(log(q[i]) / w) *
36  +   dclass[c[i]])
```

The bin width of the empirical diameter distribution is set in line 4 and in line 5, maximum tree diameters are calculated for all four potential top heights. Then the empirical stem diameter distribution is set up in line 6. The indices of the maximum classes, c, of the equilibrium distributions are computed in lines 7–9, followed by the calculation of the class mid-diameters in line 10. Provisional n_0 can now be calculated in lines 11–13 from Eq. (6.13) with $n_{max} = 1$ and the diameter class corresponding to the d_{max}. Based on Eq. (6.24), the unitary or provisional numbers of trees per hectare, n'_i, are calculated in lines 14–16. The variable name used for them is ni.u. Note that we applied a list object here (cf. line 14) instead of a vector, as the n'_i relating to different h_{pot} are of different lengths. Equilibrium stand basal area is computed in line 17. Unitary/provisional and final class basal areas are initialised in line 18. Then provisional g'_i (coded as gi.u) are calculated in lines 19–21 followed by the calculation of the sums of provisional g'_i in lines 22–25. With this information, final g_i and n_i can be computed in lines 26–32. Finally, model parameters λ and final n_0 are calculated in lines 33–36.

6.3.2.4 Sterba Criterion

Sterba (pers. comm.) suggested taking the difference in stand basal area or stand basal area increment between two survey periods, ΔG, calculated as mean annual difference, and the proportion of stand basal area to be harvested, L, into account for estimating a suitable q factor:

$$L = \frac{n_{max} \times d_{max}^2 \times \frac{\pi}{4}}{\Delta G} \tag{6.25}$$

In Eq. (6.25), $L = \frac{G_{max}}{\Delta G}$, i.e. the proportion of basal areas per hectare in the largest diameter class in terms of 'stand increment'. For a stand in equilibrium, ΔG should be approximately equal to the basal area harvested. In most cases, L can be set to a value of 1.0 assuming that all trees (= 100%) in the largest diameter class (corresponding to the target diameter) are to be felled (Roloff, 2004). Using Eq. (6.13) and rearranging Eq. (6.25) yields

$$q = e^{-\frac{\log_e\left(\frac{L \times \Delta G}{n_0 \times d_{max}^2 \times \frac{\pi}{4}}\right)}{d_{max}} \times w}. \tag{6.26}$$

Based on stand basal area growth and corresponding harvesting proportion, Eq. (6.26) allows an optimised q factor to be calculated. The equation implies that for the same values of L, d_{max}, ΔG and w, different n_0 lead to different q. There is also the possibility to derive a similar equation based on the BDq approach and Eqs. (6.14) and (6.16). In both cases, q is much less arbitrary than other choices reviewed in this section and the q factor model thus takes on board elements of dynamic equilibrium models.

6.4 Dynamic Equilibrium Models

Demographic processes in tree populations that are expressed in empirical stem-diameter distributions involve growth, mortality and resulting migration throughout the classes (Schütz, 2006), see Figure 6.3. The theory of static equilibrium models (cf. Section 6.3.1) does not explicitly take these processes into account, and it is therefore interesting to consider a theory that does. The first truly demographic equilibrium modelling approach involving growth and mortality was proposed by Prodan (1949). Hanewinkel (1998) provided a good account of the history of dynamic equilibrium modelling. Kärenlampi (2019) reported using a model approach similar to the Prodan model, however, based on different growth and mortality functions.

6.4.1 Model Theory

Demographic equilibrium is achieved, if the number of trees growing into any size class (ingrowth) equals outgrowth plus losses due to natural mortality and forest interventions. Essentially, tree migration through size classes is a function of growth and mortality (Schütz, 2006; Schütz and Pommerening, 2013). Mathematically, these key processes can be expressed as follows:

(a) The *outgrowth rate* is denoted as p_i and n_i, as before, is the number of trees in class i. p_i describes the rate of trees moving on from class i to class $i + 1$ and directly depends on absolute growth rate (AGR) of arbitrary growth variable y, in the most common application to tree stem diameters on mean annual stem-diameter AGR $\overline{\overline{(\delta y)}}_i$. The outgrowth rate can be calculated as

$$p_i = \frac{\overline{\overline{(\delta y)}}_i}{w}, \tag{6.27}$$

i.e. mean annual AGR divided by the width of the stem diameter distribution. Mean annual AGR is defined as

$$\overline{\overline{(\delta y)}}_i = \frac{1}{n_i} \sum_{j=1}^{n_i} \frac{y_{j,t} - y_{j,t-\Delta t}}{\Delta t}, \tag{6.28}$$

where t and $t - \Delta t$ describe two consecutive survey years in which the trees of a stand were measured. y denotes the size variable, i.e. stem diameter in our application. Index i denotes size class i and index j denotes tree j, i.e. the mean is calculated both in terms of the time interval and the number of trees in each diameter class. However, it is also possible to use individual-tree growth rates in the modelling process rather than averaging over the trees in each class. Thus migration into and out of class i can be quantified as $n_{i-1} \times p_{i-1}$ (ingrowth) and $n_i \times p_i$ (outgrowth), respectively.

(b) The rate of tree mortality in class i in any given year is m_i. Thus, the number of trees dying of natural causes and/or those that are felled corresponds to $n_i \times m_i$.

In reality, both p_i and m_i result from model functions as we will see in Sections 6.4.3 and 6.4.4. An empirical stem-diameter distribution is balanced or in equilibrium, if ingrowth into any diameter class equals outgrowth plus losses from natural mortality and/or forest management. Based on this equilibrium concept, Prodan (1949) proposed the following equation for calculating the number of trees in successive stem-diameter classes:

$$\underbrace{n_{i-1}}_{\text{Number of trees in class } i-1} = \underbrace{\frac{n_i \times m_i \times w}{\overline{\overline{(\delta y)}}_{i-1}}}_{\text{Mortality in class } i} + \underbrace{\frac{n_i \times \overline{\overline{(\delta y)}}_i}{\overline{\overline{(\delta y)}}_{i-1}}}_{\text{Outgrowth towards class } i+1} \tag{6.29}$$

Eq. (6.29) answers the question of how many trees are necessary in class $i - 1$ to compensate for the loss of trees in class i due to outgrowth and mortality. Schütz (2001b) essentially arrived at the same results but based his equilibrium model on the following general demographic steady state equation for class i, cf. Figure 6.3:

$$\underbrace{n_{i-1} \times p_{i-1}}_{\text{Ingrowth from class } i-1} = \underbrace{n_i \times m_i}_{\text{Mortality in class } i} + \underbrace{n_i \times p_i}_{\text{Outgrowth towards class } i+1} \tag{6.30}$$

Brzeziecki et al. (2016) introduced reduction terms to account for the mortality of trees which moved out of classes $i - 1$ and i during the observation period, i.e. the time between two successive tree surveys, but then died before the end of that period, see also Brzeziecki et al. (2021):

$$\underbrace{(1 - m_{i-1})}_{\text{Mortality reduction term}} \times n_{i-1} \times p_{i-1} = \underbrace{(1 - m_i)}_{\text{Mortality reduction term}} \times n_i \times m_i + n_i \times p_i \tag{6.31}$$

For simplicity and since the difference between the last two approaches is small, we will continue with the original Schütz approach. To establish a demographically sustainable stem-diameter distribution, a simple transformation of Eq. (6.30) leads to Eq. (6.32), which can be used to calculate the number of trees in successively larger diameter classes:

$$n_{i+1} = \frac{n_i \times p_i}{p_{i+1} + m_{i+1}} \tag{6.32}$$

Alternatively, the number of trees in successively smaller diameter classes can be computed from Eq. (6.33).

$$n_{i-1} = \frac{n_i \times (p_i + m_i)}{p_{i-1}} \tag{6.33}$$

Starting with n_{min} or n_{max}, the equilibrium tree numbers of every stem-diameter class can be modelled according to Eqs. (6.32) or (6.33). The equilibrium has been identified when the number of trees in the smallest diameter class corresponds to field observations and ensures appropriate demographic proportions across the range of classes (Schütz and Pommerening, 2013), see Sections 6.4.2 and 6.4.3.3. The result is not a smooth decreasing curve, but one with different rates of decline for different stem-diameter ranges (Schütz, 2006), cf. Figure 6.11. An R implementation of this model frame is provided in Section 6.4.4.

6.4.2 Ultimate Equilibrium Conditions

In addition to the condition of demographic sustainability ensured by Eq. (6.30), there are two more equilibrium conditions which in the practical model application markedly help identify suitable guideline curves from among many candidates (Schütz, 2001b):

(1) The number of trees n_{min} in the smallest stem-diameter class needs to be realistic in the sense that it reflects currently observed and achievable numbers without being unnecessarily high. Naturally, n_{min} also depends on the size of the mid-diameter of the first class. Information on n_{min} is often hard to come by and should be partly derived from current observations at the given site and partly derived from similar woodland communities, where the trees have already formed an equilibrium or are close to one. For situations where observations from multiple plots in the same forest are available, Schütz (2006) presented a simulation method for determining n_{min}.

(2) The total amount of mortality, which can be derived from mortality rate m, basal area g and tree numbers n, should approximately equate to the cumulative growth rates, see Section 6.4.3.3:

$$\sum_{i=1}^{c} m_i \times g_i \times n_i \approx \sum_{i=1}^{c} p_i \times n_i \times (\delta g)_i \tag{6.34}$$

In Eq. (6.34), g_i is the basal area corresponding to the mid-diameter of each diameter class, i.e. $g_i = \pi \times \left(\frac{d_i}{200}\right)^2$, n_i is the number of trees per hectare in class i, whilst m_i is the corresponding mortality rate. $(\delta g)_i$ is the difference in basal area between classes $i + 1$ and i, i.e. between the next larger and the current class. If there are larger differences between both terms of Eq. (6.34), the parameters of the mortality function and the number of trees in the largest stem-diameter class are not in harmony and need to be iteratively adapted, see Section 6.4.4. Instead of basal area g it is also possible to work with stem volume v, but since this involves another modelling step we stick with basal area here to be closer to the observed data.

6.4.3 Modelling Growth, Mortality and n_{min}

The model theory presented in Section 6.4.1 provides only the frame of the Schütz and other dynamic models. This frame, although undoubtedly important, crucially depends on separate, underlying functions of stem-diameter growth δd and mortality m which are mainly dependent on tree size (Figure 6.6), particularly on stem diameter, but can also include other explanatory variables.

In principle, almost any growth or mortality function known from other work in tree and forest modelling (cf. Burkhart, 2012; Weiskittel et al., 2011) including those that depict eco-physiological processes (cf. Mäkelä and Valentine, 2020) are suitable. Important, however, is that the models involved have the same temporal resolution, i.e. that they provide rates that apply to the same time period, e.g. one year or five years. Both growth and mortality can be modelled based on continuous explanatory variables without any reference to class widths of the empirical stem-diameter distribution. Finally, the number of trees per hectare in the smallest stem-diameter class, n_{min}, needs to be determined to firmly anchor the equilibrium model in the empirical stem-diameter distribution. In the following sections, we review a few functions that have been published in conjunction with demographic modelling.

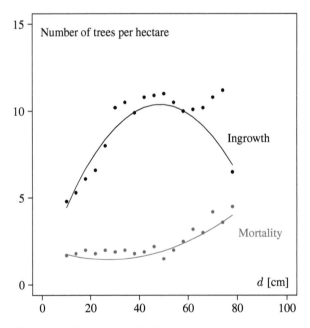

Figure 6.6 Typical example of observed and simulated annual ingrowth and mortality and their dependency on tree size, here stem diameter d. This information forms a crucial part of dynamic equilibrium models. Source: Adapted from Sterba and Zingg (2001).

6.4.3.1 Modelling Growth

Depending on the data available and the species composition of the forest stand, growth functions can be species specific or general. In the former case, separate growth functions are modelled for each species or species group. Otherwise, no distinction between different species is made.

Basic growth functions include those that are solely dependent on tree size, i.e. $\delta d = f(d)$. Brzeziecki et al. (2016), for example, estimated the mean annual absolute stem-diameter growth rate by means of the first derivative of the Chapman–Richards growth function replacing time/age with stem diameter d:

$$\delta d = a_1 + b_1 \times c_1 \cdot e^{-b_1 \times d} \times (1 - e^{-b_1 \times d})^{c_1 - 1} \tag{6.35}$$

In Eq. (6.35), a_1, b_1 and c_1 are model parameters. There are many alternative size-dependent growth functions, cf. Pommerening and Grabarnik (2019, Chapter 5). Schütz (2006) considered *basal area of larger trees* (BAL), also referred to as overtopping basal area, which is an expression of the relative size dominance of a tree in a population and is closely related to the concept of basal area percentile (Wykoff et al., 1982; Wykoff, 1990) as a non-spatial competition index. Schütz and Pommerening (2013) additionally tested a spatial competition index as explanatory variable to be included in the growth function against BAL. In these publications, the authors used both BAL and tree size for estimating tree diameter growth, i.e. this is an example of the relationship $f(d, \text{competition})$:

$$\delta d = a_2 + b_2 \times \log_e(d) + c_2 \times \text{BAL}^u \tag{6.36}$$

In Eq. (6.36), a_2, b_2 and c_2 are again model parameters, $\log_e$ is the natural logarithm and u is an exponent. Schütz (2006) used $u = 3$, whilst Schütz and Pommerening (2013) decided to work with $u = 2$. It is probably best to treat u as another model parameter and to identify its value through nonlinear regression. In a different context, Quicke et al. (1994) suggested a similar model:

$$\delta d = a_3 \cdot e^{b_3 \times d + c_3 \times \text{BAL}} \tag{6.37}$$

The sum of tree height segments (SAFC) within a search cone constructed at the tip of a subject tree using a base angle of $\alpha = 55°$ was used as a spatial competition index (Schütz and Pommerening, 2013). No size term was required in that growth estimation, i.e. this is an example of the relationship $f(\text{competition})$:

$$\delta d = a_4 + b_4 \times \text{SAFC} \tag{6.38}$$

Also here, a_4 and b_4 are model parameters. There are many more possibilities which include using existing growth functions from tree growth simulators. It is also possible to consider relative growth rates (RGR), see Pommerening and Grabarnik (2019, Chapter 6), instead of AGR.

Basal Area of Larger Trees Since the competition index basal area of larger trees has traditionally played a crucial role in the Schütz model, this measure is explained in greater detail in this paragraph. BAL is based on the *basal-area percentile*, r_i, which is a non-spatial measure of relative dominance of a tree in the basal area distribution of a forest stand and is calculated for tree i according to Eq. (6.39).

$$r_i = \frac{1}{G} \sum_{j=1}^{n} g_j \text{ for } g_j \leq g_i \tag{6.39}$$

Here G is the sum of all individual-tree basal-areas g_i. Finally, competition index BAL for tree i is calculated as

$$\text{BAL}_i = G \times (1 - r_i). \tag{6.40}$$

Thus, BAL is the sum of basal areas of all trees with a stem diameter larger than that of subject tree i (Wykoff, 1990). Usually, the measure is calculated in m^2 ha^{-1} for better comparison between stands. BAL is related to available light, water and nutrients, since with increasing basal area of larger, i.e. with more dominant trees, there are less resources available for smaller trees. BAL is a simple and effective measure that simultaneously considers relative dominance of a tree and density. It is very flexible and can easily be modified to a spatially explicit measure of competition by calculating it specifically for an influence zone around a tree. Basal area of larger trees is also very suitable for trees in small-sized sample plots. Naturally, it is also possible to distinguish between species and inter- and intraspecific versions of BAL.

Considering the example data in Table 6.6, total basal area can be calculated as $G = \frac{0.752 \, m^2}{0.025 \, ha} = 30.08 \, m^2 \, ha^{-1}$. Using Eq. (6.39), the basal-area percentile of tree #4 can be computed as $r_4 = \frac{0.031 + 0.033 + 0.042 + 0.057 \, m^2}{0.752 \, m^2} = \frac{0.163}{0.752} = 0.217$. Finally, BAL for tree #4 is calculated as $\text{BAL}_4 = 30.08 \cdot (1 - 0.217) = 23.55 \, m^2 \, ha^{-1}$.

Table 6.6 Hypothetical example of a small tree arrangement of ten trees within an area of 0.025 ha. d_i – stem diameter of tree i, g_i – basal area of tree i.

i	1	2	3	4	5	6	7	8	9	10
d_i [cm]	20	21	23	27	31	32	33	37	37	41
g_i [m²]	0.031	0.033	0.042	0.057	0.075	0.080	0.086	0.108	0.108	0.132

Source: Adapted from Gadow et al. (2021).

In contrast to the previous example, Schütz (pers. commun.) preferred calculating BAL and fitting Eq. (6.36) based on summary or tabular data, i.e. data classified according to the empirical stem-diameter distribution where the tree data are summarised in diameter classes. This preference was based on the fact that demographic modelling deals with summary data and that calculations based on individual-tree measurements produce much statistical noise which would obscure important trends.

Table 6.7 gives an impression of how BAL is calculated for tabular data: For example, BAL for the trees in diameter class 5 is calculated as the sum of g_i from class 10 to class 35, i.e. $2.04 + 3.53 + 4.83 + 5.81 + 6.43 + 6.73 = 29.38$. Consequently, for BAL pertaining to class

Table 6.7 Tree-number (n_i), basal-area (g_i) and BAL per hectare by stem diameter classes of a simulated equilibrium according to Pretzsch (2009, P), cf. Table 6.3.

Class	Sim. $\hat{n}_i$ (P)	Sim. $\hat{g}_i$ (P)	BAL
5	338	0.66	29.38
10	260	2.04	27.34
15	200	3.53	23.81
20	154	4.83	18.98
25	118	5.81	13.17
30	91	6.43	6.73
35	70	6.73	0
Total	1230	30	–

Diameter classes are in cm, basal area, and BAL in m².

10 we consider the cumulative sum of g_i from class 15 to 35. There are no trees larger than those in class 35, therefore BAL = 0 here. In R, BAL can be simply calculated using the cumsum() function and applying it to the basal area vector.

These calculations can quickly be carried out for any forest stand where the stem diameters have been re-measured at least once so that the absolute annual growth rate can be established according to Eq. (6.28). Two examples for two different species are given in Figure 6.7. The relationship between basal area of larger trees and stem-diameter AGR has been quantified by Schütz (2001b), Schütz (2006) and Schütz and Pommerening (2013) based on the model

$$\delta d = a_s + b_s \times \text{BAL}^{c_s}. \tag{6.41}$$

Model parameter a_s is the intercept, i.e. here the model curve intersects with the ordinate. This value also represents the maximum growth rate as far as the model is concerned.

In analogy to the light compensation point, the *growth compensation point*, i.e. the basal area density, where $\delta d = 0$, can be calculated from $\sqrt[c_s]{\frac{a_s}{b_s}}$. In some cases, the relationship described in Eq. (6.41) is so strong that this equation is sufficient to explain δd without the inclusion of stem diameter d.

Figure 6.7 tells us that given the same BAL *P. sitchensis* grows more prolifically than *F. sylvatica*, something that people who know the two species would expect. It becomes also clear that *P. sitchensis*, a species with intermediate light requirements, responds more to a density increase than shade-tolerant *F. sylvatica*. This is expressed by model parameter c_s, i.e. 1.069 versus 1.413. The larger c_s, the more convex is the curve and the more retarded is the decrease in AGR, and the stronger is the saturation effect towards the lower range of BAL. However, the growth compensation point is higher for *P. sitchensis* than for *F. sylvatica*.

If data from more than one re-measurement are available, the model curves of Eq. (6.41) and Figure 6.7 can be computed separately for each re-measurement and the resulting data points can then be pooled in a single nonlinear regression before using it in the

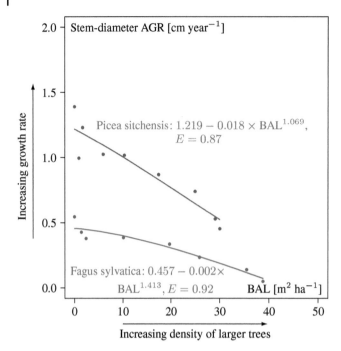

Figure 6.7 The relationship between BAL and stem-diameter AGR in a 56-years old plantation stand of *Picea sitchensis* (BONG.) CARR. at Clocaenog Forest (UK, Tyfiant Coed plot 1) and in a 142-years old, semi-natural *F. sylvatica* L. stand at Embrach (Switzerland, plot 41-007). The corresponding growth compensation points are 50.8 and 43.9 m² ha⁻¹, respectively. BAL was calculated from 4-cm diameter classes as shown in Table 6.7. *E* – efficiency[2].

equilibrium model, unless it is intended to estimate equilibrium models separately for each survey period.

6.4.3.2 Modelling Mortality

In a way similar to modelling stem-diameter growth (cf. in Section 6.4.3.1), any appropriate mortality function published in the literature (Burkhart, 2012; Weiskittel et al., 2011) can in theory be selected as long as the data available can satisfy the required independent explanatory variables. Like with growth functions, mortality models can be general or species specific. If cause of death was recorded in the original data, separate models for natural mortality and human interventions can be useful. Mortality models typically are of the form $m = f(d)$, i.e. the death process depends on tree size. However, it is also possible to include competition and environmental factors as explanatory variables. Observed mortality rates are usually measured in terms of the ratio of dead and total number of trees per stem-diameter class.

---

2 Efficiency is defined as $E = 1 - \frac{\sum_{i=1}^{n} (\hat{y}_i - y_i)^2}{\sum_{i=1}^{n} (y_i - \bar{y})^2}$, where $\hat{y}_i$ is the ith prediction of AGR, y_i is the ith observation and $\bar{y}$ is the mean observation of AGR. E approaches one with improving model performance. A value of zero indicates that the model explains no more variation than the mean value of the observations alone and negative values highlight biased estimates.

Schütz (2006) and Schütz and Pommerening (2013) decided to model mortality based on a polynomial to degree 3, where all tree deaths regardless of cause were pooled:

$$m = a_6 + b_6 \times d + c_6 \times d^2 + g_6 \times d^3 \tag{6.42}$$

In Eq. (6.42), a_6, b_6, c_6 and g_6 are again model parameters. Another good model for describing mortality rates was considered by Brzeziecki et al. (2016) whilst working on their paper:

$$m = c_7 \times (d - a_7)^2 + b_7 \tag{6.43}$$

Model parameters a_7 and b_7 determine maxima and minima whilst parameter c_7 controls the slope of the curve. Exponent 2 can potentially also be turned into a model parameter. Eventually, Brzeziecki et al. (2016) used a logistic regression model in their published study instead:

$$P(m) = \frac{e^{a_8 + b_8 \times d + c_8 \times d^2}}{1 + e^{a_8 + b_8 \times d + c_8 \times d^2}} \tag{6.44}$$

Here $P(m)$ is the mortality probability and e is again the base of the natural logarithm, whilst a_8, b_8 and c_8 are model parameters. As with modelling growth, there are, of course, many more possibilities.

6.4.3.3 Modelling n_{min}

To model the equilibrium stem-diameter distribution correctly, it is crucial to determine the number of trees in the smallest size class, i.e. n_{min}. This number should correspond to realistic, ideally observed recruitment conditions (Schütz, 2001b, 2006). This is one of the ultimate sustainability conditions, see Section 6.4.2. Later, when assembling the final model, input variable n_{max} is chosen so that Eq. (6.33) eventually delivers the correct, pre-determined n_{min}. Therefore, n_{min} must be based on a realistic estimate.

For determining n_{min}, Schütz (2006) proposed simulating different pairs of BAL and n_{min} (from Eq. (6.33)) by inputting a number of arbitrary n_{max} into the Schütz model after

Figure 6.8 Observed (red) and simulated (blue) relationships between tree numbers in the smallest size class, n_{min} and stand density measured in terms of basal area of larger trees (BAL) for *F. sylvatica* L. selection forests in Thuringia (Germany). Source: Adapted from Schütz (2006).

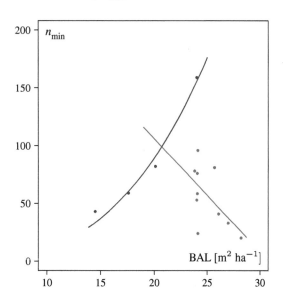

the finalised growth and mortality models have been implemented. These simulations are represented by the blue data points and the blue exponential trend curve in Figure 6.8. The simulation results are overlaid by BAL–n_{min} data from observations in different plots (purple data points) or from one plot and different survey years in the same area and with the same species composition for which the equilibrium is modelled. In this context, BAL is the basal area of larger trees that – according to model theory – all trees in class d_{min} are collectively exposed to. Such data often show a linear trend and using linear regression, a trend line is put through the data points. Extrapolating that purple trend line (in Figure 6.8) allows the point of intersection with the simulated curve to be identified.

This point of intersection (here at BAL = 21.99 m^2) gives n_{min} = 99.16 trees per hectare, i.e. the value to be eventually achieved by the equilibrium model. In our experience, values of n_{min} typically are between 70 and 250 trees per hectare.

6.4.4 Example Application of the Schütz Model

Since the Schütz model is more complex than the q factor model a detailed example is offered in this section. The example data were taken from the *Pseudotsuga menziesii* selection forest at Artist's Wood, see Section 2.2.1. In contrast to the study reported in Schütz and Pommerening (2013), for simplicity, we limited the application to plot 1 and survey periods 2002 and 2007.

First, we calculated annual AGR based on Eq. (6.28) as the arithmetic mean of all trees in each 4-cm stem-diameter class. In the same procedure, we determined the trees that were alive in 2002 but had died or were felled by 2007. In another step, mean AGR was modelled using Eq. (6.41), cf. Figure 6.9a. As shown in Section 6.4.3.1, summarised data were used in the modelling process and efficiency (E = 0.43) was comparatively low due to the high variability of the data.

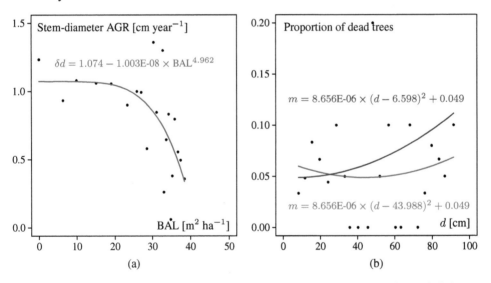

Figure 6.9 Models of mean absolute growth δd (a) and mortality m (b) rates relating to Artist's Wood (plot1, North Wales, UK) based on tree measurements in 2002 and 2007. BAL is basal area of larger trees and d is stem diameter. The fitted models refer to Eqs. (6.41) and (6.43).

Next, we modelled mortality. The observed data showed a typical U shape with high mortality among both small and large trees, whilst mortality was low in the mid-range of the empirical stem-diameter distribution (Figure 6.9b). Statistical noise was considerable in these data. We chose the model described in Eq. (6.43). Efficiency was extremely low here ($E = 0.01$), but the data and relationships in other studies appeared to be represented reasonably well by the resulting model (Schütz, 2001b; Schütz and Pommerening, 2013). Whilst we can possibly assume that the competition-growth patterns will not change much with ongoing selection-system forest management, mortality is likely to change markedly.

Recall that mortality includes both natural mortality and human interventions. Since a target diameter of 76 cm was set for the equilibrium model, in the future, for larger trees approaching 76 cm, mortality rates must be much higher than they were observed in 2002. At the same time, the mortality rate for the smallest trees should be lower than observed in 2002, since more shade-casting, nutrients and water exploiting overstorey trees are removed in future interventions. This trend is well known from previous demographic research (Schütz, 2006; Schütz and Pommerening, 2013).

In Eq. (6.43), in anticipation of a future equilibrium, the model obtained from the data observed in 2002 can be easily modified by reducing model parameter a_7. The reduction factor is determined iteratively by testing a number of candidate factors whilst checking the equilibrium conditions, cf. Section 6.4.2. From the publication by Schütz and Pommerening (2013), we knew that n_{min} should be approximately 70 trees per hectare which allowed us to narrow down n_{max} and the reduction factor. In the case of this example, a_7 in Eq. (6.43) was multiplied by 0.15 resulting in the blue curve in Figure 6.9b. As a result of the modelling work presented here, there is only a 6% difference in the sums on both sides of Eq. (6.34), i.e. basal area growth exceeds mortality by 6%. Starting with an initially arbitrary number of trees in the largest stem-diameter class, n_{max} and then applying Eq. (6.33), gives the Schütz model (see the red line in Figure 6.10). For its supposedly higher accuracy, since it is based on explicit growth and mortality functions, we treated this equilibrium model as reference here, hence, the abbreviation R in Figure 6.10 and Table 6.8. As expected, q_i calculated from successive, simulated n_i according to Eq. (6.3) was not constant but ranged from 1.205 to 1.338 with mean $\bar{q} = 1.247$. Within this range, q_i increased exponentially from the smallest to the largest diameter class.

We also applied nonlinear regression to the results of the Schütz model to obtain parameters n_0 and λ of the de Liocourt model (Eq. (6.2)). Strictly speaking this should not be done, since the Schütz model is completely independent of the de Liocourt model and even does not necessarily need to have a (negative) exponential shape. We did this out of curiosity to see how close the Schütz guideline curve would be to an equivalent de Liocourt guideline curve. In this case, the fit turned out to be perfect and the results of the Schütz model could be summarised by parameters n_0 and λ. A q factor could then be derived from λ according to Eq. (6.8), see Table 6.8. However, for other stands with a different woodland structure the attempt to express the Schütz model in terms of the q factor model may not be as perfect or even fail, and it is definitely a simplification involving a potential loss of information, see also Section 6.6.

q factor models M1–M4 were applied to see how close their curves were to or how much they deviated from that of the Schütz model. As input values for the various estimates involved in the q factor models (see Section 6.3), we used information provided by the

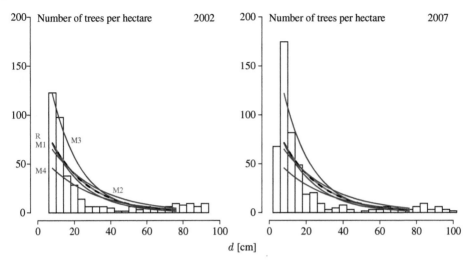

Figure 6.10 Equilibrium guide curves for the mixed *Pseudotsuga menziesii* (MIRB.) FRANCO selection forest at Artist's Wood (plot 1, North Wales, UK) compared to the observed empirical stem-diameter (*d*) distributions in 2002 and 2007. The curves are the same for both survey years and the Schütz model (R) serves as reference for the supposedly less accurate *q* factor models (M1-4). The model equations used and the model parameters are given in Table 6.8.

Schütz reference model, namely n_{max}, n_{min}, d_{max}, d_{min}, B and $\bar{d}$. Normally, as previously discussed, such 'ideal' information about the equilibrium state is not available, but in these example calculations, we deliberately wanted to give the *q* factor models ideal starting conditions to see how they would perform in comparison to the reference model. Using the

Table 6.8 Equations used for the equilibrium guide curves relating to the mixed *Pseudotsuga menziesii* (MIRB.) FRANCO selection forest at Artist's Wood (plot 1, North Wales, UK), cf. Figure 6.10.

Model	Equations	de Liocourt model parameters	k	q
R	6.30, 6.41, 6.43	$n_0 = 105.245$, $\lambda = 0.049$[a]	–	1.216
M1	6.18, 6.13	$n_0 = 109.607$, $\lambda = 0.055$	–	1.246
M2	6.22, 6.8, 6.21	$n_0 = 0.057$, $\lambda = 0.039$[b]	–	1.168
M3	6.20, 6.3, 6.13	$n_0 = 201.89$, $\lambda = 0.063$	–	1.287
M4	6.22, 6.8, 6.14, 6.16, 6.17	$n_0 = 62.165$, $\lambda = 0.039$	7.448	1.168

The equation numbers were arranged in the order of computation. The colours in the first column correspond to the colours of the curves in Figure 6.10. *k* – auxiliary model parameter of the BDq approach (Eq. (6.14)), *q* – *q* factor (Eq. (6.3)). The de Liocourt model is given in Eq. (6.2).
a) Note that the Schütz model (R) by definition is independent of the de Liocourt model and therefore **cannot** be naturally expressed by the corresponding model parameters. In this case, we applied nonlinear regression to the Schütz model results $n_i \sim d_i$ to simplify comparison. The regression results are indicated by the dashed black line in Figure 6.10.
b) This combination of model parameters n_0 and λ produces a probability density function. Therefore, the function values need to be multiplied by the total number of trees per hectare, *N*, and the class width of the empirical diameter distribution, *w*, when plotting it against the other equilibrium curves.

Schütz model results, the arithmetic mean q factor was 1.247, the geometric mean was 1.246, whilst the q factor calculated from Eq. (6.26) using the Sterba criterion was 1.996 assuming $L = 1.0$, i.e. a 100% mortality of trees in class 76 due to target-diameter fellings. Model M1 has the q factor closest to the arithmetic and geometric means, whilst the q factor of the 'q-adapted' Schütz model (R) is closer to that derived from the Sterba criterion (Eq. (6.26)).

We tried a number of combinations of equations from Section 6.3.2 in an attempt to identify the q factor models closest to the Schütz model. These are assembled in Table 6.8 and illustrated in Figure 6.10. Apparently models M1 and M2, loosely following the Pretzsch and Poznański-Rutkowska approaches and detailed in Section 6.3.2, turned out to be closest to the chosen reference. Compared to the Schütz model, M1 anticipates less trees in the mid-storey classes of the diameter distribution, whilst M2 requires slightly less trees in the first two diameter classes but then prescribes more trees from class 20 onwards. Models M3 and M4 deviate rather much from the reference, particularly for small stem-diameter classes, and the trends of these two models can be recognised from the magnitude of the corresponding n_0 parameters. Apparently even under ideal conditions, it is not easy for the static q factor model to match the dynamic Schütz model when using the standard (indirect) parameter-prediction method, although at least two combinations of estimators come very close. The best option to reproduce the Schütz model with the q factor model is direct parameter prediction by applying nonlinear regression to the $n_i \sim d_i$ vectors produced by the reference model. This should, of course, not be confused with applying nonlinear regression to empirical diameter distributions. Correctly applied regression fails to produce appropriate results, if the equilibrium diameter structure is far from exhibiting any (negative) exponential shape. It is also interesting to see that the q factor varies between the models in Table 6.8 – even for those models whose curves are graphically close.

From a silvicultural perspective, guideline models R, M1 and M2 imply that all trees larger than the target diameter of 76 cm should be felled in due course. There was a heavy intervention in 2004, but from the empirical diameter distribution recorded in 2007, we can see that there are still far too many large trees in Artist's Wood (plot 1). These cast shade and take up water and nutrients that are required particularly by the mid-storey trees where we can clearly see a gap that eventually needs filling through ingrowth from lower size classes. Regeneration/ingrowth in the small diameter classes looks satisfactory, the models even suggest less trees here.

The Schütz model arrived at an equilibrium basal area of 24.3 m^2 ha^{-1} and 386 trees per hectare. In 2002, observed stand basal area was 39.1 m^2 ha^{-1}, and there were 400 trees per hectare. By 2007 these numbers had changed to 33.6 m^2 ha^{-1} and 496 trees per hectare. Apparently, the fellings helped reduce basal-area density markedly and the stand responded not only with much regeneration/ingrowth but also with increased basal-area growth.

The key part of the Schütz model algorithm is provided in the following listing. The vectors (= columns in spreadsheet philosophy) n_i, BAL_i, p_i, δd_i and g_i are initialised in line 1. We used variable name p instead of `pi` to avoid confusion with π, which in R is coded as pi. The mid-diameters of the diameter classes are calculated from d_{min}, d_{max} and w in line 2 using the seq() function. Then the index of the largest stem-diameter class is established in line 3. The most important input to the Schütz model, the number of trees per hectare in the largest diameter class n_{max}, is provided in line 4. As discussed before, this value needs to be changed iteratively so that the number of trees per hectare in the smallest diameter class n_{min} eventually matches the number anticipated from field observations and/or simulations. Additionally, the equilibrium condition laid out in Eq. (6.34) needs to be fulfilled, see Section 6.4.2.

```
1   > ni <- bal <- p <- id <- g <- c()
2   > d <- seq(d.min, d.max, w)
3   > c <- length(d) # index of largest class
4   > ni[c] <- 1.67 # n.max: model input
5   > bal[c] <- 0
6   > m.fitted <- mort(c(a * abdn.mort$par[1], b *
7   + abdn.mort$par[2], abdn.mort$par[3]), d) * 100
8   > id[c] <- inc(abdn.L2$par, bal[c])
9   > p[c] <- id[c] / (w / 100)
10  > g[c] <- (pi * (d[c] / 200)^2) * ni[c]
11  > for (i in (c - 1): 1) {
12  +     bal[i] <- sum(g[c: (i + 1)])
13  +     id[i] <- inc(abdn.L2$par, bal[i])
14  +     p[i] <- id[i] / (w / 100)
15  +     ni[i] <- ((p[i + 1] + m.fitted[i + 1]) / p[i]) *
16  +     ni[i + 1]
17  +     g[i] <- (pi * (d[i] / 200)^2) * ni[i]
18  +}
19  > sum(ni)
20  [1] 386.473
21  > sum(g)
22  [1] 24.3334
```

In lines 6 and 7, the percentage mortality rates m_i are calculated from the mortality model (Eq. (6.43)) for each diameter class. abdn.mort$par is a vector containing the parameters of the mortality model and a and b are modifiers of parameters a_7 and b_7, in this case 0.15 and 1.0, respectively. In line 8, stem-diameter AGR is calculated for the largest diameter class from the parameters of the growth model given by vector abdn.L2$par and from BAL in the largest class which is zero (line 5). Here inc() is the growth function implementing Eq. (6.41). This growth rate is then converted to the percentage outgrowth rate p_i of the largest diameter class in line 9 following Eq. (6.27). The final step of the model initialisation involves the calculation of basal area per hectare as represented by the largest diameter class in line 10. The simulation of the Schütz model now proceeds from the largest diameter class to the smallest as indicated by the definition of the for loop in line 11 which extends to line 18. Consequently, lines 12 to 17 are repeatedly calculated starting with the second largest diameter class and eventually proceeding towards the smallest diameter class. Within the for loop, first the competition index BAL is calculated in line 12. Then stem-diameter AGR is computed in the same way as in line 8 followed by the conversion to percentage outgrowth rate in line 14. Using outgrowth rate p_i and mortality rate m_i the number of stems per hectare n_i is projected for the next smaller diameter class i in lines 15f. based on Eq. (6.33). This is followed by the calculation of the corresponding basal area per hectare in line 17, which is required for the BAL calculation of the next simulation step (line 12). Finally, after leaving the `for` loop, the total number of trees and basal area per hectare implied by the Schütz model are calculated in lines 19 and 21.

The implementation of the equilibrium condition defined in Eq. (6.34) is shown in the next listing. Individual-class basal area is calculated in line 1. This is not basal area per hectare like in the previous listing; hence, we used a different variable (`g`) here.

```
1  > g <- pi * (d / 200)^2
2  > delta.g <- rev(-diff(rev(g), lag = 1, differences = 1))
3  > delta.g <- c(delta.g, 0)
4  > sum(ni * m.fitted / 100 * g) # mortality
5  [1] 1.53124
6  > sum(delta.g * ni * p / 100) # growth
7  [1] 1.62831
```

Then $(\delta g)_i$ is calculated in line 2 using the `diff()` function. In line 3, 0 is added at the end of the `delta.g` vector for the largest diameter class. Finally, the mortality term on the left-hand side of Eq. (6.34) is calculated in line 4, whilst the growth term on the right-hand side of the same equation is computed in line 6. The results indicate that there is 6% more growth than mortality. Ideally both numbers should approximately be the same, but if they cannot be, there should be a small growth surplus rather than a mortality surplus. If the two numbers are markedly different, this is an indication that something went wrong with the modelling.

6.5 Quantifying Deviations

As the analyses in Sections 6.3.2 and 6.4.4 have shown, a graphical comparison of empirical diameter distribution and equilibrium curves has a high didactic value and is highly informative for researchers and practitioners alike. However, some equilibrium curves can lie at close proximity (cf. Figure 6.10) or differences between observed numbers of trees in diameter classes with a large range (i.e. large value of c) and the corresponding guideline model curve are not so easy to discern visually. There may also be a need to process such differences quantitatively for further analysis, for example, as a part of a continued monitoring. Three summary characteristics have been suggested for quantifying such differences. They are generic statistics applicable to any empirical size distribution and equilibrium model. The number of size classes, c, is the maximum of observed and model classes. Both measures presented in this section can be used for quantifying deviations before and immediately after interventions (or just after trees proposed for felling were marked in the field) but also for computing deviations between different survey years.

6.5.1 Mean Quadratic Difference

Sterba and Zingg (2001) and Sterba (2004) suggested the mean quadratic difference (MQD), M, between the classes of the empirical and equilibrium diameter distribution:

$$M = \sum_{i=1}^{c} \frac{\left(\log_e n_i - \log_e \hat{n}_i\right)^2}{c} \tag{6.45}$$

In Eq. (6.45), n_i is the observed number of trees per hectare of class i of the empirical size distribution whilst $\hat{n}_i$ is the corresponding number of the equilibrium model. The use of logarithms reduces the weight of differences in size classes with large numbers of trees, i.e. in small size classes. Values of $M < 0.5$ indicate equilibrium conditions, values $0.5 \leq M \leq 1$ would suggest a critical state whilst forest stands with $M > 1$ are not at equilibrium at all. A weakness of MQD is that the use of logarithms is problematic for $n_i = 0$ or $\hat{n}_i = 0$, i.e. for cases that are theoretically possible and interesting. One way of tentatively fixing this is to set such cases to $n_i = 1$ or $\hat{n}_i = 1$ when calculating M, as $\log_e 1 = 0$.

6.5.2 Modified Absolute Discrepancy

Roloff (2004) introduced the modified absolute discrepancy (MOD), denoted here as M'. This measure is a modification of the *genetic distance* defined by (Gregorius, 1974):

$$M' = \frac{1}{2} \sum_{i=1}^{c} \left| \frac{n_i}{\hat{N}} - \frac{\hat{n}_i}{\hat{N}} \right| \tag{6.46}$$

As before, n_i and $\hat{n}_i$ are the observed and simulated number of trees, respectively, while $\hat{N}$ is the total number of trees per hectare according to the model. The original genetic distance measure describes the difference between the relative frequencies of two empirical distributions and the values of this measure lie between 0 and 1. In Eq. (6.46), both absolute numbers of trees, i.e. n_i and $\hat{n}_i$, are divided by the same total number of model trees, $\hat{N}$, to additionally take differences in absolute numbers into account. Typically, $\sum_{i=1}^{c} \frac{n_i}{\hat{N}} < 1$ for $\hat{N} > N$, otherwise $\sum_{i=1}^{c} \frac{n_i}{\hat{N}} > 1$. A disadvantage of M' is that differences in smaller size classes (which usually are larger than in larger classes) much influence this measure, whilst important differences in larger classes are not well represented (Roloff, 2004).

6.5.3 Homogeneity Index

The homogeneity index, i.e. the reciprocal of the Gini index, G^*, (Eq. (4.8)) was already briefly introduced in Section 4.4.2.1. This index can also be estimated for empirical distributions, i.e. for summarised data, using Eq. (6.47).

$$G' = \frac{\sum_{i=1}^{c-1} \sum_{j=1}^{i} \frac{n_j}{N}}{\sum_{i=1}^{c-1} \sum_{j=1}^{i} \frac{n_j}{N} - \frac{g_j}{G}} \tag{6.47}$$

Here G is stand basal area per hectare. Eq. (6.47) simply states that the cumulative proportions of tree numbers, $\frac{n_j}{N}$, summed over all stem-diameter classes but one, $c - 1$, are divided by the sum of the differences of cumulative tree and basal-area proportions (de Camino, 1976). It is, of course, possible to use other size measures such as stem volume or biomass instead of basal area. Since this variant of the homogeneity index depends on the bin size, w, chosen for the empirical size distribution used, it is important to apply approximately the same w and the same c when making comparisons. G' can be calculated for the equilibrium model for which it should take values not far from 1.0 (often between 1.3 and 2.8 according to de Camino (1976)). The same characteristic can also be calculated for the observed, empirical size distribution. The more the observation deviates from the equilibrium, the larger G' becomes relative to the corresponding index of the equilibrium. With ongoing transformation of a forest stand towards an equilibrium state, G' should continuously decrease. de Camino (1976) found that homogeneity increased in selection forests with increasing adversity of environmental conditions, particularly with decreasing site fertility.

6.5.4 Application to Artist's Wood

Eq. (6.45) applied to the data from Artist's Wood (plot 1) and the corresponding Schütz model quantified in Section 6.4.4 yielded $M = 1.72$ for 2002 and $M = 2.12$ for 2007. Both values lie outside the equilibrium range and the situation worsened between 2002 and 2007. The modified absolute discrepancy gave a similar trend, i.e. $M' = 0.34$ for 2002 and $M' = 0.46$ for 2007. Although a heavy intervention took place in 2004 which was intended to bring the empirical diameter distribution closer to the equilibrium, the growth dynamics up to the year 2007 have removed the size structure further away from the equilibrium. This is strategic information for management and silvicultural monitoring, as it means that even heavier and more frequent interventions were required. On the other hand, the homogeneity index, G', relating to the Schütz model took a value of 0.97, whilst the same index gave a value of 0.98 for the observed empirical diameter distribution of 2002. According to the homogeneity index, the size structure of Artist's Wood was already very close to an equilibrium in 2002. However, by 2007, this index had changed to $G' = 2.55$ emphasizing like the other deviation measures that the situation had much worsened within the short time of five years.

6.6 Critique and Concluding Remarks

Originally developed as a tool to ensure timber sustainability by comparing the current state of a forest stand with a supposedly ideal norm, equilibrium and guideline modelling has fascinated many forest researchers. As part of the evolution of model approaches, this modelling strategy increasingly involved more and more environmental factors and underlying processes, such as growth and mortality. As the main models are essentially only computing frames (cf. Section 6.4.1), it is likely that future efforts may also involve physiological processes thus improving the scientific underpinning. It should also be pointed out that not only empirical stem-number distributions but also any size distribution can be considered. For example, Lundqvist (1995) studied the dynamics of *Picea abies* saplings in height classes. Equilibrium and guideline models currently constitute a modelling step between descriptive statistics of the current state of a forest stand and projecting its future (Hanewinkel, 1998).

A clear advantage of equilibrium and guideline models is that they are comparatively easy to apply – both in terms of model parametrisation and model application. This puts forest practitioners with a university or college education into a position to actively use them as references for their own CCF management. As can be understood from Figures 6.4 and 6.10, such reference curves have a high educational value, as they clearly highlight potential problems of the current size structure of a forest stand. Quantifying deviations, as explained in Section 6.5, can further help to tease out even subtle differences and facilitate a long-term silvicultural monitoring. And yet, equilibrium and guideline models can also be studied from a theoretical point of view, so that they are also of interest to researchers. For example, Brzeziecki et al. (2016) applied the Schütz model to the unmanaged virgin forest of Białowieża (Poland) for evaluating the demographic changes of the species populations in this forest.

6.6.1 *q* Factor Model

The calculations and discussions in this chapter have shown that it is possible and practical to represent empirical stem-diameter distributions of uneven-aged forests by means of the q factor model. This approach has been successfully used in many publications, has a low data demand and is helpful in indicating what general type of trees need to be felled to sustain complex CCF structures. The appeal of the method is the simplicity with which size structure, growing space allocation and stocking control are connected in one simple function (O'Hara and Gersonde, 2004). Only one parameter, the q factor, is essentially needed to describe the shape of a diameter distribution and the numbers of trees per size class are easy to calculate (Hansen and Nyland, 1987). On the other hand, it is a disadvantage of the dynamic equilibrium models that they cannot be summarised by a few model parameters. Given the fact that usually hardly any dynamic information such as growth and mortality rates are used to determine the model parameters as shown in this section, any results naturally need to be treated with the necessary caution. The q factor modelling family is very popular in the CCF management of Anglo-American countries (Smith et al., 1997; Nyland, 2002; Kerr, 2001, 2014). These equilibrium models have also been successfully applied in forest planning such as the Susmel model (Susmel, 1956, 1980) that is used for that purpose in Northern Italy. As the q factor model family allows a better summary of equilibrium curves compared to dynamic equilibrium models, they thus offer the opportunity to relate the model parameters to environmental variables and to stand height as a summary of these. One research strategy could be to first apply dynamic equilibrium models to different woodland communities and then to simplify and summarise these using the q factor approach. This would help establish reliable q factors that can be related to environmental conditions. Sterba and Sterba (2018) showed that the Gini index (cf. Section 4.4.2.1) is closely related to the parameters of the q factor model and vice versa.

The q factor model has two main disadvantages: (i) The fact that it relies on a constant factor q, whilst the Schütz model and the size structure of stands that in all likelihood are at equilibrium suggest that q can indeed vary from diameter class to diameter class. (ii) The q factor model is not applicable to stands with a size structure different from a (negative)

Table 6.9 Example of varying *q* factor for different ranges of the empirical stem-diameter distribution.

Diameter class [cm]	*q*
1.0–4.9	10.0
5.0–14.9	1.8
15.0–22.9	1.5
30.0–49.9	1.2

Source: Adapted from Hansen and Nyland (1987).

exponential distribution. However, many CCF woodlands do in fact not have exponential size distributions and still are likely to form some kind of equilibrium.

Drawing on earlier publications, e.g. Hansen and Nyland (1987), and on their own research, Buongiorno et al. (2000) stated that dividing the size distribution into 2–4 sections with different, specific *q* factors could be a way to improve the *q* factor model (see Table 6.9 for an example). Hansen and Nyland (1987) suggested exceptionally large *q* factors for smaller size classes, where tree mortality is high, smaller *q* factors in the medium-size range with vigorous growth and low mortality followed by even lower *q* factors for the mature overstorey with less vigorous growth and high mortality. In North American conifer forests, Hett and Loucks (1976) also found evidence for *q* factors that incrementally decreased with diameter class from small to large, whilst our results in Section 6.4.4 suggested the opposite. Such a division of the equilibrium curve is particularly necessary when the whole range of stem diameters occurring in the forest stand is considered in the model. The Sterba criterion (Eq. (6.25)) may additionally help to make *q* factors dependent on different growth and mortality patterns.

O'Hara and Gersonde (2004) commented on the second disadvantage of the *q* factor model, i.e. the inflexibility of the approach for structures that do not have negative exponential stem-diameter distributions, but may be practical candidates for CCF. In this situation, Schütz-model type approaches are the better choice, because they do not assume any particular model function. However, Hett and Loucks (1976) intriguingly proposed to superimpose the *q* factor model with a sine wave to account for multi-modal equilibrium structures in natural forests. If the necessary growth and mortality data are available, dynamic models are nearly in all situations imaginable a good choice and can also act as reference or validation model for the *q* factor approach. Chevrou (1990) presented the truncated-law guideline model as an alternative to and a generalisation of the *q* factor model. It appears to share dynamic properties with the Schütz model. This approach, at least in parts, depends on age and has not been pursued much since the time it was published.

6.6.2 Schütz Model

The comparison of a Schütz model and a *q*-adapted Schütz model curve clearly shows important differences (Figure 6.11). As explained in Section 6.4.4, a *q*-adapted Schütz model

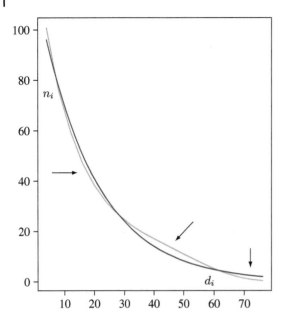

Figure 6.11 Differences between a Schütz (orange) and a q-adapted Schütz (blue) guideline curve highlighted by arrows. n_i and d_i are the number of trees in stem diameter class i, respectively.

curve can be obtained when applying nonlinear regression to the $n_i \sim d_i$ vectors produced by the Schütz model. We can clearly see that the Schütz model suggests less trees in small diameter classes, but markedly more trees in the medium range and again less trees near the target diameter compared to the q factor model. Schütz (2001b) attributed the marked 'hump' in the medium stem-diameter range to better growing conditions of the trees represented here whilst interventions in this tree cohort are low. These trees usually dominate their neighbourhoods, but have not yet reached the target diameter. The sharp decline of the curve towards the target diameter is associated with high mortality due to tree fellings in this part of the stem-diameter range. These differences are typical of the Schütz model and are the result of the stand-specific growth and mortality patterns used in assembling the model. It is for these characteristics that equilibrium curves produced with dynamic models much more than those derived from static models exhibit a *rotated-S structure* that becomes even more apparent when using a logarithmic scale for the number of trees per hectare or tree density when plotting them (Schütz, 2001b).

6.6.3 Marking Guides

Attempts have been made to derive concrete *marking* or *cutting guides* from the comparison of the current tree-size structure and the corresponding guideline curve. For example, diameter distributions with large bin widths such as 15 cm were formed for this purpose, as this bin size apparently facilitates the identification without measurement of trees of different sizes in the field (Table 6.10). For a mixed broadleaved woodland in Southern England, Kerr (2001) estimated a q factor equilibrium curve and compared it with the observed number of trees in five broad stem-diameter classes. There was a surplus of trees with stem

Table 6.10 Marking guide for Plashetts Wood (England, UK).

Diameter class [cm]	Observed n_i [Trees ha^{-1}]	Equilibrium $\hat{n}_i$ [Trees ha^{-1}]	Difference [Trees ha^{-1}]	To be removed [%]	Marking guide
<15	1326	671	655	49	1 in 2
15–29	125	230	<0	0	0
30–44	22	79	<0	0	0
45–59	48	27	21	44	3 in 7[a]
>60	21	12	9	43	3 in 7[a]
Totals	1542	1019	685		

a) If natural regeneration of *Quercus* spec. is required, removing these trees should be delayed, as they are the main seed source.
Source: Data from Kerr (2001).

diameters <15 cm and >45 cm. Table 6.10 suggests removing half the trees with a stem diameter <15 cm, these would be mainly *Acer pseudoplatanus* and up to 3 in 7 of the larger trees, mainly *Quercus* spec. Any decision to remove large trees may also affect the regeneration potential of the site. Kerr (2001) emphasised that any such system as in Table 6.10 should be considered a guide to management rather than a firm prescription.

Such summaries are then similar to the historical check-method distributions discussed in Section 6.2. The final column of such simplified and tabled diameter distributions states how many trees should be removed in each 15-cm class (Kerr, 2001). However, in practice, there is no quick way to mark trees so that the residual stand will conform to a specific equilibrium model curve, since this usually involves several interventions and years, i.e. the forest manager should definitely not be tempted to try to achieve the equilibrium in a single operation (Schütz, 2001b). For that reason, any such attempts of producing marking guides are always applied approximately at best (Buongiorno et al., 2000). This is the reason why we did not include such efforts much in this chapter, as the necessary thinnings or silvicultural systems cannot be easily derived from non-spatial data, and they involve many more considerations than the equilibrium models and the current size structure can offer. The practitioner or research manager using the guideline models is better off relying on their silvicultural knowledge and experience when attempting to gradually move the existing stand structure closer to the equilibrium in the field. Among other important considerations, an appropriate spatial strategy needs to be adopted in marking so that any felling also benefits regeneration and the development of the mid-storey trees, see Section 5.4.

6.6.4 Adjustment of Allowable Cut

Occasionally, the equilibrium curve exceeds the number of trees of the empirical diameter distribution, i.e. the numbers of trees in some classes of the observed empirical diameter distribution (also termed *deficit classes*) are below the equilibrium curve (cf. Figure 6.2b). This is often the case for classes accounting for mid-storey trees, see Figure 6.10b

(d range 20–60 cm). Guldin (1991) suggested that for determining the allowable cut in these instances the corresponding basal area of the 'deficit' is redistributed into the diameter classes where the number of trees in the observed empirical diameter distribution exceed the equilibrium curve (also referred to as *surplus classes*). This compensation for deficit implies that the allowable cut is reduced in the surplus diameter classes (Figure 6.12). The redistribution of basal area does not affect the equilibrium curve; it only serves to correct the allowable cut. Cancino and Gadow (2002) proposed an iterative approach for such a redistribution of basal area and adjustment of allowable cut. This approach (i) produces additional auxiliary equilibrium curves by increasing the original equilibrium basal area in small, iterative steps, e.g. 0.5 or 1 m² ha⁻¹ per step. (ii) The adjusted cutting line is defined for each of the additional auxiliary equilibrium curves as the smaller values of either the empirical distribution or the auxiliary equilibrium curve. Finally, (iii) that new equilibrium basal area, B, and corresponding auxiliary model are selected that produce a total basal area of the adjusted cutting line equalling the original equilibrium basal area (cf. Figure 6.12).

The following listing provides the R code for optimising the adjusted line of allowable cut. In line 1, variable `B.iter` is set to equilibrium basal area B stored in variable `B`. In line 2, model parameter λ is calculated which is not affected by the optimisation. A `repeat` loop then runs from line 3 to line 15 with a `break` condition and corresponding statement in lines 13f. This condition and statement ensure that the loop is terminated when the sum of class basal area of the adjusted line of allowable cut is about to exceed $B = 30$ m² ha⁻¹. At the beginning of each iteration, variable `B.iter` is increased by 0.1. Then k is calculated according to Eq. (6.14) in line 5. Here, vector `classes[1: c]` represents the mid-points

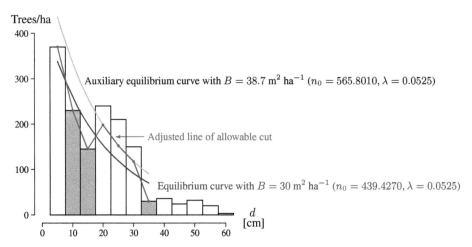

Trees/ha

Auxiliary equilibrium curve with $B = 38.7$ m² ha⁻¹ ($n_0 = 565.8010, \lambda = 0.0525$)

Adjusted line of allowable cut

Equilibrium curve with $B = 30$ m² ha⁻¹ ($n_0 = 439.4270, \lambda = 0.0525$)

d
[cm]

Figure 6.12 Example of an adjustment of allowable cut based on the data provided by Cancino and Gadow (2002), see Table 6.3 and Figure 6.4. According to the method developed by Cancino and Gadow (2002), by means of an iterative process, an auxiliary equilibrium curve with basal area $B = 38.7$ m² ha⁻¹ was identified which allowed to derive an adjusted line of allowable cut to compensate for the deficit in the number of trees in the observed diameter classes highlighted in grey. Stem diameter is denoted as d.

of the diameter classes up to class 35 which has index c. In line 7, n_{max} is calculated from iterated B and k as in Eq. (6.16). In line 8, the remaining numbers of trees n_i are computed according to Eq. (6.11) from n_{max}. Model parameter n_0 is calculated in line 9 using Eq. (6.17). As a centrepiece of the Cancino–Gadow approach, the minimum values of the numbers of trees of either the auxiliary equilibrium function or the observed numbers of trees are computed in lines 10f. and overwritten vector ni now gives an iteration of the adjusted line of allowable cut.

```
1   > B.iter <- B
2   > lambda <- log(q) / w
3   > repeat {
4   +   B.iter <- B.iter + 0.1
5   +   k <- pi / 40000 * sum(q^(c - seq(1, c, 1)) *
6   +   classes[1: c]^2))
7   +   n.max <- B.iter / k
8   +   ni <- q^(c - seq(1, c, 1)) * n.max
9   +   n0 <- ni[3] / exp(-lambda * classes[3])
10  +   for (i in 1: c)
11  +   ni[i] <- min(ni[i], obsClasses[i])
12  +   gi <- pi / 40000 * ni * classes[1: c]^2
13  +   if (sum(gi) > B)
14  +   break
15  + }
```

For the break condition and statement in lines 13f., the basal area of each class is now calculated in line 12 using the number of trees of the adjusted line of allowable cut. Finally 0.1 needs to be subtracted from variable B.iter to give the equilibrium basal area before $\sum g_i$ exceeded B. Then variables k, n_{max}, n_i, n_0 and n_i need to be recalculated for that last but one value of B.iter (not shown in the listing). Following that, variables n0 and lambda provide the model parameters n_0 and λ of the final auxiliary equilibrium model (Eq. (6.2)). For those who wish to reproduce the calculations, the rounded integer values of $\hat{n}_i$ relating to the auxiliary model with $B = 38.7 \text{ m}^2 \text{ ha}^{-1}$ are 435, 335, 258, 198, 152, 117 and 90. The corresponding $\tilde{n}_i$ of the adjusted cutting line are 370, 230, 145, 198, 152, 117 and 30.

6.6.5 Species Representation

Both static and dynamic models can theoretically be modelled separately for tree species populations, see, for example, Brzeziecki et al. (2016). This makes good sense, if the species involved in one forest stand or larger entity occur in sufficiently large numbers resulting in empirical diameter distributions where all classes within a certain range are more or less represented by trees of the species under study. Often, however, there are only one or two dominant species with large numbers of trees and several minor species with just a few trees each. These can then be pooled into one collective species group to result in a meaningful diameter distribution. Alternatively, the equilibrium curve for a stand in question is modelled regardless of species and the desired species composition is considered in the marking strategy (Table 6.10).

6.6.6   Identifying Model Parameters

The parameters of both the static and the dynamic models are estimated using a mix of current and future, anticipated stand characteristics and any of the choices made have a substantial influence on the model result. This implies that at least some details of the desired equilibrium stand structure should be known when the modelling begins, although logically they should rather be the result of the modelling process (Hanewinkel, 1998). This is a dilemma which is difficult to resolve. It can partly be compensated for by experience with such data analyses and forest types. An iterative modelling approach as suggested by Schütz (2006) and described in Section 6.4.4, where the model parameters are gradually adapted until general silvicultural criteria are matched, is clearly an option.

However, even if growth and mortality data are available for modelling the equilibrium structure, these data are historical, and it is plausible to assume that the patterns of growth and mortality will change as part of ongoing transformation of the forest stand towards the equilibrium and particularly, once the forest stand has finally arrived at equilibrium. Bachofen (1999) pointed out that equilibrium curves need to be adapted from time to time, since forest structure and, as a consequence growth and mortality, keep changing. Data from re-measurements should be used for updating the equilibrium. As a precaution to avoid excessive updating, equilibrium modelling is often restricted to forest stands that are not too far from an ideal size structure (Roloff, 2004). For early transformation, where equilibrium guidelines may even be more required and useful, crude assumptions based on the q factor are perhaps sufficient to indicate the broad direction of required interventions. Many model users have found examples for the model input parameters in natural stands (O'Hara and Gersonde, 2004). Given the general uncertainty of the exact location and shape of equilibrium distributions, it is reasonable to assume that there is not just one equilibrium curve that is valid for a given forest stand, but a number of them forming a field or region of confidence (Sterba, 2004). Within the boundaries of such a region, any curve can be selected and applied. Naturally, there is greater uncertainty for small diameter classes than for larger ones (Figure 6.13). After a few years with fresh observations, another curve from the same equilibrium region may need to be chosen for future reference. Mitscherlich (1952) was probably the first to suggest equilibrium regions as opposed to only one, definite curve.

6.6.7   Silvicultural Monitoring

Although some doubt remains about what exact shape the demographic equilibrium curve of a given forest stand should have and where in the graph it should be situated, the model curves shown in this chapter have demonstrated that they can indeed be used as references to compare observed empirical size distributions with. Considering these references in relation to changes in size distributions over time can help estimate how close or how far away a forest stand is from an ideal forest structure. As such demographic equilibrium models along with the deviation measures of Section 6.5 constitute useful characteristics for silvicultural monitoring.

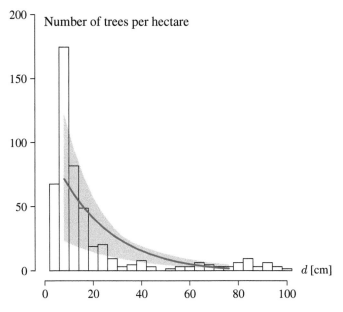

Figure 6.13 Equilibrium guide curve for the mixed *Pseudotsuga menziesii* (MIRB.) FRANCO selection forest at Artist's Wood (plot 1, North Wales, UK) along with an example of a confidence region (grey) and compared to the observed empirical diameter distributions in 2007. The Schütz model (R, red) serves as reference and the supposedly less accurate *q* factor models M2, M3 and M4 form a region of confidence that can be accepted for empirical future stem-diameter (*d*) distribution to be close enough to the equilibrium. The model equations used and the model parameters are given in Table 6.8.

6.6.8 Software

As we have seen in this chapter, specialised software for calculating equilibrium models is usually not necessary, since the calculations required are comparatively straightforward. Spreadsheets have been used much in the past (see, for example, Clarke, 1995; Kerr, 2001), but R scripts can handle such models including the identification of model parameters much more elegantly. `WestProPlus` and `SouthPro`, for example, were add-on tools developed by the US Forest Service for implementing the BDq approach in different forest types that can be combined with spreadsheet software and are usually coupled with growth simulators. Another software package is `PEP` developed by the Baden-Württemberg Forest Research Station at Freiburg in Germany (see `https://www.fva-bw.de/daten-und-tools/programm-pep-30`). This software has implemented a number of different equilibrium models including the Schütz model and uses growth and mortality data for setting up localised matrix models that form the basis of the equilibrium models (Klädtke and Yue, 2003). Matrix models share similarities with equilibrium models and project growth and mortality on the basis of transition probabilities (Virgilietti and

Buongiorno, 1997). Understanding the equilibrium models reviewed in this section is a good preparation for exploring matrix models. They are the next logical step in refining equilibrium models, however, they have a greater demand on data and modelling expertise. Individual-tree growth models provide even more details and allow much more fine-tuning whilst data requirements and modelling expertise reach a high level (Sterba, 2004).

7

Putting it All Together: Implementing CCF for Different Management Purposes

Abstract

How can we integrate the different tools and treatment strategies presented in the general silvicultural toolbox of this book for the consistent management of a forest stand or estate? Forest stands have a long lifetime and the sequence of possible CCF interventions needs to form a consistent treatment in pursuit of one or several management objectives. One way to encapsulate and describe such a consistent treatment system is *forest development types*. These have recently been adopted in several countries and the generic principles of forest development types are discussed in this chapter. Management objectives can greatly differ, since it is possible to carry out CCF for many different reasons and purposes. Traditionally silviculture, including CCF management, mainly focussed on timber production but in many countries, other ecosystem goods and services have now assumed a much more prominent role. Part of the appeal of CCF is that this type of management is believed to offer many alternatives and solutions for delivering a diverse range of ecosystem goods and services. In this chapter, the main CCF-related management implications and considerations for greatly differing forest management objectives are presented and discussed.

7.1 Introduction

In Chapter 5, we referred to the fact that silvicultural systems in the Anglo-American literature are perceived as being broad management programmes defining and encapsulating all treatments, whilst in Central Europe, they are mainly considered regeneration methods with the notable exceptions of the selection system and the two-storeyed high forest. But what integrates continuous cover forestry (CCF) management then if not the silvicultural systems? This is an important question that has been partly addressed by various forest planning concepts in different countries, i.e. national or regional solutions. From a theoretical point of view, it is noteworthy that a generic system, the *forest development types*, has evolved which attempts to encapsulate CCF treatments in a consistent way and also includes transformation/conversion to CCF (Larsen and Nielsen, 2007; Haufe et al., 2021). The less experience with CCF there is in a country, the more

Continuous Cover Forestry: Theories, Concepts, and Implementation, First Edition. Arne Pommerening.
© 2024 John Wiley & Sons Ltd. Published 2024 by John Wiley & Sons Ltd.

a system like forest development types is required to support the introduction of this management type.

Another important question relates to forest management objectives. In the past, all silvicultural teaching and research assumed that commercial forestry and timber production formed the dominant purpose of any forestry activity. To justify this dominance, the claim has been upheld that many other ecosystem goods and services would be accomplished and delivered in the wake of commercial forestry more or less automatically. Here and there perhaps some additional management like determining habitat trees would be necessary but not much else or anything special would be required. The division of forest land into *productive* and *non-productive* forests also stems from the domination of timber production (Bončina, 2011b). This domination has recently been much relaxed. Many methods described in this book can be employed for quite different types of forestry, and commercial forestry has no supremacy any more. As discussed in Section 1.9, the most important fields of application of CCF methods are

- Timber production including pulp,
- Climate-change mitigation including carbon forestry,
- Biodiversity and conservation,
- Water management,
- Urban and recreation forestry,
- Sustainable energy production,
- Forest cemeteries,
- Protection forests.

The traditional idea that commercial forestry should have the highest priority to a large degree is a legacy of the historical trauma of the devastated forest resources in Europe of the eighteenth century. Related to this, there has also been for a long time the aforementioned, widespread perception that other objectives are a by-product of commercial forestry and that several management goals require similar treatments. Instead of maximising a single management objective, the intention was to optimise several relevant objectives at the same time. The number and priorities of the management objectives involved depend on societal demands and on the suitability of the forest area. This concept has been termed the *integration approach* to forest management (Bončina, 2011b). For example, Figure 7.1 gives the simplified structure of deriving silvicultural prescriptions based on a 10-year forest management plan. Prescriptions are short-term interpretations of the guidance given in general plans and programmes. The current state of a stand is described using a qualitative text component and a quantitative component which includes the results of a very basic forest stand inventory. Starting with this general description of the forest stand, a long-term vision of how this forest stand is envisaged to look like when mature is provided next. Part of this vision and the following 10-year plan is a weighting of the three most important ecosystem goods and services relevant to the area for which the plan is prepared, i.e. production, conservation and recreation in this example. Conservation in this case also includes protection, e.g. coastal protection near the

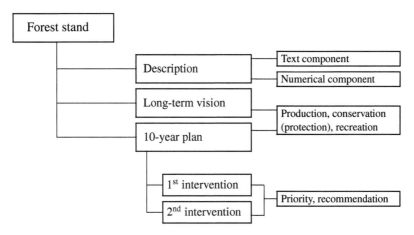

Figure 7.1 Basic principle of defining silvicultural prescriptions as part of an integrated approach to forest management. Adapted from Guest (2008).

sea or the protection of settlements from rockfall in the mountains. This weighting is to be considered in any prescription which in Figure 7.1 is subdivided into two interventions. To allow for options, the management prescriptions can be labelled as either priority or recommendation.

This integrated approach is contrasted by the *segregation approach* where forest land is divided into areas of single purpose only. The segregation approach presents a straightforward way to ensure multiple-objective management in the whole area by using single-objective management in parts of the area. Nature conservation, for example, is often carried out in forest reserves or some other type of strictly protected forest area, where usually no other management objective is pursued. Outside these reserves, conservation is not considered (Bončina, 2011b).

Both concepts have been implemented using many different variants and compromises. In Central Europe, mostly the integration approach prevails, whilst in North America, the segregation method tends to dominate. Both approaches have advantages and disadvantages. Integration has limitations and can often lead to conflicts or 'foul' compromises that the stakeholders involved find difficult to support. No single stand structure can maximise the production of all ecosystem goods and services (O'Hara, 2014). The segregation approach on the other hand, can give rise to extreme forms of RFM outside nature reserves that are very close to industrialised agriculture. CCF is often perceived as a way of taking the 'rough edges' off RFM so that the integration of multiple objectives is easier to achieve. Incidentally, this reasoning can be even detected in the documentation associated with forest development types and in British forest design plans (Haufe et al., 2021). May this be as it is, given the rapid development of challenges, such as climate change and energy crises, there are always situations where certain objectives have to be prioritised or even have to be carried out as sole objective. And yet, both forest practitioners and forestry students are often left alone with the legitimate question of how to carry out CCF silviculture to meet these objectives in isolation. Theoretically, anybody with a good basic education in ecology, soil science and silviculture should be able to select and assemble the necessary tools from the general silvicultural toolbox. However, given the rapid changes of environmental, societal and management conditions along with the fact that most silviculture teaching and associated textbooks still focus on timber production and commercial forestry, the time has come to offer more explicit details and primers on how to meet alternative management objectives. This is the plan for the majority of this chapter.

7.2 Forest Development Types

The term and concept of forest development types originated in plant sociology and was used several times before it was formalised by Otto (1994), Palmer (1994) and Perpeet (2000) beginning in the mid-1990s. Originally, forest development types were a result of the ongoing debate about and strong move towards CCF in Germany. They arose from the need to aid the practical implementation of CCF in the country. At the time, particularly in forest practice, it was felt that existing forest planning concepts poorly reflected the dynamic nature of forest ecosystems and the important role of natural disturbances. The increased frequency and unpredictable nature of natural disturbances started to become more apparent in the 1990s.

Based on natural woodland communities and potential natural vegetation, forest development types (FDT) are a framework of management programmes for forest stands with comparable management potentials and objectives. They are templates and guidelines for forest stands that according to their environmental conditions belong to the same woodland community. The centrepiece of FDTs is a long-term vision of forest structure, species composition and main deliverables in terms of ecosystem goods and services. This is accompanied by a management programme detailing how to achieve the long-term vision. FDTs should not be confused with descriptions of stands at the current point in time (Haufe et al., 2021). For example, as an extreme case, a current stand can be entirely made up of non-native *Picea abies* but is deliberately allocated to a mixed *Quercus* FDT, because this type more adequately describes the ecologically most suitable species mixture. Forest development types can therefore include forest stands that currently are – in terms of species composition and forest structure – comparatively far removed from the vision, but with time and appropriate management, can and should come much closer to it.

FDTs have now been introduced in several European countries including Britain, Denmark, Germany and Slovakia. An important element of forest development types is to actively involve the semi-natural development of forest ecosystems and as a consequence to reduce the impact of human interventions, which is an important objective of CCF (see Section 1.5). The concept of forest development types also includes the possibility that the eventual outcome of managing a forest stand according to these principles may differ from what was originally stated in the FDT vision, since natural disturbances and successional processes are explicitly included.

In traditional (commercial) forest management, such potentially large deviations from specified targets were considered failures (Puettmann et al., 2009); therefore, this 'allowance' and explicit inclusion of alternatives constituted a considerable paradigm shift. Foresters historically often attempted to reduce uncertainty by preventing or controlling disturbances rather than incorporating the reality of such events into their plans. In many situations, these policies resulted in increased vulnerability and loss of resilience in the face of the inevitable occurrence of disturbances (Franklin et al., 2018). The adherence to natural development emphasises the commitment to CCF and leaves an element of open-ended processes (Bartsch et al., 2020). Therefore, in stark contrast to earlier forest planning approaches, in the description of FDTs, the term 'target' is often softened, and several possible alternative outcomes are given. Instead of emphasising targets, a higher weight is placed on processes.

This explicit tendency towards management that accepts uncertainty and risk, tries different methods and learns by comparison between the results of management and predictions is called *adaptive management* (Kimmins, 2004).

Adaptive management is the integration of design, treatment and monitoring to systematically test assumptions in order to adapt and learn (Newton, 2007). This process helps avoid repeating the same mistakes in the future and allows moving forward in iterative steps in complex situations (Messier et al., 2013), particularly where an immediate solution is elusive. More specifically, Newton (2007) offered the following details of adaptive management:

Testing assumptions. Systematically trying different management strategies to achieve a desired outcome. Depends on first thinking about the situation at the specific forest site, developing a specific set of assumptions about what is occurring and considering what actions could be taken to affect these events. These strategies are then implemented, and the results are monitored to assess how they compare to the ones predicted at the outset, on the basis of the assumptions.

Adaptation. If the expected results were not obtained, the assumptions were wrong, the strategies were poorly executed, the conditions at the forest site have changed or the monitoring was faulty. Adaptation involves changing assumptions und interventions in response to the information obtained as a result of monitoring.

Learning. Process of systematically documenting the management process and the results achieved.

For adaptive management to work well, it is desirable to undertake management interventions in the form of experiments (*experimental silviculture*). Ideally, this requires appropriate controls and replications, something that is often difficult to achieve in practice, although this dramatically increases the potential for drawing the right conclusions (Newton, 2007).

Following the general tenets of CCF as reviewed in Section 1.5, forest development types usually explicitly promote mixed-species stands and diverse forest structure, particularly for enhancing resilience to climate change and the recreational value of forests. Since forest development types are based on environmental conditions and on the corresponding woodland communities as an expression of the natural vegetation to be expected on a given site, a suitable FDT can be reliably identified based on simple *ecograms* (see Figure 7.2). These are usually founded on vegetation-classification systems such as the work by Ellenberg and Leuschner (2010) in Europe. Particularly, for forest practice such diagrams are a quick way of identifying the FDT guidelines they need to turn to when managing a given site and ensure that a species composition is selected that leads to site-adapted resilient forest stands. Each forest stand is assigned to one FDT. Large differences between current species composition and/or forest structure constitute a need for transformation/conversion (cf. Section 2.4).

Although starting with the potential natural vegetation (Ellenberg and Leuschner, 2010) on any given site and using this as a reference, forest development types can also

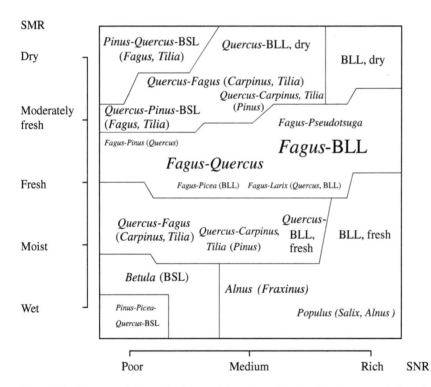

Figure 7.2 Ecogram of forest development types based on the Ellenberg vegetation classification (Ellenberg and Leuschner, 2010) for lowlands in South-West Germany. The font size of letters indicates the dominance of an FDT relative to others. BLL – long-lived broadleaves other than *Fagus* or *Quercus*, BSL – short-lived broadleaves other than *Betula, Alnus, Populus* and *Salix*. SMR - soil moisture regime, SNR - soil nutrient regime. Source: Adapted from Perpeet (2000).

include locally non-native, introduced species (Larsen and Nielsen, 2007). As long as these are not invasive and the conservation status of a given site is not too high, non-native and even exotic species can continue to be included. Such species are often a reality in many countries and help meet the societal demand for timber production or for carbon sequestration. In Central Europe, introduced *Pseudotsuga menziesii* plays an important role in climate change mitigation (Isaac-Renton et al., 2014). The documentation of forest development types acts as a guide for management activities. In most European countries, forest development types follow the integration approach explained in Section 7.1, i.e. they make an effort to include particularly the broad groups of ecosystem goods and services described as production, conservation and recreation. However, sometimes specific priority weights are assigned to the different ecosystem goods and services. Danish forest development types, for example, have the following structure (Larsen and Nielsen, 2007):

Structure of the Danish FDT documentation

Name. The name includes the dominant (*primary*) and co-dominant (*secondary*) species.
Structure. A verbal description of how forest structure could appear when fully developed.
Species composition. The long-term species composition and their relative importance.
Dynamics. Expected successional dynamics including spatial patterns of species and size diversity.
Ecosystem goods and services. Indication of broad groups of management objectives relating to forest functions, i.e. *production*, *conservation* and *recreation*.
Occurrence. Suggested application in relation to environmental conditions (climate and soil). For this purpose, the country is divided into four sub-regions with typical climatic characteristics. In addition, specific FDT variants are defined in terms of soil nutrient and soil water regimes.

More recently, forest development types have been modified to also include climate-change adaptations (Witt et al., 2013). Since all relevant environmental information in terms of soils and climate are available in digital formats, forest development types have been mapped for larger regions, which in addition to tools like ecograms (Figure 7.2) is of great help to forest practice. German and British FDTs include a detailed description of the environmental conditions and their range (Landesbetrieb Forst Baden-Württemberg, 2014; Haufe et al., 2021). There is also often an appraisal of the ecological and conservation context along with an assessment of adaptability to climate change.

The structure of FDT descriptions vary from country to country. For example, what is labelled as 'structure' in the Danish system is referred to as 'target' in the British and 'vision statement' in the German systems. Both in Britain and Denmark, there was a detailed consultation and active involvement of forest practice when designing forest development types to ensure that this framework would be a useful tool for managers

Table 7.1 Main British FDT categories.

Number	Description
1	FDTs dominated by *Picea* spp.
2	FDTs dominated by light demanding conifers
3	FDTs dominated by conifers from the Pacific Northwest of America
4	FDTs dominated by other conifers
5	FDTs dominated by *Quercus* spp.
6	FDTs dominated by *Fagus sylvatica*
7	FDTs dominated by *Betula* spp.
8	FDTs dominated by long-lived broadleaves other than *Quercus* spp. and *F. sylvatica*
9	FDTs dominated by short-lived broadleaves other than *Betula* spp.

Source: Adapted from Haufe et al. (2021).

and planners (Haufe et al., 2021; Larsen and Nielsen, 2007). The FDT framework usually is designed as an open expandable system where new FDTs can be added when required.

The Danish FDTs are divided into three broad groups, i.e. *broadleaved-dominated*, *conifer-dominated* and *historic* FDTs. The last group includes coppice forests, forest pastures, forest meadows and unmanaged forests. The first two groups are broad species containers (Larsen and Nielsen, 2007). Haufe et al. (2021) defined the main FDT categories in Table 7.1. Dominance and co-dominance is usually defined in terms of volume or basal area proportions. For example, according to the British FDT system, there is a primary species usually contributing in excess of 50% of stand basal area, whilst the secondary species contributes less than 50%. The primary species also makes a major contribution to the FDT name. These boundaries are less firm in the Danish and German systems that usually involve more species in the definition. Using the preposition 'with', the Danish FDT system adopted a wording that expresses species dominance indirectly, e.g. *Fagus sylvatica* with *Fraxinus excelsior* and *Acer pseudoplantanus* or *P. menziesii* with *P. abies* and *F. sylvatica*.

In Germany, there are separate FDT systems for each federal state. In anticipation of climate change and transformation/conversion, the South-Western state of Baden-Württemberg explicitly defined *transitional* FDTs, e.g. *P. abies* → Mixed *Quercus* spp. or *Pinus sylvestris* → Mixed *F. sylvatica* for their importance in the state so that the transitional character is flagged up in the name by the arrow operator.

Perpeet (2000) also emphasised the role of tree and forest models in simulating treatment options for stands allocated to forest development types. The same author also suggested that each FDT should have a number of management demonstration sites and nature reserves that can help develop intervention strategies for other forest stands of the same FDT. These reference stands are then also used to assess how remote or how close a management unit is from or to the potential natural vegetation. They also help decide how much management input is really necessary on any given site to meet a certain objective compared to leaving a stand completely to nature. The South-West German forest development types, i.e. the area of Europe in which FDTs originated, possibly have the most sophisticated structure (Landesbetrieb Forst Baden-Württemberg, 2014) which is summarised in the following box.

Structure of the South-West German FDT documentation

Name. The name includes the names of dominant and several minor species, sometimes the mixture type or an indication of transition to other species and forest structures.

Vision. A verbal description of how forest structure could appear when fully developed including species composition and types of species mixtures. Species and structural changes are also noted here.

Basics. Using sub-sections, this section includes the geographic occurrence/relevance of this FDT, its application in relation to environmental conditions (climate and soil), the typical origin and history of stands belonging to this FDT, a description of typical site types and the position of the FDT within the sequence of natural succession, an ecological appraisal of the FDT, its climatic adaptability and a description of its contribution to nature conservation.

Treatments. Using again several sub-sections, this part of the FDT indicates expected long-term species composition, vertical and horizontal forest structure. This is followed by a definition of expected ecosystem goods and services pertaining to the FDT. The management details are then spelled out including transformation strategies, frame tree numbers, target diameters. Special attention is given to individual-based thinning regimes (see Chapter 3) in different development stages. Target-diameter harvesting and regeneration (silvicultural systems, cf. Section 5.3) methods are also specified.

Natural disturbances. This section specifies what measures should be taken when large-scale disturbances (e.g. gales, insect calamities) occur and the management specified in the treatment section cannot go ahead as planned. This includes potential replanting schemes and the use of woody debris and open areas for nature conservation.

The most remarkable elements of the south-west German FDT structure is the integration of management details including those required for transformation/conversion to CCF and plans for how to deal with the aftermath of natural disturbances. This provides helpful guidance to forest practice, and it is always good to have a plan B in times of great uncertainty when climate change may render anticipated pathways of management impossible. Treatments can be specified in a number of ways and basic concepts are explained in Section 5.2. They should include information on thinning regime, thinning type, thinning intensity

Table 7.2 Example of a complete individual-based treatment guideline for mixed *Picea sitchensis* (Bong.) Carr. (SS) – *Betula* spp. (BI) woodlands at Coed y Brenin in North Wales (UK) based on the W4 *Betula-Molinia* woodland community, the frame-tree concept and on results from a chronosequence.

THT < 10–12 m	THT = 12–20 m	THT = 20–30 m	Regeneration
Reduction to 1500–2500 trees/ha.	Installation of extraction racks.	Selective target diameter harvest of SS > 40 cm, BI > 25 cm every 5–7 years but not before advance regeneration is established.	Final overstorey removal after 5–15 years
Selective removal preferred.	Permanent marking: 50 SS frame trees every 15 m, 30 BI frame trees/ha in clusters.		Allow for 10–15 BI standards/ha.
Removal of BI wolf and whip trees.	Removal of 2–3 do-SS competitors per frame tree every 3-5 years or at least in two separate heavy selective crown thinnings.	Regeneration method: Group/strip system.	
		Target basal area: 25 m²/ha	
	Target basal area: 30–35 m²/ha No more than 5–8 m² BI and 50–80 m² SS to be removed in any intervention.		
	SS: Max h/d = 80, min c/h = 50; BI: Promote clusters. Keep at least 1.5 m free space between BI and SS crowns.		

THT – top height (mean height of the 100 largest trees per hectare), h/d – height-diameter ratio, see Eq. (2.2), c/h – crown ratio, see Eq. (2.3). See also the schematic visualisation of a mature stand in Figure 7.3.
Source: Adapted from Pommerening and Grabarnik (2019).

and thinning cycle. As they need to cover the whole lifetime of a forest stand belonging to a certain FDT, such treatment guidelines also need to cover harvesting and regeneration methods. Using top height as reference ensures that the treatment recommendations are comparatively independent of site (Table 7.2), i.e. they can be applied across the variations of environmental variables occurring in the forest area allocated to the FDT under study.

All this information usually is the synthesis of practical experience, many field trials and model simulations. Combining multiple information sources usually leads to robust, generalisable results. In many European countries, such treatment guidelines are based on the principles of individual-tree forest management, see Section 3.5, because these principles help with the implementation and through their focus on frame trees achieve a rationalisation effect that benefits conservation and recreation interests at the same time. Biological rationalisation is also important, since climate change and the increased frequency and severity of disturbances that come with it suggest reducing financial investments in forest management. The amount of information provided in the FDT documentation differs and more may be necessary in countries where CCF is new and detailed silvicultural advice is lacking. In every case, the treatment information does not constitute firm prescriptions but guidelines and deviations or detours may sometimes be necessary depending on local conditions and unforeseen disturbances.

Already Larsen and Nielsen (2007) reported that schematic visualisations of vertical profiles of mature stands of FDTs have been useful for defining them and for communicating long-term objectives. They also help with the implementation of FDTs, particularly with realising the envisaged stand structure and are usually shown at the top of the FDT documentation sheet. The Danish FDT visualisations are more detailed and artistic than the German and British ones, which tend to be more schematic. The *P. sitchensis – Betula* spp. FDT described in Table 7.2 and in Figure 7.3, for example, includes two species with very different growth dynamics and therefore the specimens

Figure 7.3 Example of a schematic visualisation of a typical vertical profile of a mature stand of the forest development type described in Table 7.2, i.e. mixed *P. sitchensis* (Bong.) Carr. – *Betula* spp. woodlands.

Figure 7.4 'Structural arrow' indicating a range of possible stand structures based on Figures 1.10 and 2.15. The word 'layer' refers to canopy layer. Options which are not suitable for the FDT described in Table 7.2 and Figure 7.3 are greyed out, all others are shown in black font. The option highlighted in bold is the recommended stand structure. Source: Adapted from Haufe et al. (2021).

of both are segregated from the other by arranging them in conspecific groups (cf. Section 4.3). The visualisation also gives an impression of the relative dominance of the species involved.

Haufe et al. (2021) additionally devised a simplified stand structure scale in the form of a 'structural arrow' (cf. Figure 7.4). This scale offers a quick orientation of possible stand structures in the wide range of CCF applications. This is another piece of help breaking CCF down into concrete applications. The particular recommendation in Figure 7.4 again relates to the mixed *P. sitchensis – Betula* spp. FDT of Table 7.2 and Figure 7.3. Given the importance of *Betula* spp. in the FDT, complex structures are unsuitable. However, there are two species involved and locally and temporarily there may be more than one canopy layer. Figure 7.4 is just an example, naturally it is possible to modify the design of structural arrows in any way that is deemed to be more suitable.

As discussed in this section, forest development types and similar approaches are general, flexible frameworks which have their foundation in site and vegetation classifications. They differ from country to country and even from region to region, since there is no single system that perfectly fits all requirements. They integrate vision statements, transformation/conversion to CCF, silvicultural systems and thinning treatments. FDTs help to make CCF work and can be applied either to implement an integrated approach to forestry including several ecosystem goods and services or to accomplish a segregation approach to forestry. As part of the latter, FDTs can also be applied to forest conservation and other, more specialised applications of forestry, which are discussed in Section 7.3.

7.3 Specialised CCF Management

Although there is a long tradition, especially in Europe, of pursuing integrated forest management where the attempt is made to deliver several ecosystem goods and services simultaneously (cf. Section 7.1), the integration concept often has its limitations and is not necessarily applicable to all forest land of a forest district or estate. Particularly CCF has traditionally much subscribed to the integration concept of ecosystem goods and services (Larsen et al., 2022). However, current challenges, the speed at which they are unfolding and other pressing issues increasingly leave little room for compromises.

Table 7.3 Examples of management objectives and corresponding sivicultural strategies.

Objective	Silvicultural strategies
Integrated production of valuable timber with minimum effort	Use of biological rationalisation (Section 2.6) as much as possible, simultaneous achievement of conservation and ecology objectives; long production periods, large timber, site-adapted species originating from natural regeneration, natural forest structure, minimum of interventions
Segregated production of valuable timber with maximum yield	Long production periods, large timber, minimising risk by investing in insect and deer control, increasing yield through fertilisation, tree breeding, the use of non-native species and RFM methods
Conservation	No interventions or restoration interventions, locally native species and provenances (from potentially natural vegetation or original species)
Protection (against noise, landslides, wind, cold air, air pollution etc.)	Use CCF in general, particularly selection forests to maintain permanent cover
Drinking water (quantity and quality)	Species with low water demand, medium-to-low tree density
Recreation	Maintaining old-growth and mixed-species forests
Energy wood	*Minimum of input*: almost no intervention, reliance on natural forest development *Maximum of yield*: Breeding of fast growing species, fertilisation, agricultural methods
Timber mining	Harvesting in existing forests, no interventions for regeneration and tending, no silviculture and no sustainable forestry

Source: Modified from v. Lüpke (2002, pers. communication).

Even if the general decision has been made to largely follow the integration approach, there are always some woodland areas in each forest district or estate that require special treatment in pursuit of just one specialised objective (cf. Table 7.3). Conservation areas, forest reserves, forest margins and streamside forests are typical examples where only one objective, conservation in this case, is pursued. Single-objective management often involves tree fellings and naturally at least some of the trees removed can be allowed to generate income which can then contribute to the management costs, even if the objective is not commercial (Binkley, 2021). Single objectives also help understand what is required for each single objective before putting them all together and trying to achieve a compromise in the integrated variant of forest management. This information is also crucial to woodland owners who wish to diversify the forest business on their estates. Specialised management objectives are already a reality in silvicultural training courses offered to practitioners, see Section 8.2. Therefore, primers are necessary that gently point practitioners, novices to CCF but also modellers, into the direction of possible strategies when charged with the task of optimising management for one important ecosystem service. Such strategies are broadly outlined in the following sections. Table 7.3 offers an overview of examples of specialised silviculture.

7.3.1 Timber Production

The production of timber meets various important societal needs and the silviculture of this objective has been abundantly discussed in many textbooks (cf. Bartsch et al., 2020; Burschel and Huss, 1997; Mayer, 1984; Nyland, 2002; O'Hara, 2014; Rittershofer, 1999). Timber production involves both the quantity and quality of timber produced (O'Hara, 2014). In this section, we will mainly focus on the management requirements and strategies for quality timber and pulp for industrial products and paper production.

7.3.1.1 Requirements

The production of quality timber for producing furniture and construction timber requires strategies where all management is directed towards a comparatively small number of economically valuable trees that fetch high prices on the timber market and as a consequence promise high financial returns. This management objective requires *maximising the stem volume of a few individual good-quality trees* generally leading to large average tree sizes (Bartsch et al., 2020).

With trees used for pulp and other industrial purposes, timber quality is a minor consideration. Therefore, in stands where pulp and paper production are the most important objective, *maximising overall stand volume production* is a requirement (Burschel and Huss, 1997). Timber volume production is maximised when stand density is maintained at high levels resulting in comparatively small individual tree sizes (O'Hara, 2014), see also Table 7.3.

7.3.1.2 Management Strategies

Individual-tree forest management is an ideal foundation of quality timber production. Using frame-tree based methods as outlined in Chapter 3, all management should be concentrated on a small number of economically valuable trees. The most common thinning type applied in the context of quality timber production is local crown thinning. As a consequence, stands tend to develop a diverse size structure, comparatively low tree densities and more open canopies. Matrix trees not only support the quality development of frame trees but can also be used for producing energy wood, pulp and other minor timber products (Wilhelm and Rieger, 2018). Quality timber is usually based on a combination of tree size, size and number of branches on the merchantable bole, stem taper and straightness (O'Hara, 2014), cf. Table 7.3.

In some countries, special silvicultural strategies exist for enhancing the quality of particular tree species. These, for example, include delayed early thinnings in broadleaves compared to conifers, because broadleaved trees only develop straight stems with few branches in high-density environments. This is contrasted by conifers that naturally tend to have straight stems due to apical dominance. However, with genera such as *Picea* spec. and *Pseudotsuga meziesii* individual-tree resilience to wind and snow has a high priority, hence, earlier thinnings than in broadleaves are required (Rittershofer, 1999). With light-demanding species, thinnings start earlier than with more shade-tolerant species, even if both occur in the same forest stand. For example, in a mixed-species forest, frame-tree based thinnings may start in *Quercus robur* and *Larix decidua* stands when they are approximately 15 years old, while for *P. abies*, frame trees thinnings may only start when they have passed approximately 25 years. In quality-timber production, methods of biological rationalisation (cf. Section 2.6) play an important role to keep the production costs as low as possible.

There are also even more specific strategies for growing species that are prone to epicormic growth or other defects along the stem axis, e.g. *Quercus* spp. and *Fraxinus* spp. In stands with such species, underplanting is carried out using shade-tolerant species (referred to as nurse, trainer or auxiliary species), e.g. *Carpinus betulus* or *Tilia* spp. (termed stand improvement underplanting, see Section 2.4.1) and Table 7.4. The underplanted trees shade the stems of the primary tree species and prevent epicormic growth or forking of the main stem. Care, however, must be taken that the underplanted trees do not start competing with the primary species. This is usually achieved by a considerable size difference due to delayed underplanting which can be maintained by pollarding the underplanted trees from time to time in order to reduce their competition pressure on the light-demanding trees of the overstorey (Pretzsch et al., 2017). In Denmark and Wales, occasionally *P. abies* is planted under *Quercus* spp., but in other countries this is considered a troublesome mix, as *P. abies* is hard and expensive to keep in check and the needle litter often causes raw humus accumulation leading to soil acidification (Figure 7.5). Underplanting of broadleaved species is also carried out for improving the litter and humus layer, e.g. *F. sylvatica* under *Pinus* spp. or under *P. menziesii* using wide spacings (e.g. 4 × 1 m).

As far as the production of low-quality timber, pulp and paper is concerned, the CCF variant used in this context is likely to lean more towards the structure of RFM stands (cf. the left range of stand structure in Figure 1.10): Tree densities tend to be higher, the main canopy

Table 7.4 The most common European tree and shrub species used in stand improvement underplanting and their silvicultural characteristics.

Species	Characteristics	Site requirements
F. sylvatica L.	In Central Europe, most common species used in underplanting, grows initially more slowly than light-demanding main canopy. Catching up later, there is the danger of F. sylvatica invading the crowns of main canopy trees. Usually planted much later than the main species.	On all sites, where the soils are not too dense, wet or dry.
Carpinus betulus L.	Can be planted at the same time as main species or a little later. Initially fast growth which slows down rapidly. Low danger of invading the light-demanding main canopy.	Can also be grown where soils are dense or water-logged, however, not where SNR is low.
Tilia spp.	Slow juvenile growth is followed by rather prolific growth. Danger of invading the light-demanding main canopy.	Suitable on dense or waterlogged soils but also on good sites.
Other species	Other possible species include *Prunus padus* L., *Crataegus* spp., *Coryllus avellana* L., *Taxus baccata* L. provided the overstorey is not too dense. These species are rarely used but have potential.	Suitable on all sites.

Source: Adapted from Burschel and Huss (1997).

Figure 7.5 *Quercus robur* L. underplanted with *P. abies* (L.) H. Karst. for preventing epicormic growth on oak. This is a typical application of a stand improvement underplanting, see Section 2.4.1. However, silviculturists often consider *P. abies* an unsuitable nurse for *Quercus* spp., as the species is too aggressive on most sites and has a tendency to produce raw humus leading to soil acidification. An example from Sjælland (Denmark). Source: Courtesy of Keith Blacker.

is fairly closed with less-size diversity than in stands managed for quality timber production. Species diversity may also be reduced. The main reason for this is the requirement for maximising overall stand volume. Global low thinning is the preferred thinning type in this scenario, cf. Table 7.3.

In both cases of timber production strategies, it is possible to grow mixed-species forest stands. This is desirable and leads to a more diverse range of products in stands grown for quality timber. Mixed species are less necessary in stands grown for low-quality timber, but they would enhance overall stand resilience and act as an 'insurance policy' in the face of climate change. Conifers have been traditionally preferred in timber production for their rapid growth, high yield and the fact that satisfactory timber quality can often more easily be achieved. Therefore, they are economically more attractive (Hart, 1991). In many countries, this species choice has, however, sadly been much exaggerated at the expense of low resilience to natural disturbances and climate change. For timber production, conifers have often been planted outside their natural range and on sites where they were not suitable even before the onset of climate change. Due to a lack of apical dominance and a tendency to forking of the stem, good timber quality is more difficult to achieve in broadleaved trees (Joyce et al., 1998), however, timber prices for quality broadleaved timber are usually higher than those for comparable conifer timber.

7.3.2 Climate-Change Mitigation

Climate-change mitigation is not an original tenet of CCF, but researchers and policymakers in various countries have recently discovered that CCF can offer management tools for achieving this objective, see Section 1.7. In the context of climate change, goal-oriented forestry activities can enhance

- tree/forest resilience,
- carbon sequestration,
- water retention and
- microclimatic conditions

compared to land-use forms in open land. In this section, we will mainly focus on the first two items, i.e. tree resilience and carbon sequestration.

7.3.2.1 Requirements

With ongoing climate change, higher average temperatures, reduced precipitation (during the growing season) and an increase in the occurrence of natural disturbances (including gales, wildfire, pests and pathogens) are expected in many parts of the world (Kirilenko and Sedjo, 2007; Binkley, 2021; Palik et al., 2021). Milder winter temperatures and longer growing seasons may increase the risk of insect calamities and various diseases that previously were a minor problem. Although likely global scenarios can be translated into possible local climate changes, the exact details are still largely unknown and trends may change at short notice. In such situations, ecological theories such as the *stress-gradient, Janzen-Connell, herd-immunity* and *insurance hypotheses* (Pommerening and Grabarnik, 2019, Chapter 2) can offer advice. An important rationale of growing mixed-species forests is that other species may take over when one or several species are in decline because of changing environmental conditions. Other researchers started to analyse experiments with provenances of native tree species that were introduced from warmer and dryer climates (Köhl et al., 2010; Brang et al., 2014; Frischbier et al., 2019) to see if transplanting such provenances is a viable option for mitigating climate change.

CO_2 is the most important greenhouse gas and forests are a significant part of the global carbon (C) cycle, because they contain the largest store of terrestrial carbon. Forest ecosystems store more than 80% of all terrestrial aboveground C and more than 70% of all soil organic carbon (Ameray et al., 2021). Natural forests are important global carbon sinks

(Ontl et al., 2019). Given the critical role forests play in carbon mitigation, there is increased interest in managing forests for their carbon benefits to offset greenhouse gas emissions. Forest carbon markets were developed in recognition of these benefits and landowners may receive payment for carbon-offset projects that increase carbon stocks (Franklin et al., 2018; Palik et al., 2021). Optimising carbon sequestration can be achieved by *afforestation*, *reforestation* and by adapting forest management, which is the subject of this section.

> Afforestation is the introduction of trees to sites that did not support forest or had no forest cover for a long period of time whilst reforestation is the re-establishment of forest cover on areas where it once occurred, even if missing for several years (Nyland, 2002). Reforestation is the re-establishment of forest cover either naturally or artificially (Helms, 1998).

For increasing carbon sequestration, it has often been suggested to plant more trees and thus to reverse forest degradation and to increase the global forest area. This is certainly an important strategy. However, there are surprisingly few suggestions for optimising the management of existing forests with a view to maximise carbon sequestration. Trees generally are long-lived organisms, but eventually they die or are felled prematurely and the fate of the carbon locked up in tree stems is also an important consideration in this context. Most forest carbon, however, is stored in forest soils, and these need to be managed in such a way that the release of gas from soils affecting our climate is minimised.

7.3.2.2 Management Strategies

A first step towards increasing resilience is to gradually remove all species that are not well adapted to a given site. Suitability of a species ensures resilience and long-term storage of sequestered carbon. In a forestry context, this often refers to species of the genus *Picea* and *Pinus* which for economic reasons have often been planted well beyond their ideal natural range and are now, with the onset of climate change, becoming a liability. However, specimens of these species when occurring in great quantities cannot be removed rapidly, as this would cause the disintegration of current forests which could make things even worse than they currently are. The underplanting techniques described in Section 2.4.1 are a good strategy for gradual conversion of unstable to resilient species compositions. At the same time, any species suitable and naturally colonising the site is encouraged. With ongoing climate change, increasing and maintaining tree species diversity is considered an important insurance policy (Brang et al., 2014; Palik et al., 2021). Even redundancies, i.e. tree species occupying approximately the same ecological niche, should – according to the insurance hypothesis – be promoted. Neuner et al. (2015) even found that the survival of species that are not well adapted to a given site because of climate change can be enhanced when grown in mixture with other species. Franklin et al. (2018) stated that increasing biodiversity is now viewed as one of the most important actions that can be taken to increase ecosystem resilience to climate change, not least as a risk-spreading strategy. For maintaining mixed-species forests, the principles described in Section 4.3 need to be considered. Species composition and diversity need to be monitored throughout the lifetime of a forest generation to secure the resilience effect that diversity can offer.

So far, it is not clear whether the introduction of species provenances from warmer and dryer climates is a sustainable strategy, since all evidence available is based on short-term experiments involving seedlings. There is some anecdotal evidence that within reason many tree species can much faster adapt to changing environmental conditions than is currently believed. The introduction of provenances from regions far away can have disadvantages that we currently do not understand well. Therefore it may be better to wait until more information will be available. In this context, O'Hara (2014) pointed out that climate change presents a 'moving target' for establishing populations of trees that are adapted to a given site. Trees adapted at one time may not be adapted at another. A species that may be adapted to exist in a climate may not be adapted to regenerate in that climate or vice versa. CCF stands offer the ability to overlap generations of trees with different levels of adaptations. If one generation fails because of a difficult regeneration environment or a climatic fluctuation, CCF stands have additional age/size cohorts to occupy the site. Likewise, if an older group of trees cannot persist because of environmental change, a younger cohort may be more resistant to these changes. Multiple development stages provide different levels of adaptations that provide resiliency in CCF stands. The implementation of CCF strategies to develop complex stands will therefore be part of the broader strategies to develop forests that are resistant and resilient to climate change (O'Hara, 2014). The increase in the frequency and intensity of natural disturbances suggests making greater use of biological rationalisation (cf. Section 2.6) whilst reducing the costly investments of traditional silviculture such as replanting windthrow and burnt areas as well as pre-commercial thinnings. If climate change is likely to make the outcomes of timber production uncertain, it is better to keep investments low and allow nature to take the lead.

Species, such as *P. menziesii* introduced to Europe more than hundred years ago, that are apparently better adapted to climate change whilst growing more prolifically (and thus sequestering more carbon) than native species, may play a greater role in the future (Isaac-Renton et al., 2014; O'Hara, 2014). Such introductions can constitute a climate-change adaptation strategy and are also termed *forestry-assisted migration* (Pedlar et al., 2012). This should not be confused with assisted migration for saving species that are threatened by rapid climate change elsewhere.

Carbon forestry has now become an established term to denote forest management with the explicit objective of increasing carbon sequestration (Pukkala, 2018). Carbon storage is an increasingly common forest management objective supported by international agreements and developing carbon markets. Forests have great potential for storing carbon. However, as with many forest management decisions, there are trade-offs associated with increasing stand-level carbon stores. For example, greater carbon storage often results in greater fire hazards, lower tree vigour and susceptibility to insects and pathogens because of the higher tree densities involved or reduced structural complexity (O'Hara, 2014).

On closer inspection, the broad strategies suggested for increasing forest carbon sequestration markedly overlap with tenets of CCF (cf. Section 1.5). Large carbon losses as suffered after clearfelling are avoided in CCF. This particularly relates to the loss in soil organic carbon (Ameray et al., 2021), since trees replanted on clearfell sites soon compensate for the biomass in living trees. The general idea of CCF is to support managed forests with structures, management practices and/or species compositions that are more akin to the potentially natural processes of tree vegetation on any particular site than those that are commonly observed in rigid plantation management (see Section 7.2). RFM forests tend to sequester and store more C in biomass than CCF forests, but the latter store more C in soils so that total ecosystem C is often higher in CCF. In addition, due to niche complementarity mixed-species forest stands (as favoured in CCF) have a higher C sequestration potential than monospecies stands. CCF usually also implies longer lifetimes of trees compared to RFM and this also has a positive effect on carbon sequestration in living biomass (Ameray et al., 2021). Lengthening tree lifetimes not only retains more C by increasing average stem size but also delays emissions that occur after harvesting. Larger tree sizes may also create the potential for a greater level of off-site carbon storage in long-lived timber products (Palik et al., 2021). CCF practices also include increasing the resilience to natural disturbance and enhancing forest recovery following disturbance (Ontl et al., 2019) thus reducing the probability of sudden, large carbon losses. Despite continued, regular interventions CCF woodlands are net carbon sinks at all times. Compared to RFM, the main advantage of CCF woodlands is that they never cease to act as net carbon sinks. The increased structural complexity and thinning frequency (as opposed to large-scale harvesting operations) of CCF woodlands further reduce the probability of large-scale disturbances and associated large carbon losses (Hanewinkel et al., 2014). On the other hand, as briefly alluded to before, overall carbon storage in tree biomass – depending on species and environmental factors – might on average be lower in CCF stands relative to RFM (Lundmark et al., 2016), because CCF requires more open forest canopies. This, however, might be compensated for by growth stimulation effects caused by repeated thinnings that has been observed for tree volume growth of forest stands (Assmann, 1970), although to date the research community knows little about the effects of selective thinnings on carbon sequestration (Franklin et al., 2018). Increasing structural complexity as part of ongoing CCF management can result in increasing niche complementarity of trees in different canopy layers and thus lead to increased overall carbon sequestration in biomass (Binkley, 2021). Increasing drought and diseases may require the reduction of tree density, as changing environmental conditions decrease the carrying capacity of a given site (Palik et al., 2021).

An unresolved key question of carbon forestry is whether it may be beneficial to concentrate sequestered carbon in a few large and long-living trees rather than maximising overall carbon sequestration by spreading carbon across many small and medium-sized trees. From a logistic point of view, the former option is perhaps more suitable to a concept of long-term insurance of carbon sequestration by converting the timber of individual large trees to long-term wood products such as furniture and construction timber along the chain of custody. Alternatively, researchers and stakeholders have considered putting large tree stems into long-term underground storage (Zeng, 2008).

Zeng and Hausmann (2022) suggested wood harvesting and storage (WHS), a hybrid nature-engineering combination method to mitigate climate change by harvesting timber sustainably and storing it semi-permanently for carbon sequestration. According to this method, woody biomass is collected and buried underground in anaerobic conditions or stored in other environments designed to prevent decomposition, thus forming an effective carbon sink. To a traditional forester, this may sound strange, since the timber is not put to 'proper' use. However, even with timber used in furniture and for construction purposes, it is not really certain for how long the carbon is safe in these products. WHS is a way of decreasing this uncertainty. The centrepiece of WHS is a *wood vault* (cf. Figure 7.6), which essentially is a (partially) underground cavity embedded in the biologically inactive subsoil. Here wood is buried in a way that ensures anaerobic, dry or cold conditions so that decomposition is prevented. For example, to ensure anaerobic conditions, the cavity is insulated with clay or other low-permeability material. After enclosure is completed, the top is backfilled with original topsoil and organic layer to allow natural vegetation to gradually reclaim the burial site. WHS is inspired by the natural process of coal formation. Compared to natural coal formation, the rate of WHS wood burial is accelerated by human intervention through timber harvesting and the construction of wood vaults. WHS reverses fossil fuel extraction and burning.

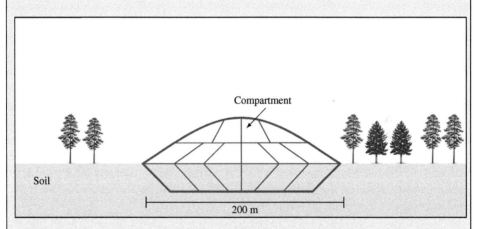

Figure 7.6 Schematic representation of a *wood vault* that is partially above and partially underground. The vault is divided into compartments (indicated by the red lines) that are sealed as soon they are filled. The seal is made of low permeability clay or similar natural material to ensure anaerobic conditions. Once the vault is completely filled, the original topsoil is put back. Source: Adapted from Zeng and Hausmann (2022).

Zeng and Hausmann (2022) proposed several variants of the wood vault including a semi-underground burial mound (cf. Figure 7.6) akin to landfills, warehouses, timber deposits at the bottom of lakes (water vaults), open storage in deserts and in permafrost. For underground and semi-underground burials, Zeng and Hausmann

(Continued)

(Continued)

(2022) recommended compartmentalisation, i.e. the division of the cavity into multiple sections or cells, each of them is sealed by a layer of clay or similar material to avoid degradation and CO_2 emissions before the whole vault is eventually closed. Compartmentalisation also minimises subsequent leakage and greenhouse gas emission in later years. Wood vaults can be of nearly any size, but for logistic reasons larger entities, e.g. 50×50 m or 100×100 m may be advisable minimum numbers.

Archaeological evidence has shown that such wood storage can preserve timber for many hundreds if not thousands of years, mainly because anaerobic bacteria are unable to digest lignin. Sensors should be used to regularly monitor the inside of wood vaults including greenhouse gases such as CO_2, O_2, CH_4. In addition, pH, and temperature, humidity and water table also need to be measured regularly.

7.3.3 Biodiversity and Conservation

Biodiversity and the conservation of habitats is an important tenet of CCF, see Section 1.5. In general, biological diversity means the variability among living organisms and its conservation is required by law (Gaston and Spicer, 2004). Indicators of tree diversity suitable for monitoring were introduced in Chapter 4. When performing a literature review of forest management strategies that enhance or maintain biodiversity, one cannot help but notice that most papers available reverse this question along the lines of what is the effect of forest management on biodiversity (Kovac et al., 2018). Therefore, biodiversity management in forests appears to be considered a bit of an inverse problem rather than a proactive strategy. However, there is agreement that the emulation of natural disturbance through forest management has an important role to play in biodiversity management (Kimmins, 2004; Binkley, 2021). Thus, human disturbances can increase biodiversity compared to strictly preserved forest ecosystems on the same site. What also needs to be considered is that biodiversity in a forest stand – as other traits – varies in time and space. The greatest biodiversity is achieved with a variety of stand structures (cf. Figure 1.10) over broad spatial scales; therefore a range of different stand structures across forests and landscapes is also important (O'Hara, 2014).

7.3.3.1 Requirements

There is a long tradition of attempting to maintain biodiversity by banning human interventions. The rationale behind this strategy is that high levels of biodiversity exist in natural forests and any previously managed forest set aside for strict conservation will eventually return to the biodiversity level typical of the woodland community under study. However, since forest biodiversity is commonly highest at intermediate intensity and frequency of natural disturbances, contemporary biological and forest sciences agree that natural and man-made disturbances are the main agents of ecosystem diversity (Kovac et al., 2018). In general, human interventions and natural disturbances share many similarities and can be characterised in similar ways, but Kimmins (2004) pointed out that the latter often involve a much larger variety of scales than the former and impacts of forest structure and density also vary much more than in human interventions. Management for biodiversity and conservation often involves the retention of mature habitat trees that can serve as roosts, nest trees or provide cavities, etc., and later naturally transform to deadwood.

7.3.3.2 Management Strategies

Forest restoration or ecological restoration is a special variant of conservation and can largely be interpreted as a variant of conversion, see Section 2.4. Broadly speaking, forest restoration engages in activities that assist in the recovery of forest ecosystems when they have been degraded, damaged or destroyed (Newton, 2007; Palik et al., 2021). In a sense, restoration is a way of counteracting the effects of concerting forests to other uses (O'Hara, 2014). The domain of forest restoration can range from situations where no forest is currently present to cases where a current forest has traits that are undesirable (Binkley, 2021). Restoration may involve the re-establishment of the characteristics of a forest ecosystem, such as composition, structure and function, which were prevalent before its degradation. Restoration can, for example, apply to situations of low tree density (overcutting), of invasive species and of uniform young forests that require old-growth structures (O'Hara, 2014). In the context of forest restoration, it is widely assumed that the promotion of native tree species will confer greater environmental benefits than the establishment of non-native tree species often employed in RFM. However, as experienced in Britain, plantation forests of exotic tree species can sometimes be of significant habitat value for wildlife and in some cases, it may be necessary to first establish non-native tree species to act as a nurse for establishing native species. Although it can sometimes involve intensive operations, the preferred method of forest restoration is often to allow the forest to recover naturally through the process of succession. This is also termed *passive restoration* (Newton, 2007). Restoration in general can involve silvicultural systems and other restoration methods including various types of planting and seeding. The group (shelterwood) system (see Section 5.3.2) creating gaps of variable size, for example, is often applied in this context, since it promotes the regeneration or re-introduction of light-demanding species (cf. Appendix B). The re-introduction of native broadleaved tree species or allowing them to colonise stands previously dominated by conifers usually is an important component of forest restoration. Halo thinnings (see Section 5.2.2) can help safeguard remnant broadleaved trees in conifer plantations. At the same time, the attempt should be made to preserve or create as many habitats as possible, e.g. by promoting deadwood and small wet areas. Nurse crops using nitrogen-fixing species (cf. Section 2.3) can be a good tool of restoration on deforested sites (O'Hara, 2014).

In North America, variable-density thinning is seen as a major way to implement and balance different habitat requirements in space and time (Kimmins, 2004; Franklin et al., 2018; Palik et al., 2021), see Section 2.4.3. The concept was originally developed in the context of conservation and biodiversity. As previously discussed, VDT can be interpreted as a special method of forest restoration fostering a fast development of structural characteristics of old-growth forests (O'Hara, 2014; Binkley, 2021). In addition to encouraging standing and down deadwood, it also encourages the active creation of '*decadence*', i.e. the deliberate killing or wounding of healthy trees or retaining defective trees. Decadent, declining and injured trees are now recognised as the near-term source of deadwood, and the decline process is considered important to biodiversity. Creating decadence can include methods such as girdling, which involves removal of the bark (phloem and cambium) in a band around the circumference of the tree. High stumping, thus leaving a stump of 3–6 m, is another technique (Palik et al., 2021). However, any CCF management aiming at increasing or maintaining a high level of species and size diversity makes a valid contribution to maintaining biodiversity, since tree diversity and more direct measures of biodiversity are correlated (Pommerening and Grabarnik, 2019). Tree lifetimes are usually extended in CCF compared to those in plantations and therefore promote large, old trees that act as important habitat trees. Since they have a wide range of tree sizes, multiple canopy strata and often mixed tree species, CCF woodlands therefore typically have a large level of habitat diversity where the longevity of trees in CCF also contributes to this (O'Hara, 2014).

There is also the option of applying coppice with standards and the coppice selection system (see Section 1.3), which are known to increase biodiversity (Vanbeveren and Ceulemans, 2019; Vollmuth, 2022). The adoption of crown thinnings (cf. Section 5.2.2) is a good strategy for introducing greater structural diversity in forests. In addition, individual-based forest management (perhaps in combination with group thinnings, see Section 3.4 and Section 5.2) and specific structural thinnings as outlined in Section 3.6 help diversify forest stands thus increasing resilience. Following from this, the regeneration through group (shelterwood) systems (see Section 5.3.2), as already mentioned, can help preserve irregular horizontal and vertical forest structure.

7.3.3.3 Forest Margins

Many authors have emphasised the importance of considering the fact that forest ecosystems are part of a greater landscape (Palik et al., 2021; Binkley, 2021). The boundaries of forests may legally define different ownerships, but from an environmental point of view, they are not firm boundaries at all but places of frequent exchange and interaction and sometimes they are transition zones. This, for example, becomes obvious when wildlife that mostly live in forests cross the forest boundaries damaging the crops in an adjacent agricultural field. Therefore forest edges and transition zones between forest and open landscape deserve special attention and have been addressed in various national and regional definitions of and guidelines for CCF, see Section 1.5. As transition zones, they constitute special habitats for many plant and animal species. In addition, the interaction between strong winds and forest margins with a steep edge often leads to strong vortices and turbulences that can have devastating effects on forest stands behind the margins (see Figure 7.7). Edge effects or edge influence relate to the reciprocal influences of adjacent patches on each other with respect to some parameters, processes or organisms (Franklin et al., 2018; Palik et al., 2021). These effects are particularly crucial when there are forest stands close to the forest margin which are in transformation to CCF, since transformation can temporarily decrease tree resilience to wind and snow. The absence of well-designed forest margins in 'wall to wall' plantations can also lead to drought or frost effects near the boundaries of forests with the open landscape and prevents forest microclimate from building up (Otto, 1994). Trees with thin bark (e.g. *F. sylvatica*) can suffer from sunburn (i.e. an overheating of the cambium layer) if unprotected by vegetation at forest margins with southern exposure. Forest margins also reduce the risk of wildfires and sustain rare plant and animal

Figure 7.7 Blowdown and associated processes that occur along an edge are part of a suite of edge-related phenomena known as edge effects (Palik et al., 2021). An example from Clocaenog Forest (North Wales, UK) where a sharp forest edge was abruptly created by clearfelling an adjacent *P. sitchensis* (Bong.) Carr. stand. Source: Courtesy of Jens Haufe.

species (Weihs, 1990). At the same time, they increase the aesthetic value of forests, particularly in the context of recreation (Figure 7.9). Similar to the management and protection of other special habitats, the design and treatment of forest margins is a central concern of CCF.

Some authors distinguish between *outer* and *inner* forest margins (Otto, 1994). While the former have already been described, inner forest margins form the transition between individual forest stands, between forest stands and forest roads, between forest stands and water courses, power lines or bogs to give just a few examples. Inner forest margins are also important, but usually have a smaller width compared to outer forest margins whilst being more strongly influenced by the surrounding forest. Particularly important outer forest margins are those that are exposed to prevailing winds and to strong sunlight, e.g. the Southern and South-Western parts of forests in Europe. Relative to other forest margins, the highest management priority should be assigned to them (Burschel and Huss, 1997). In plantations, outer forest margins are often horizontally straight and have a steep, vertical ascent, i.e. there is no gentle gradient towards the open landscape (Figure 7.7). Good-practice outer forest margins can vary in width between 15 and 30 m (cf. Figure 7.8). The higher the wind risk at a particular point, the wider the forest margins should be (Weihs, 1990). They are often subdivided into three different zones that can have irregular boundaries, (i) a zone of low vascular plants (approximately 5 m in width), (ii) a zone including shrubs and secondary tree species of low total height (up to approximately 15 m; ~5–15 m in width) and (iii) a transition zone involving secondary and primary tree species (total height >20 m; ~15 m in width). The trees here should be allowed to approach the characteristics of open-grown trees with exceptionally low *h/d* and large crown ratios, see Section 2.2. Outer forest margins should be included in the (re)planting of stands on bare land. They can be retrospectively introduced by heavy thinnings of mature trees in the margins. Shrubs can be planted in zone 2 or their colonisation can be encouraged by loosely piling up deadwood and shrub cuttings in a linear or slightly wavy fashion along the forest edge. Sturdy wooden posts on either side may help hold the deadwood in place. Such dry or dead hedges create wind-still habitats; however, light, water and air should still be able to reach the ground so that

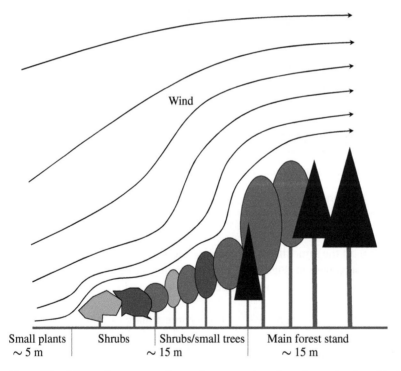

Figure 7.8 Schematic structure of outer forest margins where forests border with the open landscape. The boundaries between the zones should be irregular and wavy. Source: Adapted from Otto (1994) and Burschel and Huss (1997).

seeds protected by the deadwood can germinate in these piles and seedlings can grow through them to eventually form a live hedge. In German-speaking countries, this concept is known as *Benjes* hedge. It is also possible to combine dead hedges with limited shrub planting. The use of dead hedges is a realisation of the principles of biological rationalisation (Section 2.6) and the aforementioned principle of passive restoration (Figure 7.9).

The zones should have irregular boundaries, particularly the outer shrub frontier should be undulating thus allowing for small inlets of 10–20 m diameter every 50–60 m (cf. Figure 7.10). Small advance groups of shrubs or small trees in front of the main line of the

Figure 7.9 Almost perfect mature forest margins enveloping a small mixed *Quercus petraea* (MATT.) LIEBL. - *F. sylvatica* L. community woodland outside the village of Sieboldshausen near Göttingen (Lower Saxony, Germany). The forest margins include species such as *Acer campestre* L., *Cratraegus* spp., *Rosa canina* L. and *Sambucus nigra* L. The unfortified road is convenient for the maintenance of the forest margins and shared with farmers and hikers. Source: Arne Pommerening (Author).

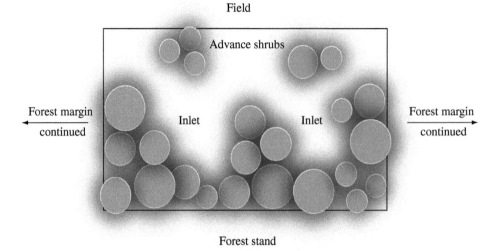

Figure 7.10 Schematic structure of an outer forest margin with inlets and advance shrub groups seen from above. Source: Adapted from Codoc (2020).

forest margin would even more enhance the diversity and resilience effects. This diversifies forest margins and increases their total length and surface area. Regional features such as *clearance cairns*, i.e. irregular and unstructured collections of stones which have been removed from agricultural fields, small ponds, branch heaps, wildlife paths and beds and similar features should not be destroyed but integrated in forest margins.

Forest margins need to be regularly maintained so that the character of gradual transition from herb layer to shrubs and tall trees is preserved. In these maintenance operations, the aforementioned inlets and density reduction require special attention. It is good practice to apply methods typical of managing coppice and coppice with standards woodlands (see Section 1.3) when maintaining forest margins, i.e. coppicing some trees and shrubs whilst leaving others alone and making use of their resprouting potential. Some of the forest stand area needs to be set aside for the creation of forest margins and they then become a long-term non-productive conservation area. Having said that, it is naturally possible to sell wood felled as part of maintenance work in the forest margins for energy wood or other purposes to pay for the conservation effort. There should, however, also be a fair share of standing and down deadwood in forest margins. Maintaining forest margins is a kind of *process conservation* similar to the interventions in selection forests (cf. Section 5.4) because the transition character and structure of forest margins is otherwise lost with time. Forest margins are also aesthetically pleasing (cf. Figure 7.9) and therefore have a positive effect on visitors.

Shrub and secondary tree species planted in forest margins should be native species and provenances that naturally occur in the region. This requirement usually is automatically taken care off when applying the aforementioned method of dead hedges. Non-native tree species colonising forest margins should be removed as part of the maintenance work. The definite species selection depends on local site conditions, particularly on the soil nutrient and the soil moisture regimes. Forest margins are also an excellent refuge of rare secondary tree species that for their light demand or for their lack of competitiveness otherwise find poor conditions for survival in commercial high forests including, for example, Central European species *Sorbus aucuparia* L., *Sorbus torminalis* (L.) CRANTZ, *Sorbus domestica* L., *Malus sylvestris* (L.) MILL., *Pyrus communis* L. and *Acer campestre* L.

Ideally, inner and outer forest margins are connected with a system of hedges running along field boundaries in the open landscape outside the forest which provides habitat corridors for plants and wildlife.

7.3.3.4 Streamside Forest Buffers

Common definitions of CCF often express the need to preserve water bodies located in forests and particularly streams, see Section 1.5. Streams, watercourses or bodies of water in general that are partly or wholly located in forests constitute important habitats for organisms that depend both on terrestrial and aquatic environments with high conservation status in many countries. Often streamside forests are biodiversity hotspots and provide critical habitats for a diverse community of plants and animals (Nyland, 2002). The sides of these bodies of water merit special attention, particularly when the adjacent forest stands are managed for commercial purposes. In addition, the combination of water and forest is highly attractive to many people and therefore plays an important role in recreation. Often, aquatic food webs in small streams and pools are based on the input of leaves and other detritus from the surrounding forest. It is usually desirable to maintain some shading over streams in order to keep the water in them cool, principally because the capacity of water to absorb oxygen and other gases decreases with warming. This is important for many fish species and other aquatic animals (Smith et al., 1997). Forest streams are crucial for flood mitigation and filtering of sediments and nutrients (Palik et al., 2021). More important than the few large rivers and lakes are often the multitude of small streams that transect forest ecosystems and form networks (Kuglerová et al., 2020).

In the past, bodies of water were frequently ignored and neglected. Non-native conifers were often planted up to the banks of streams. As a consequence, the deficit of light reduced the growth of algae and interrupted the food chain of insects. In addition, decomposing conifer needles increased the acidification of the water in such streams.

Buffer, buffer strip or *leave strip* is a term used to describe the vegetated area of land in the riparian zone between the watercourse and agricultural land or other land use. *Riparian zones* include the land along permanent and intermittent streams, rivers, ponds and wetlands (Nyland, 2002). Buffer strips have the potential to conserve, enhance and protect the water environment. Buffer strips can also slow down flood flows as well as providing bank stabilisation and habitat. A buffer can consist of grassland, wetland, scrub or trees (Smith et al., 1997; Scottish Environment Protection Agency, 2009).

Streamside buffer strips enveloping watercourses have therefore been introduced to mitigate the negative effects of commercial forestry or other nearby land use on water quality and aquatic habitats. Buffer strips can filter diffuse pollutants, sediments and prevent particulate matter and dissolved nutrients from reaching streams (Smith et al., 1997). They act as place of refuge for wildlife and provide connections to other habitats (Scottish Environment Protection Agency, 2009). Streamside forest buffers are in principle special variants of forest margins (see the Section 7.3.3.3). In countries where RFM dominates, streamside forest buffers include retained parts of forest plantations that were exempted from clearfelling with the objective to protect the stream and its water (Tiwari et al., 2016). Commonly the trees in such buffers have not previously been prepared for a situation where they are all of a sudden exposed to the elements, which often leads to high mortality. This is contrasted by CCF management, where no clearfelling occurs and streamside buffers have the purpose of conservation areas where only strictly natural vegetation is allowed at low densities and management impact is reduced to a minimum (Figure 7.11). The selective removal of trees near the banks of streams increases the amount of sunlight hitting the water, increasing photosynthesis among water plants and algae and raising water temperatures where this is desirable.

When cuttings are necessary, trees should always be felled away from water bodies. Streamside buffers usually have a fixed (e.g. 2–5 m or 15–20 m) or variable width. Ecologically, though, a riparian zone spreads laterally across the floodplain and up onto adjacent slopes that drain directly into the water (Nyland, 2002). As a rule of thumb, the width of the buffer zone on each side should be at least equal to the width of the stream (McNulty et al., 2021). Recent research indicates that variable buffer width that takes site-specific differences and localised sensitivities into account yields greater conservation success (Tiwari et al., 2016). Buffer width should depend on local watercourse width, topography, steepness of slope and soil type (Nyland, 2002). In a CCF context, streamside buffers of comparatively small, fixed width are recommended in conifer forest stands. In native broadleaved forests, no streamside buffer is explicitly defined, but instead there is often a recommendation for managing the whole surrounding area as a floodplain forest and to adapt forest management accordingly (Figure 7.11). This seems to match North-American recommendations (Palik et al., 2021). The management of floodplain forests can also be designed and documented as a forest development type (cf. Section 7.2). In the context of streamside forests, common details of advice include:

- Gradual removal of exotic species, particularly of invasive species,
- Promotion of native broadleaves, particularly of species adapted to semi-aquatic conditions, e.g. *Salix* spp. and *Alnus* spp.,
- Heavy machinery should not be used in streamside forests,
- Coppicing, pollarding (where appropriate) and selective thinnings near watercourses can support biodiversity,
- Allowing for a higher proportion of deadwood than in purely terrestrial forest stands,
- Encouragement of a mosaic of small, open tree groups.

Figure 7.11 Small stream running down from Mecklenbruch bog in the Solling Mountains towards the Weser Valley near Holzminden (Lower Saxony, Germany) where the author grew up. The stream with a width between 2 and 3 m does not require a buffer, since it runs through a mixed broadleaved CCF woodland where native *Alnus glutinosa* (L.) GAERTN. is encouraged to colonise the banks. Source: Arne Pommerening (Author).

7.3.4 Water Catchment Management

The need for clean, abundant water supplies is likely to increase as the climate changes and the human population continues to grow. Due to its low-input, nature-oriented land use, the quality of water coming from forest land is almost always higher than from other land uses that expose the soil and therefore ideal for the supply of drinking water. Much of the water that sinks into the soil emerges later in springs of clear, filtered water (Smith et al., 1997). Compared with other land uses, forests generally produce less surface and subsurface water runoff due to their relatively high rates of transpiration (McNulty et al., 2021). Forest soils filter water so that the leaking of pollutants into groundwater is greatly reduced. Many forest management practices affect water services and, if well designed and implemented, can contribute to water management objectives.

7.3.4.1 Requirements

The areas of soil within a stand that are most crucial for watershed management are the source areas that are so nearly saturated that water can flow from them beneath or on top of the soil surfaces. Source areas often vary with season (Smith et al., 1997). Any removal of plants and impacts on soils can affect forest water. Undisturbed forests often have the highest water quality. Trees are the primary source of leaf litter which through decomposition contributes to forest soil formation and water filtration. Tree roots help mitigate soil movements, e.g. landslides and soil erosion by holding soils on slopes. Tree species have different water-use efficiencies (WUE) describing the relationship between plant productivity and water use. A species with a low WUE uses more water to produce the same volume of growth compared to a species with a higher WUE, see Table 7.3. Tree species with low WUE such as *P. abies* and *Betula* spp. should be avoided in catchments. Faster tree growth leads to a higher absolute use of water per unit of time (McNulty et al., 2021). Tree species with a high potential of intercepting precipitation and high resilience should be preferred. An important concern in this context is also the widespread acidification of rivers and lakes by the deposition of sulphur and nitrogen compounds derived from the combustion of fossil fuels. Aluminium and other industrial depositions are also a concern. Forest canopies capture more of these pollutants from the atmosphere than shorter types of vegetation (Forestry Commission, 2014). Evergreen stands intercept and transpire substantially more water than stands composed of deciduous trees. Cuttings in evergreen forest may therefore have a greater effect than similar cuttings in deciduous forests (Smith et al., 1997).

7.3.4.2 Management Strategies

Operations such as harvesting, soil preparation and fertilizing can increase the quantity of suspended sediments and nutrients in water bodies. Fertilising and the use of pesticides can even contribute chemicals to water bodies. Thinning and harvesting operations should therefore be always selective as this is typically the case in CCF and thus help reduce surface water runoff. Trees should be felled away from water bodies, since pulling out slash or logs can do damage to stream banks (Smith et al., 1997). Any use of chemicals in catchments should be completely avoided. Tree fellings temporarily reduce the leaf area of a forest and increase albedo, thus reducing evapotranspiration and potentially increasing forest water tables and water yield (McNulty et al., 2021; Smith et al., 1997). Regeneration cuttings as part of silvicultural systems cause larger and less temporary increases of water tables. Soil compaction reduces water infiltration and should therefore be avoided as much as possible. Permanent extraction racks (see Figure 1.9) at large distances (e.g. 40 m) and fortified by brash mats help reduce soil compaction. Soil erosion should be minimised and this may require avoiding or limiting tree fellings and extraction at steep slopes. Thinning intensities should be generally moderate whilst keeping the main canopy open, particularly in proximity to surface water. Narrow openings in the main canopy can accumulate wind-created drifts that provide useful ways of storing water and delaying run-offs (Smith et al., 1997). Maintaining more or less undisturbed buffers (see the previous section) between actively managed forest areas and surface water reduces erosion and sedimentation. Particularly mature trees with cavities and deadwood are invaluable in buffers. No heavy machinery should operate in these buffers. Forest operations can also be limited to periods where the soils are dry or frozen. They should not be scheduled shortly before imminent peak events of water accumulation (rain seasons, snowmelt). Organic surface matter formed by decomposing litter and other biotic components is essential for water infiltration and therefore for reducing rapid overland waterflows. Raw humus accumulations of needle litter of some species, e.g. of *Picea* spp., can acidify the water passing through organic soil horizons during the infiltration process. Variability of tree size as introduced by local crown thinnings and mixed species can greatly help secure high water quality. Locally adapted native tree species are often the best choice for catchments, as WUE and resilience under local conditions are high. There should be a good balance between slower and faster, minimal-water and high-water using tree species (McNulty et al., 2021). It is also good practice to retain some large woody debris within streams unless they form a barrier to fish or pose a flood risk (Forestry Commission, 2014).

7.3.5 Urban and Recreation Forestry

Woodlands in or near towns and settlements are a vital component in the daily routine of local populations. It is the local forest close by which is of utmost importance to the residents of a village or town. Urban, peri-urban and community forests are increasingly used for a wide range of activities including dog walking, jogging, cycling, berry picking and hiking (Hart, 1991). These activities can be summarised as *forest recreation*, which is defined as recreational activities and experiences conducted in forest and associated wild-land environments that are dependent on the natural resources of these areas (Hammitt, 2004). Local people not only actively use forest resources but increasingly also wish to get involved in the question of management. There is, for example, a strong movement in favour of community forests in Britain. In Britain, trees and woodlands managed predominantly for aesthetic and recreation purposes, are referred to as *amenity* trees and amenity woods (Hart, 1991).

> *Urban forestry* is generally defined as the art, science and technology of managing trees and forest resources in and close to urban areas for their value in the landscape, for their physiological, sociological, economic and aesthetic benefits including recreation and includes trees in streets, urban parks, on land reclaimed from previous industrial use as well as those in urban woodlands and gardens (Hart, 1991; Konijnendijk et al., 2006).

7.3.5.1 Requirements
The recreation pressure on woodlands in the vicinity of urban centres and other settlements can be considerable. Potential user conflicts need to be avoided or mitigated by a clever system of different forest trails and forest structure in order to segregate visitors in an

appropriate way (Arnberger, 2006). Not only trail properties, signs and forest rangers can help with that but also differences in forest structure including tree density. Infrastructure such as footpaths, log-fire places, information boards are of great importance to visitors. In a study in Finland, Koivula et al. (2020) showed that forest structures typical of CCF are clearly more attractive to visitors than forest sites with medium-to-large group and patch clearfelling. Jephcott (2002) found that the majority of forest visitors in North Wales disliked sudden visual change, preferred mixed-species forest stands and a mix of large (old) and small (young) trees, cf. Table 7.3. Species differences were often only perceived in broad categories, such as conifers and broadleaves. Low tree densities and light woodland environments were usually preferred to dense, dark forests with closed canopies. The research carried out by the aforementioned authors generally suggested that high tree diversity, particularly in terms of tree size (stem diameter and height), is much appreciated. Amenity and recreation forests should be resilient, as traces of decline (e.g. windthrow) are often perceived negatively by visitors. Straight boundary edges should be avoided and harvesting or thinning areas should blend into the landscape. The visibility of special features such as rocks and veteran trees should be enhanced. Signs of active forest management such as gaps in forest stands, stumps, crown residue and timber stacks at road sides are often perceived negatively by visitors from urban centres. Therefore, management activities need to be carried out sensibly and not too often (Franklin et al., 2018). Forest margins and moderately sized gaps can positively contribute to the recreation experience. In Britain, the safety of visitors parking their cars on car parks inside the forest or at the forest edge is a matter of government concern (Forestry Commission, pers. comm.). It is also part of management to preserve and appropriately present natural and historic monuments inside forests, e.g. scenic old tree groups, rock formations and burial mounds. Specialised thinnings are often necessary to emphasise such features, make them more visible and to present them in a suitable context.

7.3.5.2 Management Strategies

The requirements of urban and recreation forestry clearly point into the direction of complex CCF structures with more than two canopy layers and mixed species, see Figure 2.15 in Section 2.4. Such structures give visitors the opportunity to see large-, small- and medium-size trees in close proximity, see, for example, Figure 2.4. At the same time, signs of active forest management such as tree stumps and crown residue disappear in the undergrowth so that within no time after an intervention it is hard to tell whether anything recently happened to the forest stand in question (see Figure 7.12).

Selection forests (cf. Section 5.4) or two-storeyed high forests (cf. Section 5.5) are ideal for urban and community forests, since they are often small in size so that selection systems are possible to apply. The complex structure typical of selection forests is not only preferred by visitors but also lends more privacy to individual visitors or groups even if the overall visitor numbers are high. Trails should be properly marked and signposted so that users with different interests know where to go. This is a possibility to separate user groups such as mountain bikers and walkers. Visitor centres help attract people but also channel crowds through the provision of exhibitions, cafeterias, toilet facilities, conference/meeting rooms and shops.

Figure 7.12 Advance regeneration of *P. abies* (L.) H. Karst. enveloping and camouflaging stumps of mature trees felled previously at the Hirschlacke stand of the Austrian Schlägl estate. Source: Arne Pommerening (Author).

In some countries where crimes are more common than in others, there are policies in place ensuring that tree densities around car parks but sometimes also throughout community and urban woodlands are low so that visitors feel safe and it is hard for potential attackers and stalkers to hide in clusters of vegetation. Obviously, this is a measure additional to constant camera surveillance and frequent police patrols, but it helps encourage people to spend their spare time in nearby woodlands. In such conditions, urban and community forests can resemble the structure of urban parks, and individual trees show the morphology of open-grown trees rather than that of forest trees. Localised low tree densities can also be a good strategy for achieving big, old-looking trees fast, see Section 2.4.3. Thinnings and other silvicultural treatments should be arranged in ways that expose attractive features and distant vistas to public view along roads or trails (Smith et al., 1997).

The results of a vast amount of research of recent years have revealed that forest visits promote both physical and mental health by reducing stress. Utilizing forests effectively in health promotion could reduce public health care budgets and create new sources of income (Karjalainen et al., 2009). Visits to forests have, for example, been successfully used in various therapies (Sonntag-Öström et al., 2015). This is an interesting, new research field providing many opportunities for helping people in need. So far little is known about how to optimise forest stand structure in order to achieve a maximum of health benefit. Among other factors this probably also depends on what kind of forest types patients were exposed to during their early upbringing. Petucco et al. (2018), for example, found that forest visitors in six different countries had quite different density preferences: Austrian, Portuguese and Romanian visitors generally favoured dense stands of young *Quercus robur* whilst Danish, British and Swedish respondents preferred much lighter and open woodland conditions. Theoretically, it should be possible to optimise forest structure for different groups of patients depending on their health conditions. In the absence of better information, it may be a good approach to aim at a forest management that the majority of members of the general public find aesthetically pleasing. This again suggests complex CCF structures with a mix of species and tree sizes interspersed by small bodies of water whilst keeping overall tree densities low so that patients feel comfortable and do not get disturbed by a lack of light or limited visibility. Managing forest stands for improving human health may also involve growing trees with ornamental foliage such as *Larix* spp. or *Acer* spp. that show pleasant spring and autumn colours.

In light of visitor sensitivities to forest management, particularly when they are not in favour of tree fellings, it may be a good strategy to inform the general public about the importance of interventions. Jensen (2000) showed that such information clearly helps win the sympathies and support of visitors. This can be achieved through information boards installed near paths leading into the forest stand in question or by providing printed leaflets. A third option is to allow visitors to scan a QR code with their mobile phones. The code can then lead them to a website where the information text is available in full. Figure 7.13 presents an example of such an information leaflet. It can help win the visitors' support for limited forest management operations and may also enhance their experience when walking through the forest stand. In the leaflet, Jephcott (2002) briefly explained what CCF is and why Artist's Wood (see Section 2.2.1) is special without using forestry jargon. For transparency, the fate of the trees felled is also described. Leaflets were preferred by the landowner at the time, the Forestry Commission Wales, since in their experience information boards were often vandalised.

7.3.6 Sustainable Energy Wood Production

Governments worldwide are striving to reduce our dependency on fossil fuels by promoting the use of sustainable and renewable energy sources. Energy from tree biomass is an important part of this strategy (Vanbeveren and Ceulemans, 2019). Sustainable energy wood production provides forest goods that can be used for heating and for the production of electricity. Energy wood can be a by-product of other types of forest management, for example, wood from regular thinnings in small-sized trees, cf. Table 7.3. A more specialised and traditional way of producing energy wood are coppice, coppice with standards and coppice selection system (Johann, 2021), see Section 1.3. An industrialised version of coppice, i.e. *short-rotation coppice* (SRC), is usually restricted to agricultural land. SRC is the practice of cultivating fast-growing trees at high densities on a rotation of 2 to 20 years. This practice is closer to agriculture than to forestry (Lindegaard et al., 2016; Vanbeveren and Ceulemans, 2019) and is outside the remit of CCF. Energy wood is processed in a number of ways including traditional logs, wood pellets and woodchips.

7.3.6.1 Requirements

Coppice with standards and the coppice selection system are clearly CCF management options, since both use selective harvesting. The coppice system alone essentially includes clearfelling (Nyland, 2002) because similar to the situation in agricultural crops all trees of a forest stand are effectively felled at the same time, and therefore, this regime is not an acceptable CCF technique unless coppicing is carried out selectively and staggered in time. Coppice with standards and the coppice selection system can potentially host more biodiversity than CCF high forests so that this form of energy wood production comes with a very desirable by-product and is often applied in conservation (Vanbeveren and Ceulemans, 2019; Vollmuth, 2022). Naturally, according to the principles of CCF, energy wood

Coed Arlunydd - Artist's Wood, nestled beside the A5, just outside Betws y Coed within the Snowdonia National Park is a magnificent woodland, enjoyed by hundreds of people each year. Locals and visitors have fallen in love with its big trees, its lush appearance and the wealth of wildlife that lives within it. It is also regarded as a gem by local foresters and researchers from the University of Wales, Bangor. For them it is a superb example of best forest practice that they are keen to preserve and enhance.

regeneration

How can you find out more?

For more information about Artist's Wood today, it's history and future and the ongoing research work being carried out or continuous cover forestry contact:

Dr Arne Pommerening
  School of Agricultural and Forest Sciences,
  University of Wales,
  Bangor, Gwynedd.
  Tel: 01248 382281 or

Forestry Commission Wales
  Forest District office address
  Tel: 01341 422289??

Other information or web sites:
  http://www.safs.bangor.ac.uk
  http://www.forestry.gov.uk

Forestry Commission

Coed Arlunydd
Artist's Wood

a woodland for wildlife and people

What makes Artist's Wood so special?

Although Artist's Wood is a comparatively small forest, it contains a wealth of trees of very different sizes. Very old big trees stand beside tiny seedlings, young and semi mature trees. The very oldest date as far back as 1921 when Douglas fir trees were planted around Betws y Coed. The youngest may be less than a year old. Such variety in size and age is rare to see within such a small area and with it has come a fantastically diverse woodland habitat of plants and animals.

How has Artist's Wood become like this?

Within Artist's Wood there is a continual cycle taking place. Seeds of the large veteran trees germinate on the forest floor and grown into small trees. As they grow the giant old trees protect them from the full force of strong winds, heavy rain and strong sunlight and suppress the growth of other plants. But as the younger trees grow older their needs change and they require more light to mature further. At this time foresters come in to harvest a small number of the big trees. They never remove them all, taking care to ensure that many are left throughout the woodland. After five or ten years the younger trees fill the space opened up by the felled giants, more young trees have grown up and require more light and new seedlings have emerged on the forest floor. The sequence carries on.

What would happen if the forest was left

Although Artist's Wood may look like a natural forest it does in fact depend on the regular cutting of the big trees. Without these interventions its appearance would change quite quickly. Within just a few decades it would be dominated by only a small number of very large trees. Little would survive in the darkness below their thick canopy of branches and the rich diversity of plants and animals would diminish. After another few decades these big trees would start to die or be blown down in heavy storms. Huge gaps would be created, empty of young trees to take their place. It would be many more decades before we would see a forest vaguely similar to the present day Artist's Wood.

How does Artist's Wood differ from other forests in Wales?

The majority of forests in Wales are quite different. Most contain trees of the same species, age and size and in the past they have been cut down all at the same time or 'clear felled.' At Artist's Wood only individual trees are removed. In general when foresters harvest only a few trees at a time and never remove them all it is known as 'near to nature' or 'continuous cover forestry.' Artist's Wood is one of the finest examples of continuous cover forestry in Wales.

Forests like Artist's Wood are more common in continental Europe where forests have been managed in this sustainable way for centuries. The use of continuous cover forestry is now being increased where possible across Wales.

So what's going to happen now?

A team of researchers from the University of Wales, Bangor and forest managers from the Forestry Commission have been studying Artist's Wood in great detail. After measuring every tree within the woodland they have worked out a way to remove the big Douglas fir trees without damaging the special character and appearance of the woodland. Each tree to be felled has been carefully selected to ensure its removal has maximum benefit on the rest of the woodland. As only single trees will be felled no large gaps will be created and the visual impact minimised. Similar operations are envisaged as necessary every five years.

What will happen to the trees removed?

The big logs from the Douglas fir trees will be removed and taken to sawmills to be sawn into large beams often needed in the restoration of historic buildings. The revenue from them will pay for the costs of preserving the woodland.

Coed Arlunydd
Artist's Wood

a unique place

Figure 7.13 Cover and inside of the leaflet explaining management at the CCF demonstration site Artist's Wood in North Wales (UK), see Section 2.2.1. Source: Jephcott (2002).

production should exclude the removal of bark, small branches (slash), crown residue and stumps from site, as these contain many nutrients that should be returned to the forest soil. In industry practice, energy wood produced from coppice forests including SRC is often combined with energy wood resulting from selective thinnings/harvesting plus residue of roadside and hedge cuttings.

7.3.6.2 Management Strategies

Where energy wood is a by-product almost any high-forest CCF management strategy is possible. In particular, energy wood is often a by-product of individual-based quality timber management, see Chapter 3. The neighbours of frame trees removed in local crown thinnings are a potential source of energy wood in these scenarios. However, it is also possible to produce energy wood in a simple-structure forest stand close to the structure of a plantation (de Jong et al., 2017) where just the clearfelling would be replaced by a shelterwood approach (cf. Section 5.3). Global crown and thinnings from below would be appropriate in this scenario. In all these scenarios, it is usually the diseased, poor quality or less vigorous trees that are felled and processed to woodfuel. In the context of this management strategy, natural regeneration is most effectively achieved by applying uniform shelterwood systems (cf. Section 5.3.1).

Coppice forests require tree species that are able to regrow from stumps or roots and climates that support this requirement. Typical coppice tree species are those that re-sprout abundantly after fellings and these include the genera *Alnus*, *Castanea*, *Eucalyptus*, *Liquidambar*, *Populus* and *Salix*, see also Section 1.3. Whilst coppice forests tend to be monospecies, coppice with standards often include a range of species. Thinnings would be carried out infrequently to encourage growth and release individual stems. The principals of individual-based forest management (cf. Chapter 3) can theoretically be applied to individual tree stems regardless whether they originate in the same or in different stools. However, there is no need to 'individualise' or single out one stem per stool. These methods are even of greater importance in coppice with standards, whilst in coppice forests thinning from below and maximising overall biomass is more appropriate. Since coppice stools reproduce vegetatively from dormant buds there is no need for a silvcultural system to produce regeneration, but some naturally occurring seedlings can from time to time replace exhausted tree stools.

7.3.7 Forest Cemeteries

Burying people in forest environments has a very long history going probably back to the beginnings of humankind (Mantel, 1990; Hasel and Schwartz, 2006; Kirkinen et al., 2022). These traditions are not limited to any particular culture, religion or continent. In recent decades, an increased interest in forest burials as an alternative to traditional cemeteries has given rise to the dedication of substantial forest land to this particular land use. As the global human population continues to grow rapidly, space for traditional cemeteries is increasingly becoming scare. At the same time in increasingly mobile societies, problems arise for descendants to maintain grave sites and forest cemeteries offer low-maintenance alternatives. Forest and tree burials have a long Buddhist tradition in East Asia and the concept of tree burials (jumokusō) emerged in Japan in the late 1990s (Nilsson Södergren, 2020). This concept can involve burying the cremated remains of a deceased person in the ground with a tree planted above the burial or nearby in what otherwise is a traditional cemetery. Another variant of this concept places biodegradable urn burials without vaults near trees in an existing woodland, otherwise known as natural burial. In both cases, trees act as grave markers and tablets inscribed with the name of the deceased are attached to these trees. In the latter variant, the woodland and forest ecosystem is kept intact, whilst relatives and other visitors are able to locate and access individual burial sites for memorialisation. Forest cemeteries in Japan are often connected with Buddhist ceremonies and temples, but not always.

Forest and tree burials were introduced to Europe only recently, i.e. towards the end of last century. Individual graves are not visibly demarcated, but trees act as grave markers (Figure 7.14). Whilst in Switzerland, the ash of the deceased person is scattered aboveground near the grave-marking tree, the cremated remains are buried in biodegradable urns near the tree markers in Germany. Ground preparation or any other manipulation of the forest is usually not permitted.

In Central Europe, the concept of forest and tree burials was initiated first by Ueli Sauter in Switzerland and later taken on by professional companies such as *FriedWald* and *Ruhe-Forst* in Austria, Germany and Switzerland. The companies rent forest land from forest owners for an extended period, e.g. 99 years, and provide cemetery services. In some cases, forest cemeteries are run by local communities. The concept has also been implemented in the Netherlands, Luxembourg, the Czech Republic and Sweden. The popularity of forest and tree burials is steadily increasing.

7.3.7.1 Requirements

Trees marking gravesites are freely selected by descendents or by the deceased themselves prior to their death. As such, the selected trees do not fulfil the 'formal' requirements of frame trees (see Chapter 3), but the forest managers involved have an obligation to keep these trees alive, which may not always be easy. For securing them, grave-marker trees and urn locations are mapped throughout the cemetery. In the case that such a tree dies, e.g. through diseases or windthrow, an alternative marker, e.g. a wooden post or a new tree planting, has to be put in place of the dead tree, but the death of grave-marker trees should generally be avoided if possible. The whole forest area has to be managed in such a way that no part of it presents a danger to visitors.

7.3.7.2 Management Strategies

Since most customers of forest cemetery companies are attracted by the prospect of being laid to rest in nature and becoming part of the perceived natural cycle of life, CCF is clearly the preferred management type in forest cemeteries. Prior to opening a new forest cemetery, the land owner has to arrange parking space, design footpaths and routes that are accessible to people with limited mobility, wheelchair users and people with sensory impairments. The landowner also carries out a preparatory thinning to reduce overall tree density.

Figure 7.14 Grave-marker tree at the burial site of the author's mentor Prof. Hanns H. Höfle at the Burg Plesse FriedWald near Göttingen (Germany). The black tablet gives a poem by Rainer Maria Rilke, the name of the deceased and his life dates. Prof. Höfle was buried in a *F. sylvatica* L. dominated mixed-species forest that he was once responsible for as Forest District manager of the Lower Saxony State Forest Service. Source: Arne Pommerening (Author).

After burials have started, forest management is mostly limited to removing dead and dying trees and to understorey clearances, i.e. largely to thinnings from below. An important part of this work involves removing all trees or parts of trees that can potentially become a danger to visitors. A more proactive approach may occasionally require releasing grave-marker trees from dominating neighbour trees following the principles of local thinnings outlined in Chapter 3 and Section 5.2. Care has to be taken to channel visitors so that they are not harmed by forest operations. Harvesting and extraction damage inflicted on grave-marking trees need to be minimised. Some forest cemeteries are simultaneously managed for nature conservation. Since this a new development, appropriate long-term forest management is still much under development, e.g. the use of silvicultural systems for natural regeneration has to be clarified across different climates and local site types. The uniform shelterwood system may be particularly appropriate for regenerating such sites where gap sizes need to be kept small.

7.3.8 Protection Forests

Using forests to protect people and their infrastructure from avalanches, debris flows, surface erosion (caused by precipitation or by wind), rockfall, floods, noise, cold/hot air, unsightly objects (e.g. rubbish tips, industrial plants, quarries) and industrial emissions (dust, gas, radiation) is almost as traditional as using forests for timber production. This forestry objective is very specialised and usually applies to certain locations only, e.g. mountains and peri-urban regions. Protection forests, which sometimes can be just small strips of forest land along major roads or landfills, have been addressed in detail in many other textbooks (e.g. Bartsch et al., 2020; Burschel and Huss, 1997; Mayer, 1984; Nyland, 2002; O'Hara, 2014; Rittershofer, 1999), and I therefore only provide a short overview here.

Forest stands need a depth of at least of at least 200–250 m to reduce noise markedly. In urban areas, depths of 80–100 m are recommended. Evergreen conifers with long crowns grown at low densities and combined with a dense shrub understorey are very efficient to absorb noise. It is good advice to select species that can tolerate salt and car fumes. Deciduous broadleaves with large leaves have an effect similar to conifers, but do not provide efficient protection in winter (Mayer, 1984).

Due to the large relative surface area of their needles, conifers have a high potential for intercepting dust. Mature conifer forests with complex forest structure are most suitable for dust interception (Mayer, 1984).

The protective ability of a mountain protection forest is mainly provided by the presence of trees acting as obstacles to mass movements. Tree stems halt falling stones and tree crowns prevent, by snow interception and by snow release, the build-up of a homogeneous snow layer that otherwise may glide as a compact blanket. Tree roots reduce the hazard of shallow landslides. The permanent input of litter reduces surface erosion and increases the water-holding capacity of the soil through the build-up of an organic layer. Tree roots can also increase the soil volume available for water storage, in particular on soils with moderate permeability. Even dead trees lying on the ground may act as barriers to downslope mass transfers (Brang et al., 2006). Tree crowns intercept dust and gases, filter air and reduce noise.

CCF plays an important role in protection forests, since in most cases, it is necessary to maintain a permanent forest with continuous tree cover to meet the protection objective at all times. Clearfellings and even cutting larger gaps are not a management option. Since there is a requirement for a permanent forest, the selection system (cf. Section 5.4) has often been proposed as the best solution, see Table 7.3. In general, growing mixed-species forests using locally native tree species in complex forest structures (cf. Figure 2.15) is a good strategy. With increasing altitude, homogeneous forest structures become increasingly rare in natural forest ecosystems and complex stand structure dominates (Brang et al., 2006). Protection forests can be situated in three different zones, i.e. the *initiation* or *release* zone, the *transit* zone and the *runout/deposition* zone of hazards. For example, in the case of rockfall, the release zone usually is of little importance whilst most protection forests mitigating avalanches are located in this initiation zone. Active forest management can help increase individual-tree resilience to disturbances and also enhances the potential for mitigating the agent or hazard in question. Crown thinnings favouring resilient trees are a prominent part of this strategy.

In European mountain forests mainly used for protection against avalanches, the inclusion of resilient upland tree species such as *Pinus mugo* TURRA, *Alnus viridis* (CHAIX) DC., *Sorbus aucuparia* (L.), *Sorbus intermedia* (EHRH.) PERS., *Salix* spp. has been beneficial. At higher elevations, it is also advisable to maintain a group structure given the heterogeneity of topography and soils (Rittershofer, 1999). Groups and clusters of trees have long inner forest margins (see Section 7.3.3) that often act as breaks to natural disturbances, preventing 'domino' effects where each fallen tree triggers the fall of adjacent trees (Brang et al., 2006). When protecting people from avalanches, high stand density is less important than in situations where rockfall is the prevailing hazard.

In upland protection forests, the availability of nurse logs, i.e. downed deadwood, has been found to be crucial to successful conifer regeneration (Figure 7.15). Decomposing nurse logs protect seedlings and saplings from predators, pathogens and drought. The sparse occurrence of nurse logs in many managed forests is therefore seen as a serious impediment to natural seedling establishment (Brang et al., 2006).

Dorren et al. (2005) concluded that tree density should not be too low and there should be a good mix of large and small trees for forests to be able to intercept large rocks efficiently. The authors also recommended that in thinning operations trees should be cut at a height in excess of 1.3 m so that tree stumps do not act as trampolines for falling rocks, and snow movement is prevented. Rammer et al. (2015) suggested cutting narrow elongated ellipses of 20–30 m length and 6–10 m width at an angle to the slope direction in the Austrian Alps when transforming even-aged forests to complex CCF. These ellipses were cut instead of traditional large gaps used in the application of group systems (cf. Section 5.3.2) to lowland sites. This gap pattern is also important for avoiding the release of avalanches (Brang et al., 2006). In conjunction with developing resilient individual trees, this design was successful in helping intercept rocks. The authors also found that using more complex forest structures (cf. Figure 2.15) in combination with biological rationalisation (see Section 2.6) reduced the costs of

Figure 7.15 Naturally established seedlings and saplings of *P. abies* (L.) H. KARST. on nurse logs in Białowieża Forest (Poland). Source: Arne Pommerening (Author).

maintaining protection forests at high elevations. Coppice forests (see Section 1.3) are also a feasible option for mitigating rockfall, avalanches and landslides. On steep slopes, no active management of forests is likely to be the best option. In cases where residual risks are high, protection forests need to be complemented by artificial defence systems, e.g. avalanche barriers, rockfall nets, terraces, poles, tripods and dams (Brang et al., 2006).

8

Training for CCF

Abstract

It is comparatively easy to manage plantations, but managing CCF woodlands requires much more knowledge and long-term experience with the interaction between trees at a particular location, the corresponding environmental conditions and human interventions. Moving from plantation management towards CCF does not only involve a transformation of forests but also a transformation of organisations (such as forest services and forest companies) and their operational staff. Knowledge and experience deficits can be addressed by specific silvicultural training and much experience in designing CCF courses for forest practitioners has been made in different countries. In the last few decades, marteloscopes have been very effective in strengthening the transition between indoor training and practical implementation. In marteloscope exercises, participants log down their tree selection preferences which are then analysed statistically. Instant feedback in the form of individual result and feedback sheets are made available to each trainee after completion but it is also possible to carry out research on the tree selection behaviour of trainees. These research results can provide important information on training requirements and training quality. Until recently, little attention has been paid to aspects of training and yet this topic is crucial to the successful uptake of CCF.

8.1 Introduction

One of the greatest obstacles to the uptake of continuous cover forestry (CCF) is organisations and their staff, since they are surprisingly inert and slow to adapt. Countries with a long RFM tradition have highly compartmentalised state forest services and forest companies, cf. Section 1.10. In such conditions, forest managers do not oversee any longer the whole process of forest management involving planting, thinnings, regenerating trees, harvesting, road building, conservation, etc. in a forest area they are responsible for, but each of these fields have separate specialists who are in charge of only one narrowly defined field. The activities in each of these fields are fairly standardised and this organisation of work is reminiscent of work processes in large car or furniture factories. Specialisation makes it easier to assign new staff to posts and to move them around, as the need arises. For example, when an employee is promoted, s/he most certainly has to change workplace and often even their place of residence. This organisation principle has in some countries given rise to a trend, where even staff completely without forestry education is employed and trained in-house so that they are able to carry out the limited range of tasks in their specialised field. This heavily compartmentalised organisation of work is much supported by the area method of control, see Section 1.8. By contrast, in CCF, many of these fields or activities actually overlap in the same forest stand and influence each other, see Figure 1.14. There is hardly any spatial separation of tasks and schedules like in RFM. Often, highly complex decisions need to be made in CCF management that draw on almost every field of forest sciences. Such decisions are much supported by a system that allows staff to gradually accumulate experience regarding the interaction between local environmental factors and tree responses to forest management. In RFM, this complexity is deliberately reduced not only by simplifying and homogenising forest structure but also by compartmentalising work processes and tasks. A clear disadvantage of this strategy is that there are often problems with communication and quality issues at the boundaries of the specialised fields, since the system does not encourage looking ahead and beyond one's field of responsibility. For example, harvesting units have their own work schedules and often do not arrive in the forest when a local forest manager (who has another separate schedule) can introduce them to the harvesting operation; then they sometimes start the harvesting work anyway, since they are on a schedule. As a result, maps and/or instructions are sometimes incorrectly interpreted by harvesting staff and stands are, for example, clearfelled that were not supposed to be clearfelled for some time to come. In addition, the specialisation and compartmentalisation principle in RFM often goes hand in hand with keeping a comparatively small total workforce, and many crucial silvicultural decisions are made remotely in the office under time pressure and without a detailed appreciation of environmental conditions in the field. Often, *standing sales* are part of this office-centred forest management where sub-contractors carry out thinning operations and are allowed to market and sell the timber they harvest themselves. A large degree of the decision which trees to remove and which to leave in the forest as well as how to extract logs is left to them. In such situations, it is natural that these self-employed machine operators tend to optimise the job in hand so that it works out best for their own business rather than for the long-term development of the residual forest stand under consideration. Although standing sales are seemingly smart management options, they can cause serious silvicultural problems for the stands affected in later years.

As we have seen in previous chapters of this book, CCF also includes many activities and skills that are uncommon in RFM, e.g. crown thinnings, underplanting, transformation, the use of silvicultural systems for regeneration and individual-based forest management to name but a few. These and other fields of knowledge require additional training, and this should be offered right from the start when CCF is introduced. Forestry education at universities and colleges should in principle be largely independent of what type of forest management is preferred in a given country. Forestry students should be taught general methods and research findings rather than only a subset of matters that currently is preferred by the industry in one country but may change in the future. Who knows what the near future holds and what adaptations will be necessary before long? Armed with general knowledge, theory and a broad range of skills, any student can then, after graduation, adapt to any changes in forest management in the same way as to climate change or to other challenges should the need arise.

Unfortunately in the reality of forest-science education and industry, things are different. The continued involvement of university forest research with industry and vice versa, the recruitment of academic staff who previously worked in industry and other factors often lead to a situation where the university curriculum of forest management is quite substantially influenced by what currently is *en vogue* in the regional or national forest industry. More specifically, when RFM dominates forestry in a given country, in my experience, this domination also influences forest science education, and the students at universities and colleges tend not to be taught much about the details of alternative forest management types. This knowledge then is missing and when forestry graduates subsequently come to work in the industry and changes or adaptations are required, they are unable to make them because they have not received the necessary general education as far as silviculture is concerned.

> With a broad, general forestry education that included CCF, it is comparatively easy for any individual to adapt to RFM. However, it is considerably less easy to adapt to CCF, when higher education exclusively focussed on teaching the principles of RFM.

This explains why training is so fundamental when CCF is introduced. Even if forestry graduates have received a broad general education at the beginning of their career, after many years of working in RFM practice, they are mentally so bound to the procedures and processes of RFM that it is nearly impossible for them to make the switch to CCF. Also here training can help and unlock memories of the general education involving CCF that these staff have previously received at university.

8.2 Training Requirements

Traditional forestry training typically involves a combination of indoor lectures, visits to research or management-demonstration plots (MDP) and general field trips (that do not include plots of measured trees). Increasingly computer labs play an important role in training including, for example, harvester and forest growth simulators (Pretzsch, 2009). Training in CCF should ideally involve all levels of staff and also include self-employed machine operators. The latter do the real work on the ground and often receive contracts from state forest services or from companies and carry out crucial thinning and harvesting operations without much supervision. A particular challenge is to find the appropriate and inclusive wording so that all participants are able to follow the training and to implement its teachings. For example, with university graduates, it is possible to include simple mathematical equations in the training and to discuss various approaches to forest management and their variants in great theoretical detail, whilst with machine operators and forest workers, a more practical and direct language needs to be adopted. Contents may need to be simplified and encapsulated in more definite statements rather than discussions about various angles and nuances of different methods. For example, instead of referring to trees with low h/d ratios the trainer may want to mention 'short, fat hairy trees' instead. Visual aids are important in this context – a good picture often delivers the clearest message (Haufe, pers. comm.). If the target audience is very heterogeneous, specialised courses need to be offered that can better address different levels of work and education, e.g. forest managers, machine operators and forest planners. It is also possible to assign different tasks to different audience groups which are closely related to their field of work. Field trips and visits to research sites and MDPs are important, because many practitioners need to see ongoing CCF management, particularly transformation to CCF, with their own eyes. Field trips to other regions or countries, particularly where transformation to CCF has only started a little earlier but is not too much advanced (see Section 1.10), can also be very helpful and should be organised. However, as mentioned before, one should not underestimate how important it is to set up or to identify existing MDPs in the region or country where CCF is currently introduced: Practitioners want to see CCF at work on soils and with species, they are familiar with and under environmental and economic conditions, they have to work with every day. In my experience, this is important and can make the difference between success and failure of training. Research sites are not necessarily always suitable as MDPs or for involvement in field trips: Some of them have quite different purposes, e.g. ecophysiology, where large groups of visitors should not trample around, and the research purpose may only be marginally or indirectly related to the main issues of training. For experimental reasons, other research plots reflect extreme variants of forest management rather than best practice. Therefore, it is fundamentally important to come up with a network of MDPs fairly soon after the political decision has been made to introduce CCF. These should include the most important woodland communities, species compositions, soil types and elevations in a given country. In the context of CCF, important course topics include

- Instant new CCF (Section 2.3),
- Transformation/conversion (Section 2.4),
- Underplanting (Section 2.4.1),
- Biological rationalisation (Section 2.6),
- Individual-based forest management (Chapter 3),
- Managing mixed-species forests (Section 4.3),
- Thinnings, particularly local crown thinnings (Section 5.2),
- Silvicultural systems (Section 5.3),
- Forest development types (Section 7.2),
- Carbon forestry (Section 7.3),
- Maintaining and increasing biodiversity in CCF (Section 7.3),
- Forest margins and streamside buffers (Section 7.3),
- Recreation and amenity forestry (Section 7.3).

Training events should be short and to the point, ideally a little less than one working day, as practitioners are often pushed for time and attending a training course is a bit of a sacrifice for them (Haufe, pers. comm.). Each of the topics listed in the previous box should be provided in a separate course, i.e. one topic, one course. Sometimes it is even necessary to offer the same topic on several occasions, for example, biological rationalisation in broadleaves and conifers, in uplands and lowlands. In addition, a range of even more specialised CCF courses are possible as the need arises. Depending on the level of education, basic courses in silviculture, soil sciences and other fields of forest science have to be provided first before it makes sense for applicants to take a course in CCF.

Indoor training can be based on slides produced with presentation software. When university lecturers act as trainers, there is often the temptation to re-cycle slides from lectures intended for an academic student audience. However, such training aids are not necessarily suitable, since the academic nature of a subject area involves a certain degree of exploration and experimentation that may confuse trainees. For training purposes, it may be better to prepare dedicated learning materials that give clear and straight statements and mostly use visual aids. Visual aids are particularly important when explaining stand structure dynamics (Haufe, pers. comm.), since these are naturally difficult to perceive by the human mind. It is also a good idea to keep the number of slides to a minimum and to involve the audience, as lengthy lectures can be tiring and thus diminish the learning success. Indoor sessions should be complemented by outdoor-sessions which are typically more enjoyed by practitioners. In Britain, a ratio of 50 : 50 between indoor and outdoor sessions with the indoor part in the mornings and the outdoor part in the afternoons has worked well (Haufe, pers. comm.). Outdoor sessions could include practical tree selection, for example, for individual-based forest management (cf. Chapter 3) or the marking needed for implementing large-scale silvicultural systems (cf. Chapter 5). In outdoor sessions and on field trips, laminated flashcards and posters using little text and mainly visual aids with carefully selected colours can help explain important concepts. A limited number of flashcards can also be handed out to the trainees as 'presents' which usually increases motivation. Field trip and presentation handouts are usually much appreciated, both as hardcopies or as digital copies (Haufe, pers. comm.).

In a country, where there is little experience with CCF, it may be a good strategy to recruit at least one trainer from a region or a country where CCF has a long tradition, or alternatively, to recruit someone who spent at least a few years in such a country studying CCF. A mixed team of training staff with different education, gender and experience is ideal. Trainers should have a forest-science background, and if such a person grew up and has received their education in a region with extensive CCF experience, s/he should be first required to properly understand the RFM background and traditions in the new country before designing courses. This perhaps only needs a few months or up to a year, but it is helpful in taking course participants gently from a world of methods and practice they know well, e.g. RFM, towards the unknown, new territory of CCF and to explain how CCF methods can help improve things.

8.3 Marteloscopes

A rather special device and method of silvicultural training is the *marteloscope*. It is frequently used in the context of CCF training. The results of a survey listing organisations using marteloscopes for training has been published by Pommerening et al. (2018). Since this method is an important part of CCF training and the author had much research and training experience with this tool, a detailed review of marteloscopes follows.

8.3.1 Origins

A curious, early example demonstrating that forestry staff started to doubt that tree-selection decisions taken by different forest managers were sufficiently similar are the data published by Heger (1955). Although the two German foresters in this example managed to produce broadly similar empirical crown-class distributions for both stands, there are marked differences between individual classes (see Figure 8.1). Particularly large are the differences in the first three classes, i.e. in the dominant part of the two forest stands. Apparently, it was easier for the two foresters to agree on the dominated trees than on the dominant trees. These classification differences were then potentially propagated in subsequent thinnings, since, as discussed in Section 5.2, the crown classes formed the basis of implementing thinning types and thinning intensity in Germany.

In the mid-1990s, Prof. Klaus von Gadow initiated a research group at Göttingen University in Germany investigating the tree selection behaviour of forest managers and machine operators in different forest ecosystems. The group designed a special survey method for data gathering which went by the names of *thinning-event inventory* and *harvesting-event inventory*. The key idea of this survey method was to schedule the data gathering at the time of tree marking prior to the actual tree removal. In contrast to traditional forest inventory methods, the thinning-event inventory captured both the initial forest stand conditions and the residual stand at the same time (Pommerening and Grabarnik, 2019). The changes in forest density and structure could then be analysed and decisions could be revised if necessary (see Figure 8.2).

Towards the end of the 1990s, a group at AgroParisTech-ENGREF at Nancy (France) led by Max Bruciamacchie realised the potential of thinning-event inventories for training

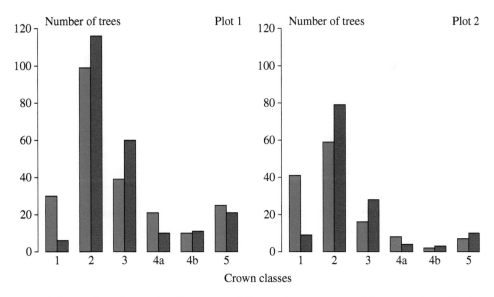

Figure 8.1 Number of trees classified separately by two forest managers (shown in red and blue) according to the crown class classification by Kraft (1884, see Section 4.2) in two forest plots of Neudorf Forest District (Germany). Both forest managers classified the same number of 224 and 113 trees, respectively, in the two plots. Source: Data taken from Heger (1955).

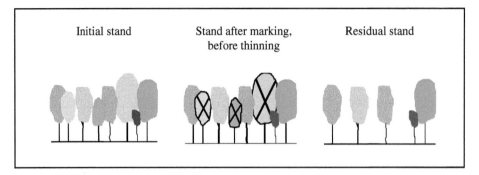

Figure 8.2 The principle of the thinning-event analysis: Human marking behaviour is analysed immediately after marking a forest stand and before trees are felled. This provides the opportunity of revising the decision-making. The initial forest stand, the stand after marking but before thinning and the residual stand are analysed simultaneously. Source: Adapted from Gadow et al. (2021) and Pommerening and Grabarnik (2019).

and education in CCF. The group at Nancy offered field-based training courses to forestry professionals and students. In these courses, the participants were asked to mark trees for thinnings on a sheet of paper similar to a questionnaire, see Table 8.1. Their choices were analysed using specialised software or macros embedded in spreadsheet software. Personalised result and feedback sheets were handed out to every participant at the end of the training session. In order to attach a new brand to this type of research plot, the group at Nancy coined the name marteloscope (from French *martelage* – marking) for these training sites. From these beginnings, the idea of marteloscopes spread through the networks of Pro Silva, AFI (Association Futaie Irrégulière) and other professional forestry organisations in France, Switzerland, Britain and Ireland (Pommerening and Grabarnik, 2019).

In the first decade of this century, the author of this book and his Tyfiant Coed project team at Bangor University used the marteloscope idea in a training project in North Wales. Poore (2011), Susse et al. (2011) and Soucy et al. (2016) described the application of marteloscopes for education and training:

> The marteloscope is a didactic tool and a permanent plot within the forest in which tree measurements and associated software are linked to provide a framework for in-forest training in tree selection as part of integrated training in silviculture.

The popularity of marteloscopes is currently on the increase, however, its original research purpose has only been pursued at few institutions.

8.3.2 Plot Design

A marteloscope is typically set up as a research plot with rectangular boundaries. Often a suitable size for long-term plots is 100×100 m, where the number of trees to be included should ideally be between 150 and 500 trees. A marteloscope of this size takes a trainee approximately three hours to complete. For short-time use or when there are constraints on the time available for tree marking, plots of 50 × 50 m or 40 × 40 m are also suitable. In general, the plot area should be sufficiently large so that the trainees do not influence each other's decision-making. In a commercial forestry context, it can also be recommended to select an area where thinnings have not occurred during the last 10 years, so that management urgency is high, which provides more opportunities for tree marking compared to recently thinned forest stands. Otherwise, the participants may struggle to find sufficient options for decision-making. For practical and statistical reasons, it is good practice to consider setting up two marteloscope plots in the same forest type at close proximity, a principle Pommerening and Grabarnik (2019) referred to as *twin plots*. As time goes by, the two plots are thinned in turns by the local forest owner (as part of routine forest management) so that one twin can be used for marteloscope training at all times whilst the other twin – i.e. the plot that has most recently been thinned – is used for management demonstration in field trips. This recently managed plot allows the trainees to better understand the silvicultural instructions in the context of a long-term vision. Good CCF management practice suggests organising one set of twin-plots for each FDT (cf. Section 7.2). Every tree in a marteloscope plot typically has a unique identity number, which is painted on the stem surface with waterproof paint as clearly as possible and should be visible from afar. If feasible one can consider painting tree numbers twice on each tree from opposite directions for easier identification. A minimum of tree measurements should include stem diameters (with a lower threshold of 5–7 cm). If more funds are available, surveying should ideally also include tree locations, total heights and heights to base of crown (at least on a sample basis), volume/biomass, habitat value and timber quality. Marteloscopes are not very different from standard research plots commonly used in silviculture and growth and yield research. In fact, plots from the latter two research fields can often be re-used as marteloscopes which offer the benefit of having past growth rates at the trainer's disposal (Pommerening and Grabarnik, 2019).

Mapping the stem centre locations of trees is useful for logistic and practical reasons, since maps can be produced (see Figure 8.3a). These help maintain the experiment but can also serve as aids for orientation when handed out to trainees, particularly when the marteloscope area is large. The map can also be used to visualise the choices made by a particular trainee, as this is the case in Figure 8.3a. The corresponding stacked empirical diameter distribution is an example of simple descriptive statistics (Figure 8.3b) illustrating a trainee's choices. In this example, an extraction rack runs through the centre of this *Picea sitchensis* marteloscope, a curious feature, since the selection of frame trees near roads and rides is generally discouraged for the potential damage they can suffer from the extraction of felled trees or from traffic.

When setting up marteloscope plots, it is always a good idea to get the local forest manager on board and to secure her or his enthusiasm and support. S/he may even help and make a nearby cabin or small office available that can be used for analysing and printing the results, which is particularly useful in a training context (Pommerening and Grabarnik, 2019). The local forest manager often benefits from the mapped marteloscope data and takes an interest in the silvicultural discussions.

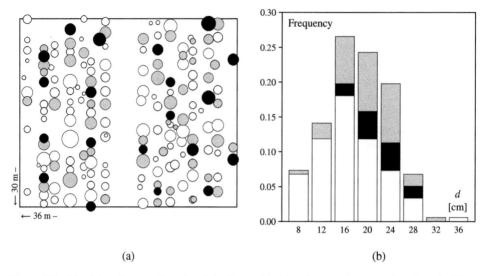

(a) (b)

Figure 8.3 Map (a) and proportions (b) of the frame (black) and competitor trees (grey) marked by one participant as part of a small temporary marteloscope experiment at Clocaenog Forest in North Wales (UK) in 2006. The remaining trees and corresponding proportions are shown in white. Stem diameter is denoted by *d*. Source: Pommerening and Grabarnik (2019).

8.3.3 Marking Sheet

Table 8.1 gives an example of a possible marking sheet. The design of the marking sheet may vary considerably depending on the purpose of the training session and can also be implemented in a software application for tablet computers. The design of the marking sheet surely has an influence on the marking behaviour. It is therefore advisable to consider the design carefully and to be resourceful in order to achieve an unbiased outcome of the training event. I often recommend to leave the marking sheet completely blank and to ask the trainees to note only those trees that they want to select, since the order of trees in the marking sheet and additional size information (cf. Table 8.1) may influence trainee decisions. Handing out empty sheets of paper to the trainees would probably reduce the influence of the marking sheet to a minimum (Pommerening and Grabarnik, 2019).

For better orientation, it can be useful advice to suggest that the trainees walk through the marteloscope stand in parallel, rectangular strips, which can be supported by using the same strategy for numbering the trees when setting up the marteloscope. Following this marking strategy is not always possible but particularly for novices in selective individual-tree management, this may be helpful to get started.

In the context of local crown thinnings (see Section 5.2), we additionally recommend to require the trainees for each tree they mark for removal to note the frame tree(s) this proposed felling is supposed to benefit. In rare situations, one tree felling can promote two frame trees simultaneously. This requirement, also termed *competitor-for-frame tree rule* by Kruse et al. (2023), provides valuable information for the subsequent analysis of the trainees' choices, but most importantly, it introduces the didactic element of encouraging the trainee to observe the principles of frame-tree based thinnings, cf. Section 3.5. This recommendation also provides an opportunity for the trainees to briefly look back and

Table 8.1 Design of a typical, basic tree-marking sheet for use in marteloscope research and training.

Tree No.	Species	DBH	Frame	Thin
1	Birch	55.4		×
2	Scots pine	60.6	×	
3	Scots pine	61.5		×
4	Birch	33.5		
5	Birch	42.1		
6	Scots pine	52.3		
7	Birch	15.6		
8	Birch	53.7		×
9	Scots pine	64.3	×	
10	Birch	24.2		
...	...	...		
...	...	...		
...	...	...		

DBH is stem diameter at breast height and measured in centimetres. 'Frame' is short for frame tree and a mark in column 'Thin' implies that a tree is selected for removal.
Source: Pommerening and Grabarnik (2019).

reflect on the last few choices they made. As mentioned before, local crown thinnings play a prominent role in CCF and are important to include in marteloscope training because they demand greater skill on the part of the forest manager (Smith et al., 1997).

Marteloscope software on tablet computers can be a great help for both data capture and analysis. However, before relying on this in the field, it is necessary to test the hard- and software under field conditions and to check whether both really meet the requirements of the training session. Even if the test turned out to be positive, it is a good idea to provide for paper and pencils as a back-up plan just in case that the software does not work after all. Hot drinks in cold or wet weather can pleasantly predispose trainees.

In the data processing, the marking sheets are digitalised unless the data were already entered with a computer tablet during the marking process. A cross or tick indicating tree selection is converted to a '1'. No selection results in '0', so that a typical trainee data column consists of a sequence of 0's and 1's.

8.3.4 Marking Exercise

Marteloscope training sessions should not include more than 20 trainees, ideally 10–12 persons, because otherwise the participants obstruct each other or influence each other's decisions (cf. Figures 8.4 and 8.7). Tree marking in groups is a possibility, but the training effect is greater when all participants make completely independent choices, even if their confidence is low, and they believe that they are inexperienced. People who mark trees in

(a)   (b)

Figure 8.4 (a) Students of Göttingen University selecting trees in a marteloscope near Reinhausen (Germany). (b) A student from the Estonian University of Life Sciences selecting trees in a marteloscope at Järvselja Forest (Estonia). Source: Arne Pommerening (Author).

groups should indicate the group affiliation on their marking sheets or in the software so that group decisions are handled correctly in the analysis. There is also the possibility to split large groups of participants and to hold two or more separate marteloscope events. Other possibilities include assigning each person to an individual starting time and/or an individual starting location at the margins of the forest stand.

Each marteloscope training event must have clear objectives, and depending on these more or less detailed instructions, example marking and training is given to the participants prior to the marking event. Instructions should give the trainees an idea of the type(s) of trees to select and of the approximate number of trees and/or basal area to be marked. This can be accompanied by a brief qualitative and quantitative description of the forest stand. Instructions typically depend on the training purpose. Vítková et al. (2016), for example, were interested in the trainees' ability to mark for crown thinnings (see Section 5.2) in Ireland and carried out one marteloscope event each before and after delivering a training workshop on crown thinning with the same participants. There must be a clear, measurable definition of the marking strategy the trainees are asked to adopt; otherwise, it is hard to give constructive feedback based on objective, quantitative criteria. It is also an option to carry out one marteloscope event without instructions and another one that gives clear instructions prior to the marking so that the trainees can see the difference between their 'old ways' and the new techniques delivered in the training. Instructions become even more useful when supported by practical marking demonstrations in an adjacent woodland (Pommerening and Grabarnik, 2019). These demonstrations should use different coloured ribbons for marking frame trees and competitors. The ribbons should be wrapped around the stems of selected trees approximately at 1.3 m above ground level. To illustrate the principle of local management units (Schädelin's 'thinning cells', see Figure 3.6) at least two adjacent frame trees and their neighbours should be included in the practical demonstration. The trainees can be actively involved in the demonstration and a short discussion can follow. An example instruction in a crown-thinning context is

---

- Identify 80–120 frame trees ha^{-1}, using timber quality as the main selection criterion.
- Thin to release frame trees, removing 1–3 competitors each as part of a crown thinning.
- Retain at least 18 m^2 ha^{-1} basal area.
- Place a tick next to trees you select for removal in the appropriate column of the marking sheet.

This instruction is a blend of information on local and global crown thinning, since there is also information on global, residual basal area. In most cases, this is not necessary, if a firm limit for the maximum number of competitors is given. For subsequent analysis into training quality and efficiency, it is also good practice to record the trainees' names, gender, work affiliation, professional and geographic backgrounds, but this information can be collected beforehand as part of the registration process. In any event, such information needs to be treated confidentially. Any additional information can potentially turn out as useful covariates or for post-stratification in the analysis and aid the interpretation of the results. A time limit may also be set to give trainees a realistic expectation of how long the marking work, should take (Haufe, pers. comm.). To avoid 'strategic behaviour', the organiser should repeatedly assure the trainees that their results and any other personal data will be treated anonymously (Pommerening and Grabarnik, 2019).

8.3.5 Analysis and Feedback

Ideally, every trainee should receive an individual result sheet for personal feedback immediately after the marteloscope session has ended. This is an important element of marteloscope training, and it is easy to achieve when tablet computers and associated software are used. Then the software can carry out the analysis and compile a result sheet automatically, which appears on the screen of the tablet computer. However, the use of such computers and standardised software/analysis is not always practical or desirable. In bad weather conditions, the use of tablet computers may be limited and different analysis variants may be necessary for different training purposes so that standardisation is not always easy. Also, it is informative to show in one graph the marking results of one trainee in the context of those of all other participants (without revealing their names), and this is only possible after pooling the results of all trainees and then sending the summarised results back to the individual devices. This may sometimes be difficult to achieve in outdoor conditions. In the worst case, the result sheets can be sent to the individuals later.

It is good practice to have a final discussion after the marteloscope session has ended, since bringing people together and exchanging professional experience is one of the many benefits of marteloscope exercises and of training in general. This could also involve a repeated walk through the marteloscope to discuss individual questions or trees that are difficult to mark (Pommerening and Grabarnik, 2019).

Pommerening et al. (2018) suggested subsetting trees of a given marteloscope in the analysis: In their research, the authors found evidence that differences in trainee behaviour can potentially be better elaborated when focussing on trees that are difficult to mark, since it is psychologically easier for humans to agree on obvious negative cases (Matonis et al., 2016). Teasing out these subtle differences between people may enhance the training effect. In practical terms, the organisers of the marteloscope training session could pre-select a subset of trees in a marteloscope that they collectively perceive as difficult to judge on. This subset is not revealed to the trainees. In the analysis, the organisers then analyse the markings once for all trees and once for the subset only. This will allow to understand how trees that are difficult to judge can influence the outcome of the training and what is necessary to improve to be understood.

8.3.6 Reference Marking

A key question often asked is what reference can be used to compare trainee decisions with. There is, of course, the possibility to compare with the marking of an established expert. This can be the trainer or the organiser of the training workshop but also a recognised senior forest practitioner. In current marteloscope exercises, using such a person as reference is actually common practice. Individual experts are asked to estimate the timber quality of all (live!) trees in the plot and similar gradings are carried out for the habitat value of trees. In addition to habitat value, other criteria such as individual-tree resilience or aesthetic value can be estimated by experts in a similar way. When the actual marking of trainees is analysed, it is compared with the timber and habitat quality assessment of such experts and some of the marteloscope software used even calculates values or points that the trainee has achieved in the marking (Mordini, 2009). This analysis capability sounds rather grand and promising; however, for logistic and financial reasons usually only one expert per subject area (economics, habitats, etc.) is invited to carry out the estimations. Experts can make mistakes or have a certain agenda (Pommerening and Grabarnik, 2019). There is always the danger that an expert's marking is biased. A more reliable approach would be to ask 5–10 experts to carry out economic and ecological gradings and then to determine individual-tree gradings as means of such expert estimations. There are also great uncertainties when it comes to placing points or grades to habitat or ecological features. In addition, these gradings are likely to change much with time and require regular updating. A more sophisticated approach to consolidating expert opinion could be to apply methods from political sciences, specifically multi-winner approval voting (Pommerening et al., 2020), but there are also other methods available as detailed in Pommerening and Grabarnik (2019, Chapter 7). One method that is similar to establishing references in point process statistics is to use *random tree selection* (Eberhard and Hasenauer, 2021). To carry out such a thinning, random numbers can be generated in R or in a similar software and the total number of trees to be thinned at random can be determined as the mean number of trees selected for thinning per trainee or as a Poisson distributed random number, where the aforementioned mean number of trees selected is the parameter of the Poisson distribution, see Pommerening and Grabarnik (2019, Section 4.4.9). Applying this method offers the possibility to formally test whether the tree marking performed by trainees differs from random tree marking. The possibility of simulating the selection of frame trees based on size and distance was already indicated and referred to in Figure 3.8.

Stem-diameter structure, spatial forest structure, basal-area or volume growth rate are simple, objective criteria that can be used as a reference. The use of growth rates is based on the idea of timber sustainability, i.e. that under normal circumstances not more should be removed than will re-grow until the next intervention is due. Under certain circumstances, equilibrium models describing ideal sustainable size structures can also act as references, see Chapter 6. Another reference can, for example, be the spatial dispersion of frame trees but also the basal area of competing neighbours marked for removal relative to the basal area of the frame tree, see Figure 3.10. These quantitative measures are preferred by the author as a basis for marking references and for the general analysis, see Pommerening and Grabarnik (2019, Chapter 7).

In most training projects, particularly in the context of introducing CCF, it is paramount that established silvicultural techniques that are known to produce intended results in terms of transformation or the diversification of forest stand structure, are correctly applied by the trainees, whilst the optimisation of the economic and ecological consequences are of secondary importance. Including economic and ecological results is clearly interesting, but they only provide short-term snapshots, whilst a detailed analysis of the silvicultural techniques applied by the trainees in my opinion is likely to be of much greater long-term relevance.

8.3.7 Analysis Methods and Strategies

There are two kinds of analysis results required in marteloscope exercises:

1. Personalised results for individual feedback,
2. Group results:
   - Results of marking *real* or *virtual* groups,
   - Results of the *whole* training class.

The first type of results is primarily for providing feedback on the performance of the individual trainee. As mentioned before, this is a very important part of the marteloscope concept and should be provided as soon as possible after the exercise. After making the marking effort, trainees usually are keen to receive direct, constructive feedback; otherwise, the purpose of the whole activity is doubtful.

Group results can involve the results of those trainees who teamed up and decided to mark trees jointly (real group) or of those that were retrospectively assigned to groups of interest (virtual groups), e.g. gender, experienced staff, novices. The term 'group results' also covers the results of analyses that involved all participants of a course or exercise as one large, single group.

Group results can serve at least two important purposes. (i) They are helpful for highlighting how one individual trainee has performed in the context of all other or of a subset of other trainees. This enhances the individual feedback and allows every participant to compare their results with those of others which usually is a natural thing for people to do. For obvious reasons, the names of the other participants should not be revealed in this process under any circumstances. (ii) Group analyses offer the possibility to carry out research on the behaviour of the whole group of participants and can reveal behavioural differences. This is of general research interest, since the interaction of people with trees has not been considered much in the past. Such group analysis can also help provide better inputs for tree growth models, since real-person and averaged behaviour can then be used in growth simulations to make them specific to the behaviour of local people. Finally, group analysis can also play an important role in quality assurance and provides important quantitative information on how successful the training has been. This is obviously a crucial concern in any training provision.

The most intuitive way to define groups is probably to include all participants in a particular course or marteloscope exercise in the analysis. However, it is also possible to consider a subset of these people for post-stratification (virtual groups) or to include participants from past exercises in the same marteloscope to provide contrasts. Depending on the purpose of analysis, it is often even possible to include the marking results of trainees from different marteloscopes in other countries (Kruse et al., 2023). The comparison of trainee behaviour superimposed on that of another can highlight important differences that are worth following up. Therefore, the analyst can be quite creative with the definition of what constitutes a group.

A detailed review of characteristics suitable for the analysis of marteloscope data can be found in Pommerening and Grabarnik (2019, Chapter 7), Pommerening et al. (2020) and Pommerening et al. (2021a). Some of these characteristics were also introduced in Sections 3.5 and 5.2.2. The most common analysis types are given and explained in the following list:

Agreement. There is always a high variability in tree marking, even if the marking objectives are clearly defined, and all participants have the same level of education. To quantify, agreement/variability helps understand the silvicultural baseline of a certain group, uncovers training deficits and can be used for monitoring training quality. This information is also important for modelling and simulation, e.g. when projecting the long-term consequences of people's marking decisions. This type of characteristic is a little less important for personalised feedbacks as opposed to group analyses.

Intervention type. Several characteristics can be applied to quantifying thinning types. This is a very important consideration in the context of CCF, since local crown thinnings are recommended and staff with experience in RFM tend to carry out global thinnings from below. This type of characteristic is useful for both personalised feedbacks and group analyses.

Intervention intensity. Characteristics that are partly related to those quantifying the intervention type and express the proportions of suggested tree removal in terms of basal area, biomass, stem volume or number of trees. Basal area is the most universally applicable and meaningful variable in this context. This type of characteristic is useful for both personalised feedbacks and group analyses.

Individual intervention intensity. If the trainees were required to indicate the frame trees which they envisaged to benefit from their tree-removal markings (*competitor-for-frame tree rule*; Kruse et al., 2023), it is possible to quantify the average number of neighbours per frame tree proposed for removal and to analyse the ratio of competitor and frame-tree basal area, see Figure 3.10. The corresponding graphs should reveal a declining trend and the residuals should not be too large. Overall the observed point cloud should show little variance. An alternative is the A-thinning index (see Eq. (3.3)). These characteristics are useful for both personalised feedbacks and group analyses.

Resilience. In local-thinning exercises, h/d (Eq. (2.2)) and c/h (Eq. (2.3)) characteristics can be calculated for each frame tree. In global thinnings, the hundred largest residual trees per hectare can be used for these calculations instead. In a second step, the mean is computed for both ratios. Resilience to wind and snow hazard is high, if $\overline{h/d} < 0.8$ and $\overline{c/h} > 0.4$. Resilience is low, if $\overline{h/d} > 0.8$ and $\overline{c/h} < 0.4$, see Figure 2.2. Otherwise, resilience to wind and snow hazard is medium (Pretzsch, 1996). This type of characteristic is useful mainly for personalised feedbacks.

Distances and dispersion. Several characteristics can be used to check up on frame-tree distances and the frame-tree dispersion patterns. Distance and dispersion rank lowest on the list of criteria for frame-tree selection (cf. Section 3.4); however, they are not unimportant and gross violations of the principles reviewed in Chapter 3 can help identify training deficits or poor learning outcomes/training quality. These calculations require mapped tree locations. This type of characteristic is useful mainly for personalised feedbacks.

Forest structure. Since the transformation of RFM to some form of CCF is much about diversifying forests and modifying spatial forest structure, measures of forest structure can be helpful and particularly the changes that are introduced when a trainee's marking solution is implemented in the field. Stem-diameter distribution is one useful characteristic (cf. Figure 8.3b) and others are introduced in Chapters 4 and 6. In addition, spatial measures of forest structure, cf. Pommerening and Grabarnik (2019, Chapter 4) are ideal for identifying subtle changes in forest structure as introduced by the marking; however, they require mapped tree locations. This type of characteristic is useful mainly for personalised feedbacks.

Growth sustainability. When stem-diameter growth rates are available, it is possible to quantify the sum of the basal area or stem volume of the trees to be removed as suggested by the trainees relative to the basal-area or volume increment of the last five or ten years. If course participants marked less than the observed basal area increment, this would lead to a net increase of stand density and otherwise to a net decrease. Ratios around one suggest that stand density would not change. A slight decrease of stand density is usually desired, since increasing stand density can result in a loss of forest structure and tree diversity required for successful CCF management. This type of characteristic is useful for both personalised feedbacks and group analyses.

The choice of characteristics naturally depends on the objectives. In mixed-forest stands, some of these characteristics could also be calculated for specific species populations. In research, it is often useful to apply a wide range of statistics and then to select those that provide the most meaningful results. Even seemingly similar or competing characteristics tend to highlight different aspects, and one might miss interesting trends and clues, if some of them were excluded before trying them. General forest-stand summary characteristics were found not to be sufficiently sensitive to discriminate between different strategies of tree selection.

Individual feed-back sheets could start with maps visualising the tree selection and stacked diameter distributions (cf. Figure 8.3). Then basic statistics including basal area per hectare, trees per hectare, mean stem diameter and minimum and maximum stem diameter for the whole forest stand as well as for the main tree species should be given for both states, i.e. before and after the marking. Following this, most of the other characteristics could be included, but most importantly, those on intervention type and intensity, on resilience, distances and dispersion should be presented.

8.3.8 Important Results so Far

For organising training involving marteloscopes, it is useful to review the results from past studies to remind ourselves of what we know already about human tree-selection behaviour. This information can then be considered for designing effective training courses.

In countries where RFM is the standard or a majority type of forest management, thinnings from below are commonly applied in forest practice, and this is what forestry staff are most experienced in. Crown thinnings are uncommon in these countries, and forestry staff have great difficulties to switch to this thinning type. The more experience trainees possess, and the longer they have worked in forestry, the more difficulties they have to follow crown-thinning principles (Vítková et al., 2016). Those with little or no forestry experience are the best in adapting to crown thinnings. This phenomenon has now been confirmed in a number of studies and is known as the *experience paradox*. Knowledge of or experience with individual-based forest management (Chapter 3) does hardly exist in RFM dominated countries and 'seeing the trees for the forest' is a real problem here, since the staff involved are only used to dealing with stands as a whole.

Anecdotal evidence (e.g. Figure 8.1) and early studies have already cast doubt on the widespread assumption that forestry staff select trees in good agreement (Zucchini and Gadow, 1995; Füldner et al., 1996; Daume et al., 1997) and in accordance with textbooks, regional guidelines and local prescriptions. This has been a widespread assumption in forest models and textbooks up to the mid 1990s. The authors of the early studies on tree-selection behaviour have found only little agreement in the marking behaviour among forestry professionals and that the aforementioned assumption was but a myth. Individual-tree agreement in forestry is generally lower than in other fields such as medicine (Pommerening et al., 2018).

Spinelli et al. (2016) studied the silvicultural results (in terms of basal area and trees per hectare) performed by a number of test persons with different professional backgrounds in mixed CCF woodlands in Northern Italy. They found no significant difference in the marking behaviour of test persons from different professional groups. This is indeed common when only yield summary characteristics are used to identify differences (Eberhard and Hasenauer, 2021). However, Spinelli et al. (2016) also identified a substantial lack of agreement in terms of the selection of individual trees. The authors speculated that different practical experience in tree marking is a possible explanation for the lack of agreement.

Vítková et al. (2016) could demonstrate that education and subsequent training can profile people's choices in terms of tree selection behaviour. The authors reported tree-marking experiments involving test persons with different experience and education. They required the test persons to perform the marking twice in the same experimental forest in Ireland, once before and once after receiving training in crown thinning methods. Apparently, seasoned experts were reluctant to adopt the new thinning method and the training led to confusion and decreasing agreement in this group. In contrast, novices, i.e. trainees without previous forestry experience, responded well to the training and the agreement in this group was significantly higher than among the experts. These findings were confirmed by Bravo-Oviedo et al. (2020) in an independent marteloscope experiment in Italy, and this phenomenon is described by the aforementioned experience paradox. The influence of gender and geographical origin of trainees on the marking results has so far not turned out to be significant. Pommerening et al. (2018) also found little agreement in tree marking when analysing 36 marteloscope experiments from all over Britain, where two different thinning types were applied, i.e. local crown and global low thinning. Overall agreement was low but particularly so in crown-thinning experiments (Figure 8.5). The authors' results suggested that there is a need to provide more training related to crown thinnings, since the adoption of this method in a country where RFM is the standard proved difficult. Interestingly, there was no correlation between measures of forest structure and Fleiss' kappa or other agreement characteristics. The authors, however, found that more complex forest structure appears to facilitate the decision process, i.e. complexity of stand structures makes choices more obvious. This potentially implies that marteloscope exercises in forests with a uniform structure are perceived as more challenging by CCF novices than in more complex woodlands further along the structural gradient shown in Figure 1.10. This finding emphasises that training in simple-structured forests that mark the beginning of transformation is very important.

Pommerening et al. (2021a) studied the tree-selection behaviour of trainees in 26 local crown-thinning and global low-thinning experiments in Britain. The same trainees

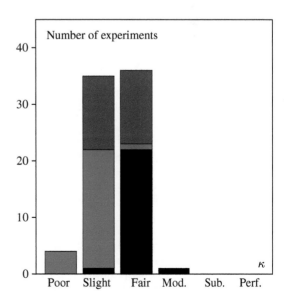

Figure 8.5 Empirical distribution of the Fleiss kappa characteristic, κ, from British marteloscopes indicating agreement in individual-tree selection. Black – global low thinning, red – local crown thinning (frame-tree competitors), blue – local crown thinning (frame trees). The κ interpretation classes are poor, slight, fair, moderate, substantial and perfect agreement (Pommerening et al., 2018). Source: Pommerening et al. (2021a).

marked for both thinning types in the same marteloscopes. The results revealed that highest agreement in individual-tree selection was achieved in the low-thinning experiments followed by frame-tree selection, whilst the lowest agreement resulted from the selection of frame-tree competitors (Figure 8.5). The results also confirmed that all frame trees were generally selected in accordance with the theory and current state of the art including the selection of mainly dominant trees and the low variability in average frame-tree size among the trainee selections.

The B ratio clearly separated between low and crown thinning. Interestingly, the frame trees even scored lower than the trees selected for the crown thinnings, i.e. their competitors (Figure 8.6a). The selected frame trees were significantly more dominant than the frame-tree competitors. Mean $\overline{h/d}$ ratios, a resilience indicator, were lowest for the frame trees followed by the trees selected as frame-tree competitors (Figure 8.6b). Apparently, trees selected as frame trees and frame-tree competitors were dominant trees with comparatively free crowns; otherwise, their $\overline{h/d}$ ratio would be higher. The frame trees had $\overline{h/d}$ ratios within the realisation space recommended by Abetz and Klädtke (2002), though close to the upper boundary, see Figure 3.12. The largest and most variable mean $\overline{h/d}$ ratios were the result of the tree selection for the low thinnings. All these findings were perfectly consistent with both theory and marking instructions (Pommerening et al., 2021a).

Apparently, difficulties arise when trainees select the competitors of frame trees among their nearest neighbours (cf. Figure 8.5). This should ideally happen in the spirit of a crown thinning, and this process does not seem to be sufficiently clear to most participants, or past experience is a hindrance to choosing the right trees (experience paradox). This suggests that participants in training courses need more support for selecting frame-tree competitors.

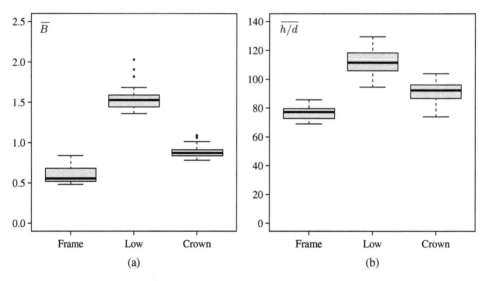

Figure 8.6 (a) Mean ratio of the proportion of number of trees selected by trainees and the corresponding basal area of the selected trees (Eq. (5.1)), $\overline{B}$. (b) Mean ratio of total height and stem diameter of selected trees (Eq. (2.2)), $\overline{h/d}$. Frame – Frame-tree selection, Low – Trees selected in global low thinnings, Crown – Trees selected in local crown thinnings (= frame-tree competitors). Source: Pommerening et al. (2021a).

Perhaps this can partly be achieved by introducing the aforementioned didactic requirement that trainees need to state clearly which frame tree the marking of trees for removal is supposed to benefit (*competitor-for-frame tree rule*; Kruse et al., 2023).

There has been a long discussion about whether tree-selection decisions should be delegated to harvester drivers. In eight *Picea abies* forest stands in Austria, Eberhard and Hasenauer (2021) compared the tree selection results of forest managers with those of harvester operators and additionally projected stand development for 50 years into the future using the growth simulator MOSES. Apparently, 70% of the trees selected by forest managers were identical to those selected by the harvester operators. In terms of forestry summary characteristics after 50 years of simulation, the authors detected no difference in future stand development. Austria is a country with a comparatively long CCF tradition and both forest managers and machine operators enjoy a high level of education in the country, which may explain these results. The forest operators were instructed by the forest managers before both of these staff groups took separately part in the experiment. That forestry summary characteristics show little difference between trainees is not a big surprise, since they are not very sensitive. However, the differences to forest structure may still be large. In the growth projection, it remained unclear whether the growth simulator was required to mimic the marking behaviour of each participant in the experiment. Still, the high number of individual-tree agreement is remarkable in this Austrian study, since they have not been confirmed elsewhere. Similar tests should be run in other countries, too.

8.3.9 Virtual Marteloscopes

Growth simulators probably were the first software applications that allowed thinnings to be carried out using computers. Apart from performing automated thinnings based on some algorithm, the simulators usually also included an option for identifying trees to be thinned manually at the computer screen by clicking on a tree map or on tree representations in 3d visualisations (Eberhard and Hasenauer, 2021).

Mordini (2009) mentioned the use of virtual marteloscopes in her MSc thesis. Once established, software enabling virtual marteloscopes naturally much reduces the considerable logistic effort required for field marteloscopes. Trainees can carry out virtual tree markings from the comfort of their home so that distance training is a real possibility. Although using tree maps produced from mapping real forest stands clearly has its merits, it is also possible to simulate forest stand maps in such software based on point process models (cf. Pommerening and Grabarnik, 2019, Chapter 5), which would reduce costs even more whilst establishing controlled and standardised conditions at the same time. The analysis of virtual marteloscope exercises can be carried out in exactly the same way as for the results from field marteloscopes. Training success is certainly influenced by the visualisation quality of the software and by the size of the computer screens used. It would, for example, clearly help, if users could switch between 2d tree maps and 3d visualisations. Although such software could theoretically offer much more individual-tree information compared to what is available in the field, providing this information would probably not be prudent, since the user should ideally be put into a situation not too different from that of a field exercise (Figure 8.7). In the future, it may be possible to use virtual-reality rooms where trainees could physically walk through a 3d projection of marteloscope trees which is close

Figure 8.7 Trainees selecting trees for a local crown thinning in the mixed *Pinus sylvestris* L. - *Picea abies* (L.) H. Karst. and *Betula* spp. marteloscope at the Svartbergets Försökspark (Vindeln, Northern Sweden), which is part of the Swedish University of Agricultural Sciences, SLU. Source: Courtesy of Lydia Kruse.

to outdoor conditions. Like with other training and simulator software, irrespective of how good the visualisation is, there will always be a difference between tree virtual markings and those in the field even when carried out by the same people for the simple reason that the surrounding conditions and environments are never the same as indoors. Still using virtual marteloscopes is an excellent option, and every training centre should include them in their programmes.

8.3.10   Limitations of Marteloscope Training

Marteloscopes are particularly efficient in the training of thinning and selective harvesting methods. As we have seen in other parts of this book, such methods assume a very important role in CCF. Therefore, field and/or virtual marteloscopes should definitely form an important part of CCF training. They also provide the opportunity for trainees to be active as opposed to just passively listen and reflect on what the trainers say and show. In addition, they are a fantastic chance to discuss approaches to management among peers enriched by the hands-on experience of active tree marking. Regeneration methods can only be applied to a limited extent in marteloscopes, since the plot size involved is often not sufficient for full-scale silvicultural systems. Such methods would, for example, include delineating strips and groups of strip and group shelterwood systems, whilst the removal of individual trees as part of uniform shelterwood or single-tree selection systems would be possible to emulate through tree marking. Marteloscopes are not limited to local thinning methods. Any kind of thinning including global methods, GDT (cf. Section 2.4.2) and VDT (cf. Section 2.4.3) can be marked for in marteloscopes.

Quantitative characteristics used in the analysis for providing feedback are a considerable help to make marking exercises and CCF methods transparent. They are crucial to the learning success and without them, the exercises would be of little value. Instead of using the characteristics presented in Pommerening and Grabarnik (2019, Chapter 7) some research and training groups derive economic and ecological values of frame trees and competing neighbours. As feedback results they provide statistics on these two groups of values in an attempt to support the integrated variant of CCF that tries to meet multiple management objectives at the same time (Mordini, 2009; Cosyns et al., 2018). Whilst this is certainly interesting, these results can only provide short-term information or perhaps a projection for the time of harvesting that is based on many quite unpredictable assumptions. Owing to my personal experience, I am not certain that such information aids the training success in the context of this book. Comparing the results of individuals with other trainees – past and present – on the basis of the statistics in Pommerening and Grabarnik (2019, Chapter 7), but also with reference markings as discussed in a previous paragraph of this section, ensures reliable information that directly relates to the training objectives.

Marteloscopes are valuable aids that should be included not only in training for CCF but also in the education of forestry students at universities. However, as any training aid, their role should not be exaggerated. Along with field trips, management demonstration sites, indoor seminars and outdoor demonstrations, they are but one component among many others.

Appendix A

Overview of the Most Common Principles of CCF

As a synthesis of Section 1.5, a list of the most common tenets and principles of continuous cover forestry is given here along with a short explanatory comment (cf. Figure 1.8).

Continuity of woodland climate and soil. Any management intervention should not significantly modify woodland climate and soil conditions typical of the local area or restore them.

Reliance on natural processes. As much as possible natural processes of regeneration, self-pruning, size differentiation and self-thinnings in tree seedlings/saplings should be encouraged and preferred to artificial, man-made processes where possible. The impact of necessary human interventions should be reduced to a minimum.

Promoting vertical and horizontal structure. A varied horizontal and vertical stand structure supports natural processes, promotes biodiversity, contributes to resilience and makes forests more appealing to visitors.

Attention to site limitations. The choice of tree species and provenances should be made dependent on local site conditions.

Species, size and genetic diversity. A mix of different tree species, mixed conifers and broadleaves as well as a mix of tree sizes and high genetic variation are encouraged.

Selective individual-tree silviculture. Trees are individually selected for thinning and harvesting in an attempt to gently steer a forest stand into a desired direction and often as a compromise between silvicultural, economic and conservation needs. There is no rotation period.

Conserving old trees, deadwood, endangered plant and animal species. To maintain biodiversity and for keeping the ecosystem intact, legacies of the old-growth phase including remnant trees and downed/standing deadwood should be retained. Endangered plant and animal species should be conserved. Old scenic trees and other natural features enhancing the recreation experience are also sustained.

Promoting native tree species/provenances and broadleaves in particular. The use of locally adapted native species and provenances contributes to resilient ecosystems offering many habitats. As part of this, the dominance of introduced conifers is reverted, and ancient woodland sites are restored.

Environmentally sensitive forest protection, thinning and harvesting operations. This principle ensures that any man-made disturbance of forest ecosystems is kept to a minimum. Soils in particular should not be affected chemically or mechanically.

Continuous Cover Forestry: Theories, Concepts, and Implementation, First Edition. Arne Pommerening.
© 2024 John Wiley & Sons Ltd. Published 2024 by John Wiley & Sons Ltd.

Environmentally sensitive wildlife management. In the absence of natural predators, the density of deer and other grazing and bark-stripping animal populations should be in balance with the requirement of using natural processes of growing trees such as natural regeneration and not reduce biodiversity.

Establishment and maintenance of forest margins, special habitats and nature reserves. To strengthen forest ecosystems and holistic forestry, transition zones between the open landscape and woodlands are created. In the same way, special habitats such as riparian and floodplain forests and unmanaged nature reserves are established and maintained.

Appendix B

Light Demand of Tree Species

One pre-requisite for successful natural regeneration is to meet the light demand of the established seedlings and saplings. Tree species may be sorted into functional groups (such as shade-tolerant, intermediate, light-demanding) or more precisely characterised by their minimum light requirement in the percentage of full illumination or on Ellenberg's scale (see Table B.1). In reality, the trade shade tolerance operates at a continuous scale and also depends on other environmental conditions, e.g. on the soil nutrient regime. Ellenberg and Leuschner (2010) developed a nonlinear, nine-level scale that is widely used among European ecologists. Level 1 corresponds to 'full shade', level 5 to 'semi-shade tolerant' with usually more than 10% of full light required. Level 8 corresponds to 'light loving' with usually more than 40% of full light required whilst level 9 denotes 'full light'.

Even some light-demanding species, such as *Fraxinus excelsior* and *Quercus petraea*, are capable of surviving under very low light levels when they are young. Light demand rises quickly with size and age and if such regeneration is not released, it usually dies. Regeneration of shade-tolerant species can start earlier than that of light demanders, which require faster release operations (Davies et al., 2008).

Very light-demanding (or shade-intolerant) tree species often act as *pioneer species* that are able to colonise open ground rapidly. These species are characterised by a relatively short lifespan and rapid juvenile growth. Their total height remains relatively low. *Sorbus aucuparia* and *Betula* spp. are typical examples.

Other light-demanding species such as *Larix* spp., *Quercus* spp. and *Fraxinus excelsior* show larger final sizes and have longer life expectancies, but still have fast early height growth which soon slows down. Shade-tolerant species (e.g. *Fagus sylvatica*, *Tsuga heterophylla* and *Abies alba*) start rather slowly but have long-lasting high growth rates and may reach very large final heights. Intermediate species, as the term implies, fit somewhere between light demanders and shade bearers: *Picea* spp., *Pseudotsuga menziesii*, *Acer pseudoplatanus*, *Tilia* spp. and *Castanea sativa* are typical examples. The group of shade-tolerant species is adapted to occupy the understorey of forests (Davies et al., 2008).

Newton (2007) reported various attempts to quantify shade tolerance which are made difficult by the fact that shade tolerance varies with age, i.e. young trees, particularly seedlings, are much more shade-tolerant than more adult stages of the same species.

Table B.1 Light demand of selected tree species.

Species	Ellenberg's scale[a]	Light demand (% of full illumination)[b]	General light demand[c]
Larix decidua, Larix kaempferi	7	20	
Sorbus aucuparia	6	12	
Populus tremula, Betula spp.	6, 7	11	
Pinus sylvestris, Pinus contorta	7	10	Very light demanding
Populus spp.	6	7–9	
Quercus robur	7	4–5	
Alnus glutinosa, Prunus avium	5, 4		
Quercus petraea	6	3–4	
Pinus nigra	7		Light demanding
Fraxinus excelsior	5	17	
Acer platanoides	4	1.8	
Ulmus spp.	4		Intermediate
Tilia cordata, Castanea sativa	5		
Picea abies, Picea sitchensis	7	2.8–3.6	
Pseudotsuga menziesii, Carpinus betulus	6, 4	1.8	Moderately shade tolerant
Acer pseudoplatanus, Tilia platyphyllos,			
Corylus avellana	4		
Fagus sylvatica	3	1.2–1.6	
Thuja plicata, Chamaecyparis lawsoniana	4, 5		
Tsugo heterophylla	6		Very shade tolerant
Abies alba, Abies grandis		1.6	
Taxus baccata	4	0.9	

Superscripts indicate figures taken from [a]Hill et al. (1999), [b]Mayer (1984) and [c]Burschel and Huss (1997). Source: Modified from Davies et al. (2008).

References

P. Abetz. Eine Entscheidungshilfe für die Durchforstung von Fichtenbeständen [A decision aid for the thinning of Norway spruce stands]. *Allgemeine Forstzeitschrift*, 30:666–667, 1975.

P. Abetz. Reaktion auf Standraumerweiterungen und Folgeerscheinungen für die Auslesedurchforstung bei Fichten [Response to release thinnings as part of frame-tree management in Norway spruce]. *Allgemeine Forst- und Jagdzeitung*, 147:72–75, 1976.

P. Abetz. Müssen wir in der waldbaulichen Behandlung der Fichte wieder umdenken? [Do we have to re-consider the management of Norway spruce?]. *Forstwissenschaftliches Centralblatt*, 109:79–85, 1990.

P. Abetz and J. Klädtke. The target tree management system. *Forstwissenschaftliches Centralblatt*, 121:73–82, 2002.

S. Adnan, M. Maltamo, L. Mehtätalo, R. N. L. Ammaturo, P. Packalen, and R. Valbuena. Determining maximum entropy in 3D remote sensing height distributions and using it to improve aboveground biomass modelling via stratification. *Remote Sensing of Environment*, 260:112464, 2021.

A. Ameray, Y. Bergeron, O. Valeria, M. Montoro Girona, and X. Cavard. Forest carbon management: a review of silvicultural practices and management strategies across boreal, temperate and tropical forests. *Current Forestry Reports*, 7:245–266, 2021.

P. Ammann. Biologische Rationalisierung, Teil 3: Biologische Rationalisierung bei Esche, Bergahorn und Buche [Biological rationalisation, part 3: Biological rationalisation in ash, sycamore and beech]. *Wald und Holz*, 3/05:29–33, 2005a.

P. Ammann. Biologische Rationalisierung, Teil 4: Baumartenmischung und Anwendungsbereich [Biological rationalisation, part 4: Tree species mixture and application range]. *Wald und Holz*, 4/05:35–37, 2005b.

P. Ammann. Waldbau auf Schadflächen: Ökonomisch und ökologisch interessante Alternativen [Silviculture after disturbances: Economically and ecologically interesting alternatives]. *Zürcher Wald*, 2/2020:4–8, 2020.

M. L. Anderson. A new system of planting. *Scottish Forestry*, 44:78–89, 1930.

M. L. Anderson. Spaced group-planting and irregularity of stand-structure. *Empire Forestry Journal*, 30:328–341, 1951.

M. L. Anderson. Plea for the adoption of the standing control or check in woodland management. *Scottish Forestry*, 7:38–47, 1953.

Continuous Cover Forestry: Theories, Concepts, and Implementation, First Edition. Arne Pommerening.
© 2024 John Wiley & Sons Ltd. Published 2024 by John Wiley & Sons Ltd.

M. L. Anderson. Norway spruce-silver fir-beech mixed selection forest. Is it possible to reproduce this in Scotland? *Scottish Forestry*, 14:87–93, 1960.

T. Anfodillo, M. Carrer, F. Simini, I. Popa, J. R. Banavar, and A. Maritan. An allometry-based approach for understanding forest structure, predicting tree-size distribution and assessing the degree of disturbance. *Proceedings of the Royal Society B: Biological Sciences*, 280:20122375, 2013.

G. H. Aplet, N. Johnson, J. T. Olson, and V. A. Sample, editors. *Defining sustainable forestry*. Island Press, Washington, DC, 4th edition, 1993.

A. Arnberger. Recreation use of urban forests: an inter-area comparison. *Urban Forestry & Urban Greening*, 4:135–144, 2006.

A. Arnberger, I. E. Schneider, R. Eder, and A. Choi. Differences in urban forest visitor preferences for emerald ash borer-impacted areas. *Forestry*, 93:225–238, 2019.

E. Assmann. *The principles of forest yield study. Studies in the organic, production, structure, increment and yield of forest stands*. Pergamon Press, Oxford, 1970.

H. Bachofen. Gleichgewicht, Struktur und Wachstum in Plenterbeständen [Equilibrium, structure and increment in selection forest stands]. *Schweizerische Zeitschrift für Forstwesen*, 150:157–170, 1999.

S. C. Baker, M. Garandel, M. Deltombe, and M. G. Neyland. Factors influencing initial vascular plant seedling composition following either aggregated retention harvesting and regeneration burning or burning of unharvested forest. *Forest Ecology and Management*, 306:192–201, 2013.

I. Barbeito, I. Cañellas, and F. Montes. Evaluating the behaviour of vertical structure indices in Scots pine forests. *Annals of Forest Science*, 66:710–710, 2009.

H. Bartels. *Gehölzkunde [Dendrology]*. Verlag Eugen Ulmer, Stuttgart, 1993.

N. Bartsch, B. v. Lüpke, and E. Röhrig. *Waldbau auf ökologischer Grundlage [Silviculture on an ecological basis]*. Verlag Eugen Ulmer, Stuttgart, 8th edition, 2020.

F. W. Bauer. *Waldbau als Wissenschaft [Scientific silviculture]*, volume 1. BLV Verlag, München, 1962.

F. W. Bauer. *Waldbau als Wissenschaft [Scientific silviculture]*, volume 2. BLV Verlag, München, 1968.

Bayerische Staatsforsten. *Waldbauhandbuch Bayerische Staatsforsten. Richtlinie für die Waldbewirtschaftung im Hochgebirge [Silviculture handbook. Guidelines for managing mountain forests]*. Regensburg, 2018.

W. J. Beese and A. A. Bryant. Effect of alternative silvicultural systems on vegetation and bird communities in coastal montane forests of British Columbia, Canada. *Forest Ecology and Management*, 115:231–242, 1999.

M. Begon, J. L. Harper, and C. R. Townsend. *Ecology: individuals, populations and communities*. Blackwell Science, Oxford, 3rd edition, 2006.

U. Benecke. Ecological silviculture: the application of age-old methods. *New Zealand Journal of Forestry*, 41:27–33, 1996.

G. Bergqvist. Wood volume yield and stand structure in Norway spruce understorey depending on birch shelterwood density. *Forest Ecology and Management*, 122:221–229, 1999.

A. R. Berkowitz, C. D. Canham, and V. R. Kelly. Competition vs. facilitation of tree seedling growth and survival in early successional communities. *Ecology*, 76:1156–1168, 1995.

M. D. Bertness and R. Callaway. Positive interactions in communities. *Trends in Ecology and Evolution*, 9:191–193, 1994.

P. Bettinger, K. Boston, J. P. Siry, and D. L. Grebner. *Forest management and planning*. Academic Press, Elsevier Inc., Burlington, MA, 2nd edition, 2017.

J. Bijak, D. Courgeau, R. Franck, and E. Silverman. Modelling in demography: From statistics to simulations. In *Methodological Investigations in Agent-Based Modelling: with Applications for the Social Sciences*, pages 167–187. Springer International Publishing, Cham, 2018.

D. Binkley. A hypothesis about the interaction of tree dominance and stand production through stand development. *Forest Ecology and Management*, 190:265–271, 2004.

D. Binkley. *Forest ecology*. John Wiley & Sons, Chichester, 2021.

D. Binkley, D. M. Kashian, S. Boyden, M. W. Kaye, J. B. Bradford, M. A. Arthur, P. J. Fornwalt, and M. G. Ryan. Patterns of growth dominance in forests of the Rocky Mountains, USA. *Forest Ecology and Management*, 236:193–201, 2006.

A. Bončina. History, current status and future prospects of uneven-aged forest management in the Dinaric region: an overview. *Forestry*, 84:467–478, 2011a.

A. Bončina. Conceptual approaches to integrate nature conservation into forest management: a Central European perspective. *The International Forestry Review*, 13:13–22, 2011b.

A. K. Bose, B. D. Harvey, S. Brais, M. Beaudet, and A. Leduc. Constraints to partial cutting in the boreal forest of Canada in the context of natural disturbance-based management: a review. *Forestry*, 87:11–28, 2014.

J. B. Bradford, A. W. D'Amato, B. J. Palik, and S. Fraver. A new method for evaluating forest thinning: growth dominance in managed *Pinus resinosa* stands. *Canadian Journal of Forest Research*, 40:843–849, 2010.

P. Brang. Resistance and elasticity: promising concepts for the management of protection forests in the European Alps. *Forest Ecology and Management*, 145:107–119, 2001.

P. Brang, W. Schönenberger, M. Frehner, R. Schwitter, J.-J. Thormann, and B. Wasser. Management of protection forests in the European Alps: an overview. *Forest Snow and Landscape Research*, 80:23–44, 2006.

P. Brang, P. Spathelf, J. B. Larsen, J. Bauhus, A. Bončina, C. Chauvin, L. Drössler, C. García-Güemes, C. Heiri, G. Kerr, M. J. Lexer, B. Mason, F. Mohren, U. Mühlethaler, S. Nocentini, and M. Svoboda. Suitability of close-to-nature silviculture for adapting temperate European forests to climate change. *Forestry*, 87:492–503, 2014.

M. Brauner, W. Weinmeister, P. Agner, S. Vospernik, and B. Hoesle. Forest management decision support for evaluating forest protection effects against rockfall. *Forest Ecology and Management*, 207:75–85, 2005.

A. Bravo-Oviedo, M. Marchi, D. Travaglini, F. Pelleri, M. C. Manetti, P. Corona, F. Cruz, F. Bravo, and S. Nocentini. Adoption of new silvicultural methods in Mediterranean forests: the influence of educational background and sociodemographic factors on marker decisions. *Annals of Forest Science*, 77:48, 2020.

L. C. Brodie and C. A. Harrington. Guide to variable-density thinning using skips and gaps. General Technical Report PNW-GTR-989, U.S. Department of Agriculture, Forest Service, Pacific Northwest Research Station, Portland, 2020.

A. Bruchwald, E. Dmyterko, J. Łukaszewicz, M. Niemczyk, and P. Wrzesiński. Wzorzec rozkładu pierśnic drzew wielowarstwowego drzewostanu jodłowego Gór Świętokrzyskich [Pattern of breast height diameter distribution in a multilayer silver fir stand in the Świętokrzyskie Mountains]. *Sylwan*, 160:741–750, 2016.

M. Bruciamacchie, J.-C. Pierrat, and J. Tomasini. Modèles explicatif et marginal de la stratégie de martelage d'une parcelle irrégulière [Explicative and marginal models for a marking strategy of an unevenaged stand]. *Annals of Forest Science*, 62:727–736, 2005.

R. D. Brunner and T. W. Clark. A practice-based approach to ecosystem management. *Conservation Biology*, 11:48–58, 1997.

B. Brzeziecki, A. Pommerening, S. Miścicki, S. Drozdowski, and H. Żybura. A common lack of demographic equilibrium among tree species in Białowieża National Park (NE Poland): evidence from long-term plots. *Journal of Vegetation Science*, 27:460–469, 2016.

B. Brzeziecki, S. Drozdowski, K. Bielak, M. Czacharowski, J. Zajączkowski, W. Buraczyk, and L. Gawron. A demographic equilibrium approach to stocking control in mixed, multiaged stands in Białowieża Forest, northeast Poland. *Forest Ecology and Management*, 481:118694, 2021.

J. Buongiorno, A. Kolbe, and M. Vasievich. Economic and ecological effects of diameter-limit and BDq management regimes: simulation results for northern hardwoods, 2000.

M. Burkhart and H. E. Tomé. *Modeling forest trees and stands*. Springer, New York, 2012.

P. Burschel and J. Huss. *Grundriss des Waldbaus [Outline of silviculture]*. Parey Buchverlag, Berlin, 1997.

J. Busse. Gruppendurchforstung [Group thinning]. *Silva*, 19:145–147, 1935.

M. A. Buzas and L.-A. C. Hayek. Biodiversity resolution: An integrated approach. *Biodiversity Letters*, 3:40–43, 1996.

B. S. Cade and B. R. Noon. A gentle introduction to quantile regression for ecologists. *Frontiers in Ecology and the Environment*, 1:412–420, 2003.

E. Cairns. How could continuous cover forestry work in New Zealand? *New Zealand Tree Grower*, 22:42–43, 2001.

A. D. Cameron. Importance of early selective thinning in the development of long-term stand stability and improved log quality: a review. *Forestry*, 75:25–35, 2002.

J. Cancino and K. v. Gadow. Stem number guide curves for uneven-aged forests development and limitations. In K. v. Gadow, J. Nagel, and J. Saborowski, editors, *Continuous cover forestry: assessment, analysis, scenarios*, pages 163–174. Springer Netherlands, Dordrecht, 2002.

R. B. Chevrou. de Liocourt's law and the truncated law. *Canadian Journal of Forest Research*, 20:1933–1946, 1990.

D. J. Churchill, A. J. Larson, M. C. Dahlgreen, J. F. Franklin, P. F. Hessburg, and J. A. Lutz. Restoring forest resilience: From reference spatial patterns to silvicultural prescriptions and monitoring. *Forest Ecology and Management*, 291:442–457, 2013.

O. Ciancio and S. Nocentini. Biodiversity conservation and systemic silviculture: Concepts and applications. *Plant Biosystems*, 145:411–418, 2011.

O. Ciancio, P. Corona, A. Lamonaca, L. Portoghesi, and D. Travaglini. Conversion of clearcut beech coppices into high forests with continuous cover: A case study in Central Italy. *Forest Ecology and Management*, 224:235–240, 2006.

G. C. Clarke. Using a spreadsheet to calculate the ideal uneven-aged structure. *Continuous Cover Forestry Group Newsletter*, 7:10–13, 1995.

J. L. Clutter, J. C. Fortson, L. V. Pienaar, G. H. Brister, and R. L. Bailey. *Timber management: A quantitative approach*. John Wiley & Sons, New York, 1983.

K. D. Coates, C. D. Canham, M. Beaudet, D. L. Sachs, and C. Messier. Use of a spatially explicit individual-tree model (SORTIE/BC) to explore the implications of patchiness in structurally complex forests. *Forest Ecology and Management*, 186:297–310, 2003.

Codoc. Checkkarten Waldbau und Ökologie [Flashcards silviculture and ecology]. Brochure, Codoc, Liess, 2020.

S. Cogos, S. Roturier, and L. Östlund. The origins of prescribed burning in Scandinavian forestry: the seminal role of Joel Wretlind in the management of fire-dependent forests. *European Journal of Forest Research*, 139:393–406, 2020.

M. Coppini and L. Hermanin. Restoration of selective beech coppices: A case study in the Apennines (Italy). *Forest Ecology and Management*, 249:18–27, 2007.

H. Cosyns, D.l Kraus, F. Krumm, T. Schulz, and P. Pyttel. Reconciling the tradeoff between economic and ecological objectives in habitat-tree selection: a comparison between students, foresters, and forestry trainers. *Forest Science*, 65:223–234, 2018.

H. Cotta. *Anweisungen zum Waldbau* [*Silvicultural prescriptions*]. Dresden und Leipzig, 1816.

A. Craig and S. E. Macdonald. Threshold effects of variable retention harvesting on understory plant communities in the boreal mixedwood forest. *Forest Ecology and Management*, 258:2619–2627, 2009.

F. Crecente-Campo, A. Pommerening, and R. Rodríguez-Soalleiro. Impacts of thinning on structure, growth and risk of crown fire in a *Pinus sylvestris* L. plantation in Northern Spain. *Forest Ecology and Management*, 257:1945–1954, 2009.

R. O. Curtis. "Selective cutting" in Douglas fir: History revisited. *Journal of Forestry*, 96:40–46, 1998.

M. R. T. Dale and M.-J. Fortin. *Spatial analysis. A guide for ecologists*. Cambridge University Press, Cambridge, 2nd edition, 2014.

C. Damgaard and J. Weiner. Describing inequality in plant size or fecundity. *Ecology*, 81:1139–1142, 2000.

S. Daume, K. Füldner, and K. v. Gadow. Zur Modellierung personenspezifischer Durchforstungen in ungleichaltigen Mischbeständen [Modelling person-specific thinnings in uneven-aged mixed stands]. *Allgemeine Forst- und Jagdzeitung*, 169:21–26, 1997.

O. Davies, J. Haufe, and A. Pommerening. Silvicultural principles of continuous cover forestry – A guide to best practice. Unpublished booklet, Bangor University, Bangor, 2008.

M. De Cáceres, S. Martín-Alcón, J. R. González-Olabarria, and L. Coll. A general method for the classification of forest stands using species composition and vertical and horizontal structure. *Annals of Forest Science*, 76:40, 2019.

R. de Camino. Zur Bestimmung der Bestandeshomogenität [On defining stand homogeneity]. *Allgemeine Forst- und Jagdzeitung*, 147:54–58, 1976.

J. de Jong, C. Akselsson, G. Egnell, S. Löfgren, and B. A. Olsson. Realizing the energy potential of forest biomass in Sweden – How much is environmentally sustainable? *Forest Ecology and Management*, 383:3–16, 2017.

F. de Liocourt. De l'amenagement des sapinières [The management of silver fir forests]. *Bulletin trimestriel, Société Forestière de Franche-Comté et Belfort*, 1898:396–409, 1898.

B. de Turckheim. Planification et contrôle en futaie irrégulière et continue [Planning and control for continuous cover, uneven-aged silvicultural systems]. *Revue Forestière Française*, 51:76–86, 1999.

S. Dedrick, H. Spiecker, C. Orazio, M. Tomé, and I. Martinez, editors. *Plantation or conversion – the debate!* Discussion paper 13. European Forest Institute, Joensuu, 2007.

J. den Ouden, B. Muys, G. M. J. Mohren, and K. Verheyen. *Bosecologie en Bosbeheer [Forest ecology and silviculture]*. Acco, Leuven, 2010.

A. Dengler. *Waldbau auf ökologischer Grundlage [Silviculture on an ecological basis]*. Springer, Berlin, 3rd edition, 1944.

J. Diaci. *Nature-based forestry in Central Europe: alternatives to industrial forestry and strict preservation*. Biotechnical Faculty, University of Ljubljana, Ljubljana, 2006.

I. A. Dickie, S. A. Schnitzer, P. B. Reich, and S. E. Hobbie. Spatially disjunct effects of co-occurring competition and facilitation. *Ecology Letters*, 8:1191–1200, 2005.

O. Dittmar. Zur Z-Baum-Entwicklung in langfristigen Kieferndurchforstungen [On the development of frame trees in long-term Scots pine management]. *Allgemeine Forst- und Jagdzeitung*, 126:121–125, 1991.

H. Döbbeler and H. Spellmann. Methodological approach to simulate and evaluate silvicultural treatments under climate change. *Forstwissenschaftliches Centralblatt*, 121:52–69, 2002.

B. T. Doi, D. Binkley, and J. L. Stape. Does reverse growth dominance develop in old plantations of *Eucalyptus saligna*? *Forest Ecology and Management*, 259:1815–1818, 2010.

L. K. A. Dorren, F. Berger, C. le Hir, E. Mermin, and P. Tardif. Mechanisms, effects and management implications of rockfall in forests. *Forest Ecology and Management*, 215:183–195, 2005.

M. Dovčiak, C. B. Halpern, J. F. Saracco, S. A. Evans, and D. A. Liguori. Persistence of ground-layer bryophytes in a structural retention experiment, initial effects of level and pattern of overstory retention. *Canadian Journal of Forest Research*, 36:3039–3052, 2006.

A. Drengston and D. Taylor, editors. *Ecoforestry: The art and science of sustainable forest use*. New Society Publishers, Gabriola Island, 1997.

R. Duchesneau, I. Lesage, C. Messier, and H. Morin. Effects of light and intraspecific competition on growth and crown morphology of two size classes of understory balsam fir saplings. *Forest Ecology and Management*, 140:215–225, 2001.

M.-S. Duchiron. *Strukturierte Mischwälder. Eine Herausforderung für den Waldbau unserer Zeit [Structured mixed-species forests. A challenge for the silviculture of our time]*. Parey Buchverlag, Berlin, 2000.

G. Duduman. A forest management planning tool to create highly diverse uneven-aged stands. *Forestry*, 84:301–314, 2011.

L. Dvorak and P. Bachmann. Plenterwälder sind sturmfester [Selection forests are more resilient to storms]. *Wald und Holz*, 9:47–50, 2001.

B. Eberhard and H. Hasenauer. Tree marking versus tree selection by harvester operator: are there any differences in the development of thinned Norway spruce forests? *International Journal of Forest Engineering*, 32:42–52, 2021.

H. Eisenreich and W. Nebe. *Waldbau. Lehrbücher für die sozialistische Berufsausbildung – Forstwirtschaft* [*Silviculture. Textbooks for socialist occupational education – forestry*]. VEB Deutscher Landwirtschaftsverlag, Berlin, 1967.

H. Ellenberg and C. Leuschner. *Vegetation Mitteleuropas mit den Alpen in ökologischer, dynamischer und historischer Sicht* [*The vegetation of Central Europe including the Alps from an ecological, dynamic and historic point of view*]. Ulmer, Stuttgart, 2010.

W. H. Emmingham, M. Bondi, and D. E. Hibbs. Underplanting western hemlock in a red alder thinning: early survival, growth, and damage. *New Forests*, 3:31–43, 1989.

B. J. Enquist, G. B. West, and J. H. Brown. Extensions and evaluations of a general quantitative theory of forest structure and dynamics. *Proceedings of the National Academy of Sciences of the United States of America*, 106:7046–7051, 2009.

European Commission. New EU forest strategy for 2030. Report, European Commission, Brussels, 2021.

J. Evans. Silviculture of broadleaved woodlands. Forestry Commission Bulletin 62, Forestry Commission, Edinburgh, 1984.

L. Fahrmeir, I. Pigeot, and G. Tutz. *Statistik: Der Weg zur Datenanalyse* [*Statistics: The road to data analysis*]. Springer, 8th edition, 2016.

D. S. Falster and M. Westoby. Plant height and evolutionary games. *Trends in Ecology & Evolution*, 18:337–343, 2003.

K. Fedrowitz, J. Koricheva, S. C. Baker, D. B. Lindenmayer, B. Palik, R. Rosenvald, W. Beese, J. F. Franklin, J. Kouki, E. Macdonald, C. Messier, A. Sverdrup-Thygeson, and L. Gustafsson. Can retention forestry help conserve biodiversity? *Journal of Applied Ecology*, 51:1669–1679, 2014.

A. Fichtner, W. Härdtle, H. Bruelheide, M. Kunz, Y. Li, and G. v. Oheimb. Neighbourhood interactions drive overyielding in mixed-species tree communities. *Nature Communications*, 9:1144, 2018.

F. Fiedler. Die Entwicklung des Vorwaldgedankens unter besonderer Berücksichtigung der Birke [The historical development of the nurse crop idea with special attention to birch]. *Archiv für Forstwesen*, 11:174–191, 1962.

R. Finkeldey and M. Ziehe. Genetic implications of silvicultural regimes. *Forest Ecology and Management*, 197:231–244, 2004.

S. Flaherty, G. Patenaude, A. Close, and P. W. W. Lurz. The impact of forest stand structure on red squirrel habitat use. *Forestry*, 85:437–444, 2012.

Forest Enterprise. Forest management plan for Gwydyr Forest. Llanrwst, Unpublished, 2000.

Forestry Commission. England's forestry strategy. A new focus for England's woodlands. Brochure, Forestry Commission, Edinburgh, 1998.

Forestry Commission. The Scottish forestry strategy. Brochure, Forestry Commission, Edinburgh, 2000.

Forestry Commission. Coetiroedd i Gymru. Y Cynulliad Cenedlaethol i Gymru strategaeth am goed a choetir [Woodlands for Wales. The National Assembly for Wales strategy for trees and woodlands]. Brochure, Forestry Commission, Aberystwyth, 2001.

Forestry Commission. Managing continuous cover forests. Operational guidance booklet 7, Forestry Commission, Edinburgh, 2008.

Forestry Commission. Managing forests in acid sensitive water catchments. Practical Guide, Forestry Commission, Edinburgh, 2014.

J. F. Franklin. Towards a new forestry. *American Forests*, 95:37–44, 1989.

J. F. Franklin and R. T. T. Forman. Creating landscape patterns by forest cutting: Ecological consequences and principles. *Landscape Ecology*, 1:5–18, 1987.

J. F. Franklin, K. N. Johnson, and D. L. Johnson. *Ecological forest management*. Waveland Press Inc., Long Grove, 2018.

N. Frischbier, P. S. Nikolova, P. Brang, R. Klumpp, G. Aas, and F. Binder. Climate change adaptation with non-native tree species in central European forests: early tree survival in a multi-site field trial. *European Journal of Forest Research*, 138:1015–1032, 2019.

K. Füldner, S. Sattler, W. Zucchini, and K. v. Gadow. Modellierung personenabhängiger Auswahlwahrscheinlichkeiten bei der Durchforstung [Modelling person-specific tree selection probabilities in a thinning.]. *Allgemeine Forst- und Jagdzeitung*, 167:159–162, 1996.

K. v. Gadow. Modelling growth in managed forests – realism and limits of lumping. *The Science of the Total Environment*, 183:167–177, 1996.

K. v. Gadow. Orientation and control in CCF systems. In K. v. Gadow, J. Nagel, and J. Saborowski, editors, *Continuous cover forestry. Assessment, analysis, scenarios. Proceedings of the international IUFRO conference on 19th -21st September, 2001 at Göttingen*, pages 211–217. University of Göttingen, Göttingen, 2001.

K. v. Gadow and B. Bredenkamp. *Forest management*. Academia, Pretoria, 1992.

K. v. Gadow and V. Stüber. Die Inventuren der Forsteinrichtung [Forest management inventories]. *Forst und Holz*, 49:129–131, 1994.

K. v. Gadow, J. Nagel, and J. Saborowski, editors. *Continuous cover forestry. Assessment, analysis, scenarios*. Kluwer Academic Publishers, Dordrecht, 2002.

K. v. Gadow, J. G. Á. González, C. Zhang, T. Pukkala, and X. Zhao. *Sustaining forest ecosystems. Managing forest ecosystems*. Springer, 2021.

C. Gamborg and J. B. Larsen. "Back to nature" – a sustainable future for forestry? *Forest Ecology and Management*, 179:559–571, 2003.

J. E. Garfitt. *Natural management of woods – continuous cover forestry*. Research Studies Press Ltd., Taunton, 1995.

K. J. Gaston and J. I. Spicer. *Biodiversity. An introduction*. Blackwell Publishing, Oxford, 2004.

V. Gavrikov and D. Stoyan. The use of marked point processes in ecological and environmental forest studies. *Environmental and Ecological Statistics*, 2:331–344, 1995.

K. Gayer. *Der gemischte Wald, seine Begründung und Pflege, insbesondere durch Horst- und Gruppenwirtschaft [The mixed forest, its establishment and tending, particularly through group management]*. Paul Parey, Berlin, 1886.

T. Gebhardt, K.-H. Häberle, R. Matyssek, C. Schulz, and C. Ammer. The more, the better? Water relations of Norway spruce stands after progressive thinning. *Agricultural and Forest Meteorology*, 197:235–243, 2014.

C. Gini. *Variabilitá e mutabilitá: contributo allo studio delle distribuzioni e delle relazioni statistiche [Variability and mutability: contribution to the study of distributions and statistical relationships]*. Studi economico-giuridici pubblicati per cura della facoltá di Giurisprudenza della R. Universitá di Cagliari. Tipogr. di P. Cuppini, 1912.

H. Gockel. Die Trupp-Pflanzung. Ein neues Pflanzschema zur Begründung von Eichenbeständen [Group planting. A new planting method for establishing oak stands]. *Forst und Holz*, 50:570–575, 1995.

H. v. Goltz. Strukturdurchforstung der Fichte. Ein Weg zu stufigem Bestandesaufbau [Structural thinning of Norway spruce. An option for increasing vertical forest structure]. *AFZ-Der Wald*, 13/91:677–679, 1991.

S. G. Göttlicher, A. F. S. Taylor, H. Grip, N. R. Betson, E. Valinger, M. N. Högberg, and P. Högberg. The lateral spread of tree root systems in boreal forests: estimates based on 15N uptake and distribution of sporocarps of ectomycorrhizal fungi. *Forest Ecology and Management*, 255:75–81, 2008.

G. Grassi and R. Giannini. Influence of light and competition on crown and shoot morphological parameters of Norway spruce and silver fir saplings. *Annals of Forest Science*, 62:269–274, 2005.

O. Greger. Begründung von Elbauewälder unter besonderer Berücksichtigung des Vorwaldgedankens [Establishment of new riparian forests along the river Elbe using nurse crops]. *Forst und Holz*, 53:412–414, 1998.

H.-R. Gregorius. Genetischer Abstand zwischen Populationen [Genetic distance between populations]. *Silvae Genetica*, 23:22–27, 1974.

J. M. Gresh and J. R. Courter. In pursuit of ecological forestry: Historical barriers and ecosystem implications. *Frontiers in Forests and Global Change*, 4:571438, 2021.

V. C. Griess, R. Acevedo, F. Härtl, K. Staupendahl, and T. Knoke. Does mixing tree species enhance stand resistance against natural hazards? A case study for spruce. *Forest Ecology and Management*, 267:284–296, 2012.

R. E. Grumbine. What is ecosystem management? *Conservation Biology*, 8:27–38, 1994.

M. Guericke. Versuche zur Begründung von Eichenbeständen durch Nesterpflanzung [Experimental series of oak-nest plantings]. *Forst und Holz*, 51:577–582, 1996.

C. Guest. Newborough forest management plan. Report, Forestry consultant, Bangor, 2008.

J. M. Guldin. Uneven-aged BDq regulation of Sierra Nevada mixed conifers. *Journal of Forestry*, 89:29–36, 1991.

J. M. Guldin. Continuous cover forestry in the United States – experience with southern pines. In K. v. Gadow, J. Nagel, and J. Saborowski, editors, *Continuous cover forestry. Assessment, analysis, scenarios*, pages 296–307. Kluwer Academic Publishers, Dordrecht, 2002.

L. Gustafsson, J. Kouki, and A. Sverdrup-Thygeson. Tree retention as a conservation measure in clear-cut forests of northern Europe, a review of ecological consequences. *Scandinavian Journal of Forest Research*, 25:295–308, 2010.

L. Gustafsson, S. C. Baker, J. Bauhus, W. J. Beese, A. Brodie, J. Kouki, D. B. Lindenmayer, A. L ohmus, G. Martínez Pastur, C. Messier, M. Neyland, B. Palik, A. Sverdrup-Thygeson, W. J. A. Volney, A. Wayne, and J. F. Franklin. Retention forestry to maintain multifunctional forests: a world perspective. *Bioscience*, 62:633–645, 2012.

S. E. Hale. The effect of thinning intensity on the below-canopy light environment in a Sitka spruce plantation. *Forest Ecology and Management*, 179:341–349, 2003.

C. B. Halpern, D. McKenzie, S. A. Evans, and D. A. Maguire. Initial responses of forest understories to varying levels and patterns of green-tree retention. *Ecological Applications*, 15:175–195, 2005.

C. B. Halpern, J. Halaj, S. A. Evans, and M. Dovčiak. Level and pattern of overstory retention interact to shape long-term responses of understories to timber harvest. *Ecological Applications*, 22:2049–2064, 2012.

G. J. Hamilton and J. M. Christie. *Construction and application of stand yield models*. Forestry Commission Research and Development Paper 96. Forestry Commission, Edinburgh, 1973.

W. E. Hammitt. Recreation – user needs and preferences. In J. Burley, editor, *Encyclopedia of forest sciences*, pages 949–958. Elsevier, Oxford, 2004.

M. Hanewinkel. Kritische Analyse von auf der Basis von Gleichgewichtsmodellen hergeleiteten Zielreferenzen für Plenterwälder im Wuchsgebiet Schwarzwald [Critical analysis of goal references for selection forests in the Black Forest region based on steady-state models]. *Allgemeine Forst- und Jagdzeitung*, 170:87–97, 1998.

M. Hanewinkel, T. Kuhn, H. Bugmann, A. Lanz, and P. Brang. Vulnerability of uneven-aged forests to storm damage. *Forestry*, 87:525–534, 2014.

G. D. Hansen and R. D. Nyland. Effects of diameter distribution on the growth of simulated uneven-aged sugar maple stands. *Canadian Journal of Forest Research*, 17:1–8, 1987.

C. Hart. *Practical forestry for the agent and surveyor*. Sutton Publishing, Phoenix Mill, 3rd edition, 1991.

C. Hart. *Alternative silvicultural systems to clear cutting in Britain: A review*. Forestry Commission Bulletin 115. HSMO, London, 1995.

K. Hasel and E. Schwartz. *Forstgeschichte: Ein Grundriss für Studium und Praxis* [*Forest history: an outline for university students and practitioners*]. Norbert Kessel, Remagen-Oberwinter, 3rd edition, 2006.

H. Hasenauer. Dimensional relationships of open-grown trees in Austria. *Forest Ecology and Management*, 96:197–206, 1997.

H. Hasenauer and H. Sterba. The research programme for the restoration of forest ecosystems in Austria. In E. Klimo, H. Hager, and J. Kulhavý, editors, *Spruce monocultures in central Europe. Problems and prospects*. EFI Proceedings No. 33, pages 43–52. European Forest Institute, Joensuu, 2000.

H. Hasenauer, E. Leitgeb, and H. Sterba. Der *A*-Wert nach Johann als Konkurrenzindex für die Abschätzung von Durchforstungseffekten [Johann's *A* value used as competition index for determining thinning effects]. *Allgemeine Forst- und Jagdzeitung*, 167:169–174, 1996.

J. Haufe, G. Kerr, V. Stokes, and S. Bathgate. Forest Development Types: a guide to the design and management of diverse forests in Britain. Version 1.0, Forest Research, Edinburgh, 2021.

A. Häusler and M. Scherer-Lorenzen. Sustainable forest management in Germany: The ecosystem approach of the biodiversity convention reconsidered. Report, German Federal Agency for Nature Conservation, Bonn, 2001.

A. Heger. *Die Begründung von Mischwäldern auf Großkahlflächen unter besonderer Berücksichtigung des Vorwaldgedankens* [*The establishment of mixed-species forests on large clearfelling sites with special attention to the nurse-crop idea*]. Neumann Verlag, Radebeul, 1949.

A. Heger. *Lehrbuch der forstlichen Vorratspflege* [*Textbook on forest stock maintenance*]. Neumann, Radebeul, Berlin, 1955.

D. R. Helliwell. Dauerwald. *Forestry*, 70:375–380, 1997.

R. Helliwell and E. Wilson. Continuous cover forestry in Britain: challenges and opportunities. *Quarterly Journal of Forestry*, 106:214–224, 2012.

N. A. Helms, editor. *The dictionary of forestry*. Society of American Foresters, Bethesda, 1998.

I. M. Hertog, S. Brogaard, and T. Krause. Barriers to expanding continuous cover forestry in Sweden for delivering multiple ecosystem services. *Ecosystem Services*, 53:101392, 2022.

J. M. Hett and O. L. Loucks. Age structure models of balsam fir and eastern hemlock. *Journal of Ecology*, 64:1029–1044, 1976.

E. Heuer. Begründung von Mischbeständen aus Laub- und Nadelbäumen unter Schirm von Kiefernaltbeständen [Establishment of mixed broadleaved and coniferous forest woodlands under the canopy of mature Scots pine stands]. *AFZ-Der Wald*, 51:724–727, 1996.

W. E. Hiley. *Conifers: South African methods of cultivation*. Faber & Faber, London, 1959.

M. O. Hill. Diversity and evenness: A unifying notation and its consequences. *Ecology*, 54:427–432, 1973.

M. O. Hill, J. O. Mountford, D. B. Roy, and R. G. H. Bunce. *Ellenberg's indicator values for British plants*. ECOFACT Volume 2. Institute of Terrestrial Ecology, Huntingdon, 1999.

P. Högberg, N. Wellbrock, M. N. Högberg, H. Mikaelsson, and J. Stendahl. Large differences in plant nitrogen supply in German and Swedish forests – Implications for management. *Forest Ecology and Management*, 482:118899, 2021.

G. Hui and A. Pommerening. Analysing tree species and size diversity patterns in multi-species uneven-aged forests of Northern China. *Forest Ecology and Management*, 316:125–138, 2014.

J. Huss. Die Entwicklung des Dauerwaldgedankens bis zum Dritten Reich [The development of the Dauerwald idea until the Third Reich]. *Forst und Holz*, 45:163–171, 1990.

J. Illian, A. Penttinen, H. Stoyan, and D. Stoyan. *Statistical analysis and modelling of spatial point patterns*. John Wiley & Sons, Chichester, 2008.

M. G. Isaac-Renton, D. R. Roberts, A. Hamann, and H. Spiecker. Douglas-fir plantations in Europe: a retrospective test of assisted migration to address climate change. *Global Change Biology*, 20:2607–2617, 2014.

N. D. G. James. *A history of English forestry*. Blackwell Publishers, Oxford, 1990.

A. Jaworski and Z. Kołodziej. Beech (*Fagus sylvatica* L.) forests of a selection structure in the Bieszczady Mountains (southeastern Poland). *Journal of Forest Science*, 50:301–312, 2004.

F. S. Jensen. The effects of information on Danish forest visitors' acceptance of various management actions. *Forestry*, 73:165–172, 2000.

R. A. E. Jephcott. *Considerations required for the development of an educational interpretative trail including an evaluation of the public preferences and perceptions of forests*. MSc thesis, University of Wales, Bangor, Bangor, 2002.

K. Johann. Der *A*-Wert – ein objektiver Parameter zur Bestimmung der Freistellungsstärke von Zentralbäumen [The *A*-thinning index – an objective parameter for the determination of release intensity of frame trees]. Tagungsbericht der Jahrestagung 1982 in Weibersbrunn, Sektion Ertragskunde im Deutschen Verband Forstlicher Forschungsanstalten, 1982.

E. Johann. Coppice forests in Austria: the re-introduction of traditional management systems in coppice forests in response to the decline of species and landscape and under the aspect of climate change. *Forest Ecology and Management*, 490:119129, 2021.

S. Johnson, J. Strengbom, and J. Kouki. Low levels of tree retention do not mitigate the effects of clearcutting on ground vegetation dynamics. *Forest Ecology and Management*, 330:67–74, 2014.

D. R. Johnston. Irregularity in British forestry. *Forestry*, 51:163–169, 1978.

E. W. Jones. *Silvicultural systems*. Clarendon Press, Oxford, 2nd edition, 1952.

L. Jost. Entropy and diversity. *Oikos*, 113:363–375, 2006.

P. M. Joyce, J. Huss, R. McCarthy, and A. Pfeifer. *Growing broadleaves. Silvicultural guidelines for ash, sycamore, wild cherry, beech and oak in Ireland*. COFORD, Dublin, 1998.

A. Kangas and M. Maltamo. Anticipating the variance of predicted stand volume and timber assortments with respect to stand characteristics and field measurements. *Silva Fennica*, 36:799–811, 2002.

P. P. Kärenlampi. Spruce forest stands in a stationary state. *Journal of Forestry Research*, 30:1167–1178, 2019.

E. Karjalainen, T. Sarjala, and H. Raitio. Promoting human health through forests: overview and major challenges. *Environmental Health and Preventive Medicine*, 15:1, 2009.

H. W. Kassier. *Dynamics of diameter and height distributions in commercial timber plantations*. PhD thesis, University of Stellenbosch, Stellenbosch, 1993.

L. Katholnig. *Growth dominance and Gini index in even-aged and in uneven-aged forests*. MSc thesis, University of Natural Resources and Applied Life Sciences, Vienna, 2012.

F. Kató. *Begründung der qualitativen Gruppendurchforstung [Justification of the group thinning method]*. Habilitation thesis, University of Göttingen, Göttingen, 1973.

W. Kausch-Blecken von Schmeling. *Der Speierling – Sorbus domestica L. [The true service tree – Sorbus domestica L.]*. Verlag W. Kausch-Blecken von Schmeling, Bovenden, 1992.

G. Kenk and S. Guehne. Management of transformation in Central Europe. *Forest Ecology and Management*, 151:107–119, 2001.

G. Kerr. The management of silver fir forests: de Liocourt (1898) revisited. *Forestry*, 87:29–38, 2014.

G. Kerr. An improved spreadsheet to calculate target diameter distributions in uneven-aged silviculture. *Continuous Cover Forestry Group Newsletter*, 19:18–20, 2001.

G. Kerr, G. Morgan, J. Blyth, and V. Stokes. Transformation from even-aged plantations to an irregular forest: the world's longest running trial area at Glentress, Scotland. *Forestry*, 83:329–344, 2010.

G. Kerr, M. Snellgrove, S. Hale, and V. Stokes. The Bradford-Hutt system for transforming young even-aged stands to continuous cover management. *Forestry*, 90:581–593, 2017.

W. B. Kessler, H. Salwasser, C. W. Cartwright, and J. A. Caplan. New perspectives for sustainable natural resources management. *Ecological Applications*, 2:221–225, 1992.

J. P. Kimmins. *Forest ecology – a foundation for sustainable management*. Pearson Education Prentice Hall, Upper Saddle River, NJ, 3rd edition, 2004.

A. P. Kirilenko and R. A. Sedjo. Climate change impacts on forestry. *Proceedings of the National Academy of Sciences of the United States of America*, 104:19697–19702, 2007.

T. Kirkinen, O. López-Costas, A. Martínez Cortizas, S. P. Sihvo, H. Ruhanen, R. Käkelä, J.-E. Nyman, E. Mikkola, J. Rantanen, E. Hertell, M. Ahola, J. Roiha, and K. Mannermaa. Preservation of microscopic fur, feather, and bast fibers in the Mesolithic ochre grave of Majoonsuo, Eastern Finland. *PLoS ONE*, 17:1–22, 2022.

J. Klädtke. Umsetzungsprozesse unter besonderer Berücksichtigung Z-Baum bezogener Auslese [Differentiation processes of frame trees in local thinnings]. *Allgemeine Forst- und Jagdzeitung*, 161:29–36, 1990.

J. Klädtke. Konstruktion einer Z-Baum-Ertragstafel am Beispiel der Fichte [Construction of a frame-tree yield table for norway spruce]. Mitteilung No 173, Forstliche Versuchs- und Forschungsanstalt Baden-Württemberg, Freiburg, 1993.

J. Klädtke. Buchen-Lichtdurchforstung [Heavy release thinning in beech stands]. *Allgemeine Forstzeitschrift*, 52:1019–1023, 1997.

J. Klädtke and Ch. Yue. Produktionszielorientierte Entscheidungshilfe für die Bewirtschaftung ungleichaltriger Fichten-Tannen-Wälder und Plenterwälder [Decision supporting tool for managing uneven-aged coniferous forests and selection forests]. *Allgemeine Forst- und Jagdzeitung*, 174:196–206, 2003.

F. Klang and P.-M. Ekö. Tree properties and yield of *Picea abies* planted in shelterwoods. *Scandinavian Journal of Forest Research*, 14:262–269, 1999.

T. Knoke. Die Stabilisierung junger Fichtenbestände durch starke Durchforstungseingriffe – Versuch einer ökonomischen Bewertung [Stabilisation of young Norway spruce stands through heavy thinnings – An attempt of an economic evaluation]. *Forstarchiv*, 69:219–226, 1998.

T. Knoke and N. Plusczyk. On economic consequences of transformation of a spruce (*Picea abies* (L.) Karst.) dominated stand from regular into irregular age structure. *Forest Ecology and Management*, 151:163–179, 2001.

H. Knuchel. *Planning and control in the managed forest*. Oliver and Boyd, Edinburgh, 1953.

N. E. Koch and J. P. Skovsgaard. Sustainable management of planted forests: some comparisons between central Europe and the United States. *New Forestry*, 17:11–22, 1999.

M. Köhl, R. Hildebrandt, K. Olschofsky, R. Köhler, T. Rötzer, T. Mette, H. Pretzsch, M. Köthke, M. Dieter, M. Abiy, F. Makeschin, and B. Kenter. Combating the effects of climatic change on forests by mitigation strategies. *Carbon Balance and Management*, 5:8, 2010.

M. Kohler, J. Sohn, G. Nägele, and J. Bauhus. Can drought tolerance of Norway spruce (*Picea abies* (L.) Karst.) be increased through thinning? *European Journal of Forest Research*, 129:1109–1118, 2010.

T. S. Kohyama, M. D. Potts, T. I. Kohyama, A. R. Kassim, and P. S. Ashton. Demographic properties shape tree size distribution in a Malaysian rain forest. *The American Naturalist*, 185:367–379, 2015.

M. Koivula, H. Silvennoinen, H. Koivula, J. Tikkanen, and L. Tyrväinen. Continuous-cover management and attractiveness of managed Scots pine forests. *Canadian Journal of Forest Research*, 50:819–828, 2020.

C. C. Konijnendijk, R. M. Ricard, A. Kenney, and T. B. Randrup. Defining urban forestry – A comparative perspective of North America and Europe. *Urban Forestry & Urban Greening*, 4:93–103, 2006.

J. Köstler. *Silviculture*. Oliver and Boyd, Edinburgh, 1956.

M. Kovac, D. Hladnik, and L. Kutnar. Biodiversity in (the Natura 2000) forest habitats is not static: its conservation calls for an active management approach. *Journal for Nature Conservation*, 43:250–260, 2018.

G. Kraft. *Beiträge zur Lehre von Durchforstungen, Schlagstellungen und Lichtungshieben [On the methodology of thinnings, shelterwood cuttings and heavy release operations]*. Klindworth's Verlag, Hanover, 1884.

H. Kramer. *Waldwachstumslehre [Forest growth and yield science]*. Verlag Paul Parey, Hamburg and Berlin, 1988.

W. Kramer. Der Anbau der Weißtanne (*Abies alba* Mill.) in Nordwestdeutschland [Growing silver fir (*abies alba* Mill.) in North West Germany]. *Der Forst- und Holzwirt*, 17:367–372, 1970.

C. J. Krebs. *Ecological methodology*. Addison Wesley Longman, New York, 2nd edition, 1999.

L. Kruse, C. Erefur, J. Westin, B. T. Ersson, and A. Pommerening. Towards a benchmark of national training requirements for continuous cover forestry (CCF) in Sweden. *Trees, Forests and People*, 12:100391, 2023.

H. Krutzsch. *Waldaufbau [Silviculture]*. Deutscher Bauernverlag, Berlin, 1952.

L. Kuglerová, J. Jyväsjärvi, C. Ruffing, T. Muotka, A. Jonsson, E. Andersson, and J. S. Richardson. Cutting edge: A comparison of contemporary practices of riparian buffer retention around small streams in Canada, Finland, and Sweden. *Water Resources Research*, 56:e2019WR026381, 2020.

D. P. J. Kuijper. Lack of natural control mechanisms increases wildlife-forestry conflict in managed temperate European forest systems. *European Journal of Forest Research*, 130:895–909, 2011.

A. v. Laar and A. Akça. *Forest mensuration*. Managing Forest Ecosystems 13. Springer, Dordrecht, 2007.

E. Lähde, O. Laiho, and Y. Norokorpi. Diversity-oriented silviculture in the boreal zone of Europe. *Forest Ecology and Management*, 118:223–243, 1999.

Landesanstalt für Wald und Forstwirtschaft Thüringen. *Naturnahe Waldwirtschaft [Near-natural woodland management]*. Brochure, Thüringen Forst, Warza, 2000.

Landesbetrieb Forst Baden-Württemberg. *ForstBW Praxis: Richtlinie landesweiter Waldentwicklungstypen [Guidelines for Forest Development Types in Baden Württemberg (Germany)]*. Brochure, Ministerium für Ländlichen Raum und Verbraucherschutz Baden-Württemberg, Stuttgart, 2014.

O. Langvall and M. Ottosson Löfvenius. Effect of shelterwood density on nocturnal near-ground temperature, frost injury risk and budburst date of Norway spruce. *Forest Ecology and Management*, 168:149–161, 2002.

J. B. Larsen. Naturnær skovdrift – hvorfor og hvordan? [Near-natural forestry – why and how?] *Dansk Skovbrugs Tidsskrift*, 90:4–32, 2005.

J. B. Larsen and A. B. Nielsen. Nature-based forest management – where are we going? Elaborating forest development types in and with practice. *Forest Ecology and Management*, 238:107–117, 2007.

J. B. Larsen, P. Angelstam, J. Bauhus, J. F. Carvalho, J. Diaci, D. Dobrowolska, A. Gazda, L. Gustafsson, F. Krumm, T. Knoke, A. Konczal, T. Kuuluvainen, B. Mason, R. Motta, E. Pötzelsberger, A. Rigling, and A. Schuck. *Closer-to-nature forest management*. From Science to Policy 12. European Forest Institute, Joensuu, 2022.

W. B. Leak. An expression of diameter distribution for unbalanced, uneven-aged stands and forests. *Forest Science*, 10:39–50, 1964.

W. B. Leak. The J-shaped probability distribution. *Forest Science*, 11:405–409, 1965.

W. K. Lee, K. v. Gadow, and A. Akça. Waldstruktur und Lorenz-Modell [Forest structure and Lorenz model]. *Allgemeine Forst- und Jagdzeitung*, 170:220–223, 1999.

H. Leibundgut. *Die Waldpflege [Forest management]*. Verlag Paul Haupt, Bern & Stuttgart, 1966.

M. V. Lencinas, G. M. Pastur, E. Gallo, and J. M. Cellini. Alternative silvicultural practices with variable retention to improve understory plant diversity conservation in southern Patagonian forests. *Forest Ecology and Management*, 262:1236–1250, 2011.

Y. Li, S. Ye, G. Hui, Y. Hu, and Z. Zhao. Spatial structure of timber harvested according to structure-based forest management. *Forest Ecology and Management*, 322:106–116, 2014.

K. N. Lindegaard, P. W. R. Adams, M. Holley, A. Lamley, A. Henriksson, S. Larsson, H.-G. v. Engelbrechten, G. Esteban Lopez, and M. Pisarek. Short rotation plantations policy history in Europe: lessons from the past and recommendations for the future. *Food and Energy Security*, 5:125–152, 2016.

M. Lindner and W. Cramer. German forest sector under global change. *Forstwissenschaftliches Centralblatt*, 121(Supplement 1):3–17, 2002.

W. Linnard. *Welsh woods and forests – A history*. Gomer Press, Llandysul, 2000.

Lord Bradford. An experiment in irregular forestry. *Y Coedwigwr*, 33:6–18, 1981.

M. O. Lorenz. Methods of measuring the concentration of wealth. *Publications of the American Statistical Association*, 9:209–219, 1905.

T. Lorey. Die mittlere Bestandeshöhe [Mean stand height]. *Allgemeine Forst- und Jagdzeitung*, 54:149–155, 1878.

C. G. Lorimer and L. E. Frelich. A simulation of equilibrium diameter distributions of sugar maple (*Acer saccharum*). *Bulletin of the Torrey Botanical Club*, 111:193–199, 1984.

H. Lundmark, T. Josefsson, and L. Östlund. The history of clear-cutting in northern Sweden – Driving forces and myths in boreal silviculture. *Forest Ecology and Management*, 307:112–122, 2013.

T. Lundmark, J. Bergh, A. Nordin, N. Fahlvik, and B. C. Poudel. Comparison of carbon balances between continuous-cover and clear-cut forestry in Sweden. *Ambio*, 45:203–213, 2016.

L. Lundqvist. Simulation of sapling population dynamics in uneven-aged *Picea abies* forests. *Annals of Botany*, 76:371–380, 1995.

A. E. Magurran. *Measuring biological diversity*. Blackwell Publishing, Oxford, 2004.

A. Mäkelä and H.T. Valentine. *Models of tree and stand dynamics: Theory, formulation and application*. Springer Nature, Cham, 2020.

D. C. Malcolm, W. L. Mason, and G. C. Clarke. The transformation of conifer forests in Britain – regeneration, gap size and silvicultural systems. *Forest Ecology and Management*, 151:7–23, 2001.

K. Mantel. *Wald und Forst in der Geschichte: Ein Lehr- und Handbuch* [*Woodlands and forests in history: A textbook and guide*]. Verlag M. & H. Schaper, Hanover, 1990.

H. Mård. The influence of a birch shelter (*Betula* spp.) on the growth of young stands of *Picea abies*. *Scandinavian Journal of Forest Research*, 11:343–350, 1996.

C. Maser. *Sustainable forestry: philosophy, science and economics*. CRC Press Inc., Boca Raton, FL, 1994.

W. Mason and G. Kerr. *Transforming even-aged conifer stands to continuous cover management*. Information note, Forestry Commission, Edinburgh, 2004.

W. Mason, G. Kerr, and J. Simpson. *What is continuous cover forestry?* Information note. Forestry Commission, Edinburgh, 1999.

M. G. Matias, M. Combe, C. Barbera, and N. Mouquet. Ecological strategies shape the insurance potential of biodiversity. *Frontiers in Microbiology*, 3:432, 2013.

M. S. Matonis, D. Binkley, J. Franklin, and K. N. Johnson. Benefits of an "undesirable" approach to natural resource management. *Journal of Forestry*, 114:658–665, 2016.

J. D. Matthews. *Silvicultural systems*. Clarendon Press, Oxford, 1989.

J. D. Matthews. *Silvicultural systems*. Oxford University Press, Oxford, 1991.

H. Mayer. *Waldbau auf soziologisch-ökologischer Grundlage* [*Silviculture on a sociological-ecological basis*]. Gustav Fischer Verlag, Stuttgart, 3rd edition, 1984.

M. Mayer, C. E. Prescott, W. E. A. Abaker, L. Augusto, L. Cécillon, G. W. D. Ferreira, J. James, R. Jandl, K. Katzensteiner, J.-P. Laclau, J. Laganière, Y. Nouvellon, D. Paré, J. A. Stanturf, E. I. Vanguelova, and L. Vesterdal. Tamm Review: influence of forest management activities on soil organic carbon stocks: a knowledge synthesis. *Forest Ecology and Management*, 466:118127, 2020.

T. J. McEvoy. *Positive impact forestry: A sustainable approach to managing woodlands*. Island Press, Washington, DC, 4th edition, 2004.

S. McNulty, A. Steel, E. Springgay, B. Caldwell, K. Shono, G. Pess, S. Funge-Smith, W. Richards, S. Ferraz, D. Neary, J. Long, B. Verbist, J. Leonard, G. Sun, T. Beechie, M. Lo, L. McGill, A. Fullerton, and S. Borelli. Managing forests for water (chapter 3). In USDA FAO, IUFRO, editor, *A guide to forest-water management*, FAO Forestry Paper 185, pages 31–74. FAO, Rome, 2021.

C. Messier, K. J. Puettmann, and K. D. Coates, editors. *Managing forests as complex adaptive systems. Building resilience to the challenge of global change*. Routledge, Oxon, 2013.

M. Metslaid, K. Jõgiste, E. Nikinmaa, W. K. Moser, and A. Porcar-Castell. Tree variables related to growth response and acclimation of advance regeneration of Norway spruce and other coniferous species after release. *Forest Ecology and Management*, 250:56–63, 2007.

H. A. Meyer. Eine mathematisch-statistische Untersuchung über den Aufbau des Plenterwaldes [A mathematical-statististical analysis of the structure of selection forests]. *Schweizerische Zeitschrift für Forstwesen*, 84:33–46, 88–103, 124–131, 1933.

H. A. Meyer. Management without rotation. *Journal of Forestry*, 41:126–132, 1943.

H. A. Meyer. Structure, growth, and drain in balanced uneven-aged forests. *Journal of Forestry*, 50:85–92, 1952.

I. Michailoff. Zahlenmäßiges Verfahren für die Ausführung der Bestandeshöhenkurven [Numerical estimation of stand height curves]. *Forstwissenschaftliches Centralblatt und Tharandter Forstliches Jahrbuch*, 6:273–279, 1943.

S. J. Mitchell and W. J. Beese. The retention system: reconciling variable retention with the principles of silvicultural systems. *The Forestry Chronicle*, 78:397–403, 2002.

G. Mitscherlich. *Der Tannen-Fichten-(Buchen)-Plenterwald* [*The silver fir - Norway spruce - (European beech) selection forest*]. Schriftenreihe der Badischen Forstlichen Versuchsanstalt 8. Badische Forstliche Versuchsanstalt, 1952.

D. Mlinšek. *From clear-cutting to close-to-nature silviculture system*. IUFRO News 25. International Union of Forest Research Organisations, Vienna, 1996.

A. Möller. *Der Dauerwaldgedanke: Sein Sinn und seine Bedeutung* [*The Dauerwald idea: its meaning and significance*]. Verlag von Julius Springer, Berlin, 1922.

M. Mordini. *Modellierung und Beurteilung der ökologischen und ökonomischen Wirkungen von waldbaulichen Eingriffen – Einrichtung zweier Marteloskope in eichenreichen Flächen* [*Modelling and evaluation of the ecological and economic consequences of silvicultural interventions – Installation of two marteloscopes in oak-dominated stands*]. MSc thesis, ETH Zürich, Zürich, 2009.

G. F. Morosov. *Die Lehre vom Waldbau* [*The science of silviculture*]. Translated from Russian to German by Selma Ruoff, Hans Ruoff and Hans Buchholz. Neumann, Berlin, 2nd edition, 1959.

H. W. Morsbach. *Common sense forestry*. Chelsea Green Publishing Company, White River Junction, 2002.

R. Mosandl, H. El Kateb, and J. Ecker. Untersuchungen zur Behandlung von jungen Eichenbeständen [Investigations of various thinning treatments in young oak stands]. *Centralblatt*, 110:358–370, 1991.

D. Mülder. Nur Individuenauswahl oder auch Gruppenauswahl? [Selection of individuals only or group selection as well?]. Schriften aus der Forstlichen Fakultät der Universität Göttingen und der Niedersächsischen Forstlichen Versuchsanstalt 96, Forstliche Fakultät der Universität Göttingen und Niedersächsische Forstliche Versuchsanstalt, Göttingen, 1990.

D. M. Murray and K. v. Gadow. Relationships between the diameter distributions before and after thinning. *Forest Science*, 37:552–559, 1991.

G.-J. Nabuurs. *European forests in the 21st century: impacts of nature-oriented forest management assessed with a large-scale scenario model*. PhD thesis, University of Joensuu, Joensuu, 2001.

S. Neuner, A. Albrecht, D. Cullmann, F. Engels, V. C. Griess, W. A. Hahn, M. Hanewinkel, F. Härtl, C. Kölling, K. Staupendahl, and T. Knoke. Survival of Norway spruce remains higher in mixed stands under a dryer and warmer climate. *Global Change Biology*, 21:935–946, 2015.

A. C. Newton. *Forest ecology and conservation. A handbook of techniques*. Oxford University Press, Oxford, 2007.

C. P. Nichols, J. A. Drewe, R. Gill, N. Goode, and N. Gregory. A novel causal mechanism for grey squirrel bark stripping: the calcium hypothesis. *Forest Ecology and Management*, 367:12–20, 2016.

Niedersächsische Landesregierung. *Niedersächsisches Programm zur langfristigen ökologischen Waldentwicklung in den Landesforsten (LÖWE)* [*Lower Saxony programme for the long-term ecological forest development of the state forest*]. Brochure, Landesforsten, Hanover, 1991.

M. Nieminen, H. Hökkä, R. Laiho, A. Juutinen, A. Ahtikoski, M. Pearson, S. Kojola, S. Sarkkola, S. Launiainen, S. Valkonen, T. Penttilä, A. Lohila, M. Saarinen, K. Haahti, R. Mäkipää, J. Miettinen, and M. Ollikainen. Could continuous cover forestry be an economically and environmentally feasible management option on drained boreal peatlands? *Forest Ecology and Management*, 424:78–84, 2018.

M. Nieuwenhuis. *Terminology of forest management*. IUFRO World Series Vol. 9-en. IUFRO Secretariat Vienna, Sopron, 2000.

L. Nilsson Södergren. *Japanese urban tree burials: Diversity and individualisation*. PhD thesis, Lund University, Lund, 2020.

S. Nocentini, O. Ciancio, L. Portoghesi, and P. Corona. Historical roots and the evolving science of forest management under a systemic perspective. *Canadian Journal of Forest Research*, 51:163–171, 2021.

T. Nord-Larsen and H. Meilby. Effects of nurse trees, spacing, and tree species on biomass production in mixed forest plantations. *Scandinavian Journal of Forest Research*, 31:592–601, 2016.

M. North, J. Chen, G. Smith, L. Krakowiak, and J. Franklin. Initial response of understory plant diversity and overstory tree diameter growth to GTR harvest. *Northwest Science*, 70:24–35, 1996.

A. Nothdurft, J. Saborowski, R. S. Nuske, and D. Stoyan. Density estimation based on k-tree sampling and point pattern reconstruction. *Canadian Journal of Forest Research*, 40:953–967, 2010.

R. D. Nyland. *Silviculture. Concepts and applications.* Waveland Press, Long Grove, 2nd edition, 2002.

R. D. Nyland. Even- to uneven-aged: the challenges of conversion. *Forest Ecology and Management*, 172:291–300, 2003.

N. O'Carroll. The nursing of Sitka spruce (*Picea sitchensis*). 1. Japanese larch (*Larix leptolepis*). *Irish Forestry*, 35:60–65, 1978.

K. L. O'Hara. The historical development of uneven-aged silviculture in North America. *Forestry*, 75:339–346, 2002.

K. L. O'Hara. *Multiaged silviculture. Managing for complex forest stand structures.* Oxford University Press, Oxford, 2014.

K. L. O'Hara. What is close-to-nature silviculture in a changing world? *Forestry*, 89:cpv043, 2016.

K. L. O'Hara and R. F. Gersonde. Stocking control concepts in uneven-aged silviculture. *Forestry*, 77:131–143, 2004.

K. L. O'Hara, H. Hasenauer, and G. Kindermann. Sustainability in multi-aged stands: an analysis of long-term plenter systems. *Forestry*, 80:163–181, 2007.

K. L. O'Hara, L. P. Leonard, and C. R. Keyes. Variable-density thinning and a marking paradox: Comparing prescription protocols to attain stand variability in coast redwood. *Western Journal of Applied Forestry*, 27:143–149, 2012.

T. O'Keefe. Holistic (new) forestry: significant difference or just another gimmick. *Journal of Forestry*, 88:23–24, 1990.

C. D. Oliver and B. C. Larson. *Forest stand dynamics.* John Wiley & Sons, New York, 1996.

T. A. Ontl, M. K. Janowiak, C. W. Swanston, J. Daley, S. Handler, M. Cornett, S. Hagenbuch, C. Handrick, L. McCarthy, and N. Patch. Forest management for carbon sequestration and climate adaptation. *Journal of Forestry*, 118:86–101, 2019.

G. Örlander and C. Karlsson. Influence of shelterwood density on survival and height increment of *Picea abies* advance growth. *Scandinavian Journal of Forest Research*, 15:20–29, 2000.

H.-J. Otto. *Waldökologie [Forest ecology].* Ulmer, Stuttgart, 1994.

B. J. Palik, A. W. D'Amato, J. F. Franklin, and K. N. Johnson. *Ecological silviculture. Foundations and applications.* Waveland Press Inc., Long Grove, 2021.

S. Palmer. Waldentwicklung auf der Mittleren Schwäbischen Alb [Forest development in the Medium Swabian Jura]. *AFZ*, 10:507–510, 1994.

E. Partl, V. Szinovatz, F. Reimoser, and J. Schweiger-Adler. Forest restoration and browsing impact by roe deer. *Forest Ecology and Management*, 159:87–100, 2002.

J. H. Pedlar, D. W. McKenney, I. Aubin, T. Beardmore, J. Beaulieu, L. Iverson, G. A. O'Neill, R. S. Winder, and C. Ste-Marie. Placing forestry in the assisted migration debate. *BioScience*, 62:835–842, 2012.

M. S. Pedro, W. Rammer, and R. Seidl. Tree species diversity mitigates disturbance impacts on the forest carbon cycle. *Oecologia*, 177:619–630, 2015.

M. J. Penistan. The alternative to extensive regular clear felling. *Scottish Forestry*, 6:1–7, 1952.

D. A. Perala and A. A. Alm. Regeneration silviculture of birch: a review. *Forest Ecology and Management*, 32:39–77, 1990.

M. Perpeet. Zur Anwendung von Waldentwicklungstypen (WET) [On the use of Forest Development Types (FDT)]. *Forstarchiv*, 71:151–160, 2000.

D. A. Perry, R. Oren, and S. C. Hart. *Forest ecosystems*. The Johns Hopkins University Press, Baltimore, MD, 2nd edition, 2008.

G. F. Peterken. *Natural woodland. Ecology and conservation in northern temperate regions*. Cambridge University Press, Cambridge, 1996.

G. F. Peterken. Ecological effects of introduced tree species in Britain. *Forest Ecology and Management*, 141:31–42, 2001.

R. Petersen and M. Guericke. Untersuchungen zur Wuchsdynamik und Verjüngung der Fichte in Plenterwaldstrukturen. [Influence of selection cutting on growth and regeneration in uneven-aged stands of Norway spruce]. *Forst und Holz*, 59:58–62, 2004.

I. C. Petriţan, B. v. Lüpke, and A. M. Petriţan. Effects of root trenching of overstorey Norway spruce (*Picea abies*) on growth and biomass of underplanted beech (*Fagus sylvatica*) and Douglas fir (*Pseudotsuga menziesii*) saplings. *European Journal of Forest Research*, 130:813–828, 2011.

H. Petterson. Die Massenproduktion des Nadelwaldes [The mass production of coniferous forests]. Reports of the Swedish Forest Research Station 45, Swedish Forest Research Station, Stockholm, 1955.

C. Petucco, F. Søndergaard Jensen, H. Meilby, and J. P. Skovsgaard. Visitor preferences of thinning practice in young even-aged stands of pedunculate oak (*Quercus robur* L.): comparing the opinion of forestry professionals in six European countries. *Scandinavian Journal of Forest Research*, 33:81–90, 2018.

M. S. Philip. *Measuring trees and forests*. CAB International, Wallingford, 1994.

K. J. Pienaar and L. V. Turnbull. The Chapman-Richards generalization of von Bertalanffy's growth model for basal area growth and yield in even-aged stands. *Forest Science*, 19:2–22, 1973.

C. D. Pigott. The influence of evergreen coniferous nurse-crops on the field layer in two woodland communities. *Journal of Applied Ecology*, 27:448–459, 1990.

M. Pilarski, editor. *Restoration forestry: an international guide to sustainable forestry practices*. Kivaki Press, Durango, 1994.

A. A. Podkopaev. Gnezdovye posevy duba na smytyh sklonah Donbassa [Nest-sowings of oak on sheet-eroded slopes in the Donbas]. *Agrobiologija*, 2:241–245, 1961.

J. Pollanschütz. Durchforstung von Stangen- und Baumhölzern [Thinning of pole and medium wood]. *Allgemeine Forstzeitung*, 82:250–253, 1971.

J. Pollanschütz. Zum Thema: Durchforstung und Schneebruch [On the topic of thinnings and snow breakage]. *Holz-Kurier*, 36:4, 1981.

J. Pollanschütz. Durchforstung ist nicht gleich Auslesedurchforstung: Forstlichen Beratern ins Stammbuch geschrieben! [NB: Thinning does not necessarily imply frame-tree thinning]. *Allgemeine Forstzeitung*, 94:40–41, 1983.

A. Pommerening and P. Grabarnik. *Individual-based methods in forest ecology and management*. Springer, Cham, 2019.

A. Pommerening and S. T. Murphy. A review of the history, definitions and methods of continuous cover forestry with special attention to afforestation and restocking. *Forestry*, 77:27–44, 2004.

A. Pommerening and A. J. Sánchez Meador. Tamm review: Tree interactions between myth and reality. *Forest Ecology and Management*, 424:164–176, 2018.

A. Pommerening and D. Stoyan. Reconstructing spatial tree point patterns from nearest neighbour summary statistics measured in small subwindows. *Canadian Journal of Forest Research*, 38:1110–1122, 2008.

A. Pommerening and J. Uria-Diez. Do large trees tend towards high species mingling? *Ecological Informatics*, 42:139–147, 2017.

A. Pommerening, B. Brzeziecki, and D. Binkley. Are long-term changes in plant species composition related to asymmetric growth dominance in the pristine Białowieża Forest? *Basic and Applied Ecology*, 17:408–417, 2016.

A. Pommerening, C. Pallarés Ramos, W. Kędziora, J. Haufe, and D. Stoyan. Rating experiments in forestry: How much agreement is there in tree marking? *PLoS ONE*, 13:e0194747, 2018.

A. Pommerening, M. Brill, U. Schmidt-Kraepelin, and J. Haufe. Democratising forest management: Applying multiwinner approval voting to tree selection. *Forest Ecology and Management*, 478:118509, 2020.

A. Pommerening, K. Maleki, and J. Haufe. Tamm review: Individual-based forest management or seeing the trees for the forest. *Forest Ecology and Management*, 501:119677, 2021a.

A. Pommerening, G. Zhang, and X. Zhang. Unravelling the mechanisms of spatial correlation between species and size diversity in forest ecosystems. *Ecological Indicators*, 121:106995, 2021b.

A. Poore. The marteloscope – a training aid for continuous cover forest management. *Woodland Heritage*, 2011:28–29, 2011.

D. C. Powell. How to prepare a silvicultural prescription for uneven-aged management. White Paper 49, USDA Forest Service, Pendleton, Pacific Northwest Region, USA, 2018.

R. Poznański and L. Rutkowska. Wskaźniki zróżnicowania struktury rozkładu perśnic [Indices of diversity of the dbh distribution structure]. *Sylwan*, 12:5–12, 1997.

H. Pretzsch. Zum Einfluss des Baumverteilungsmusters auf den Bestandeszuwachs [On the effect of the spatial distribution of trees on stand growth]. *Allgmeine Forst- und Jagdzeitung*, 166:190–201, 1995.

H. Pretzsch. Erfassung des Pflegezustandes von Waldbeständen bei der zweiten Bundeswaldinventur [Monitoring forest management in the second national forest inventory of Germany]. *AFZ-Der Wald*, 15:820–823, 1996.

H. Pretzsch. *Forest dynamics, growth and yield. From measurement to model*. Springer, Heidelberg, 2009.

H. Pretzsch. *Grundlagen der Waldwachstumsforschung [Principles of forest growth and yield science]*. Springer, Heidelberg, 2019.

H. Pretzsch and P. Biber. Tree species mixing can increase maximum stand density. *Canadian Journal of Forest Research*, 46:1179–1193, 2016.

H. Pretzsch, D. I. Forrester, and J. Bauhus. *Mixed-species forests: Ecology and management*. Springer, Heidelberg, 2017.

M. Price and C. Price. Creaming the best, or creatively transforming? Might felling the biggest trees first be a win-win strategy? *Forest Ecology and Management*, 224:297–303, 2006.

D. Pringle. *The first 75 years: a brief account of the history of the Forestry Commission, 1919–1994*. Forestry Commission, Edinburgh, 1994.

Pro Silva Europe. Pro Silva. Brochure, Pro Silva Europe, Truttenhausen, 1999.

Pro Silva Europe. Pro Silva principles. Brochure, Pro Silva Europe, Truttenhausen, 2012.

M. Prodan. Die theoretische Bestimmung des Gleichgewichts im Plenterwalde [The theoretical determination of equilibrium in selection forests]. *Schweizerische Zeitschrift für Forstwesen*, 100:81–99, 1949.

S. Pryor. Practical aspects of irregular silviculture for British broadleaves. In S. Pryor, editor, *Silvicultural systems*, pages 147–166. Institute of Chartered Foresters, Edinburgh, 1990.

K. Puettmann, K. D. Coates, and C. Messier. *A critique of silviculture*. Island Press, Washington, 2009.

K. J. Puettmann, S. M. G. Wilson, S. C. Baker, P. J. Donoso, L. Drössler, G. Amente, B. D. Harvey, T. Knoke, Y. Lu, S. Nocentini, F. E. Putz, T. Yoshida, and J. Bauhus. Silvicultural alternatives to conventional even-aged forest management - what limits global adoption? *Forest Ecosystems*, 2:8, 2015.

T. Pukkala. Carbon forestry is surprising. *Forest Ecosystems*, 5:11, 2018.

T. Pukkala and K. v. Gadow, editors. *Continuous cover forestry*. Springer, Dordrecht, 2nd edition, 2012.

H. E. Quicke, R. S. Meldahl, and J. S. Kush. Basal area growth of individual trees: A model derived from a regional longleaf pine growth study. *Forest Science*, 40:528–542, 1994.

J. M. Quinton, J. Östberg, and P. N. Duinker. The importance of multi-scale temporal and spatial management for cemetery trees in Malmö, Sweden. *Forests*, 11(1):78, 2020.

J. I. Ramirez, P. A. Jansen, and L. Poorter. Effects of wild ungulates on the regeneration, structure and functioning of temperate forests: A semi-quantitative review. *Forest Ecology and Management*, 424:406–419, 2018.

W. Rammer, M. Brauner, H. Ruprecht, and M. J. Lexer. Evaluating the effects of forest management on rockfall protection and timber production at slope scale. *Scandinavian Journal of Forest Research*, 30:719–731, 2015.

H. Reininger. *Das Plenterprinzip* [*The selection principle*]. Leopold Stocker Verlag, Graz, 2001.

République et Canton de Neuchâtel. *Waldbauliche Grundsätze* [*Silvicultural principles*]. Brochure, Service des Forêts, Neuchâtel, 2001.

F. Rittershofer. *Waldpflege und Waldbau für Studium und Praxis* [*Forest management and silviculture for university students and practitioners*]. Gisela Rittershofer Verlag, Freising, 2nd edition, 1999.

G. Robinson. *The forest and the trees: A guide to excellent forestry*. Island Press, Washington, DC, 1994.

A. P. Robinson and J. D. Hamann. *Forest analytics with R. An introduction*. Use R! Springer, New York, 2010.

T. J. D. Rollinson. Thinning control of conifer plantations in Great Britain. *Annals of Forest Science*, 44:25–34, 1987.

A. Roloff. *Biometrische Analyse unterschiedlicher Gleichgewichtsmodelle für Plenterwälder* [*Biometric analysis of different equilibrium models for selection forests*]. MSc thesis, Göttingen University, Göttingen, 2004.

C. R. Rose and P. S. Muir. Green-tree retention: Consequences for timber production in forests of the Western Cascades, Oregon. *Ecological Applications*, 7:209–217, 1997.

R. Rosenvald and A. Lohmus. For what, when, and where is green-tree retention better than clear-cutting? A review of the biodiversity aspects. *Forest Ecology and Management*, 255:1–15, 2008.

S. Saha, C. Kuehne, U. Kohnle, P. Brang, A. Ehring, J. Geisel, B. Leder, M. Muth, R. Petersen, J. Peter, W. Ruhm, and J. Bauhus. Growth and quality of young oaks (*Quercus robur* and *Quercus petraea*) grown in cluster plantings in central Europe: A weighted meta-analysis. *Forest Ecology and Management*, 283:106–118, 2012.

S. Saha, C. Kuehne, and J. Bauhus. Lessons learned from oak cluster planting trials in central Europe. *Canadian Journal of Forest Research*, 47:139–148, 2017.

T. T. Salk, L. E. Frelich, S. Sugita, R. Calcote, J. B. Ferrari, and R. A. Montgomery. Poor recruitment is changing the structure and species composition of an old-growth hemlock-hardwood forest. *Forest Ecology and Management*, 261:1998–2006, 2011.

H. Salwasser. Ecosystem management: can it sustain diversity and productivity? *Journal of Forestry*, 92:6–120, 1994.

S. Sarkkola, H. Hökkä, H. Koivusalo, M. Nieminen, E. Ahti, J. Päivänen, and J. Laine. Role of tree stand evapotranspiration in maintaining satisfactory drainage conditions in drained peatlands. *Canadian Journal of Forest Research*, 40:1485–1496, 2010.

P. S. Savill. *The silviculture of trees used in British forestry*. CABI International, Wallingford, 1st edition, 1991.

P. S. Savill. *The silviculture of trees used in British forestry*. CABI International, Wallingford, 2nd edition, 2019.

M. Scaranello, L. Alves, S. Vieira, P. Camargo, C. Joly, and L. Martinelli. Height-diameter relationships of tropical Atlantic moist forest trees in southeastern Brazil. *Scientia Agricola*, 69:26–37, 2012.

W. Schädelin. Bestandeserziehung [Stand management]. *Schweizerische Zeitschrift für Forstwesen*, 77:1–15, 33–44, 1926.

W. Schädelin. *Die Durchforstung als Auslese- und Veredlungsbetrieb höchster Wertleistung [Frame-tree thinning for quality timber production]*. Verlag Paul Haupt, Bern & Leipzig, 1934.

C. Schirmer. Überlegungen zur Naturnähebeurteilung heutiger Wälder [Reflections on assessing the naturalness of present forests]. *Allgemeine Forst- und Jagdzeitung*, 170:11–18, 1998.

G. D. Schmidt. Die Weißtanne in Ostfriesland. Ein ertragskundlich-biologischer Beitrag zum Weißtannenvorkommen außerhalb des natürlichen Verbreitungsgebietes [Silver fir in Eastern Frisia. a contribution to the growth and the biology of silver fir outside its natural area of occurrence]. *Forstwissenschaftliches Centralblatt*, 70:641–665, 1951.

R. Schober. Zur Bedeutung des Umsetzens von Waldbäumen für die Z-Baum-Durchforstung. [On the importance of differentiation processes for frame-tree based thinnings]. *Allgemeine Forstzeitschrift*, 45:826–828, 1990.

J.-Ph. Schütz. Kosteneffiziente Waldpflege [Cost-effective forest management]. *Wald und Holz*, 11/2000:47–50, 2000.

J. P. Schütz. Modelling the demographic sustainability of pure beech plenter forests in Eastern Germany. *Annals of Forest Sciences*, 63:93–100, 2006.

J.-Ph. Schütz. Opportunities and strategies of transforming regular forests to irregular forests. *Forest Ecology and Management*, 151:87–94, 2001a.

J.-Ph. Schütz. *Der Plenterwald und weitere Formen strukturierter und gemischter Wälder [The selection forest and other forms of structured and mixed forests]*. Parey Buchverlag, Berlin, 2001b.

J.-Ph. Schütz. Silvicultural tools to develop irregular and diverse forest structures. *Forestry*, 75:329–337, 2002.

J.-Ph. Schütz. Intensität der Waldpflege und Baumartendiversität im Wald – oder: Naturautomation contra Entmischung [Intensity of tending and forest tree species diversity: Natural automation vs. outmixing]. *Schweizerische Zeitschrift für Forstwesen*, 156:200–206, 2005.

J.-Ph. Schütz. Waldbau I. Die Prinzipien der Waldnutzung und der Waldbehandlung. Skript zur Vorlesung Waldbau I [Silviculture I. The principles of forest exploitation and forest management. notes accompanying the lectures in silviculture I]. Unpublished manuscript, ETH Zürich, Zürich, 2003.

J.-Ph. Schütz and A. Pommerening. Can Douglas fir (*Pseudotsuga menziesii* (MIRB.) FRANCO) sustainably grow in complex forest structures? *Forest Ecology and Management*, 303:175–183, 2013.

Schweizerischer Forstverein. *Die forstlichen Verhältnisse der Schweiz [Forestry in Switzerland]*. Kommissionsverlag von Beer & Cie, Zürich, 2nd edition, 1925.

Scottish Environment Protection Agency. *Engineering in the water environment good practice guide – Riparian vegetation management*. Brochure, SEPA, Edinburgh, 2009.

R. Seidl, G. Klonner, W. Rammer, F. Essl, A. Moreno, M. Neumann, and S. Dullinger. Invasive alien pests threaten the carbon stored in Europe's forests. *Nature Communications*, 9:1626, 2018.

O. Seitschek. Waldbauliche Möglichkeiten auf Kahlflächen unter besonderer Berücksichtigung der Vorwaldbaumarten [Silvicultural options on windthrow sites with special attention to nurse crop species]. *Forst und Holz*, 46:351–355, 1991.

R. S. Seymour and M. L. Hunter. Principles of ecological forestry. In M. L. Hunter, editor, *Maintaining biodiversity in forest ecosystems*, pages 22–62. Cambridge University Press, Cambridge, 1999.

C. E. Shannon and W. Weaver. *The mathematical theory of communication*. University of Illinois Press, Urbana, 1949.

I. Short and J. Hawe. Ash dieback in Ireland. A review of European management options and case studies in remedial silviculture. *Irish Forestry*, 75:44–72, 2018.

H. Siebenbaum. 175 Jahre Aufforstung im Küstenraum [175 years of coastland afforestation]. *Allgemeine Forstzeitschrift*, 20:113–120, 1965.

E. H. Simpson. Measurement of diversity. *Nature*, 163:688, 1949.

M. J. Sjölund and A. S. Jump. The benefits and hazards of exploiting vegetative regeneration for forest conservation management in a warming world. *Forestry*, 86:503–513, 2013.

J. P. Skovsgaard. The UMF-index: an indicator to compare silvicultural practices at the forest or forest estate level. *Forestry*, 73:81–85, 2000.

M. F. Smidt and K. J. Puettmann. Overstory and understory competition affect underplanted eastern white pine. *Forest Ecology and Management*, 105:137–150, 1998.

W. R. Smith, R. M. Farrar Jr., P. A. Murphy, J. L. Yeiser, R. S. Meldahl, and J. S. Kush. Crown and basal area relationships of open-grown southern pines for modeling competition and growth. *Canadian Journal of Forest Research*, 22:341–347, 1992.

D. M. Smith, B. C. Larson, M. J. Kelty, and P. M. S. Ashton. *The practice of silviculture: Applied forest ecology*. John Wiley & Sons, New York, 9th edition, 1997.

E. Sonntag-Öström, T. Stenlund, M. Nordin, Y. Lundell, C. Ahlgren, A. Fjellman-Wiklund, L. Slunga Järvholm, and A. Dolling. "Nature's effect on my mind" – Patients' qualitative experiences of a forest-based rehabilitation programme. *Urban Forestry & Urban Greening*, 14:607–614, 2015.

M. Soucy, H. G. Adégbidi, R. Spinelli, and M. Béland. Increasing the effectiveness of knowledge transfer activities and training of the forestry workforce with marteloscopes. *The Forestry Chronicle*, 92:418–427, 2016.

H. Spellmann and W. v. Diest. Entwicklung von Z-Baum-Kollektiven [Development of frame-tree collectives]. *Forst und Holz*, 45:573–580, 1990.

H. Spiecker, J. Hansen, E. Klimo, J. P. Skovsgaard, H. Sterba, and K. v. Teuffel, editors. *Norway spruce conversion – options, and consequences*. Koninklijke Brill NV, Leiden, 2004.

R. Spinelli, N. Magagnotti, L. Pari, and M. Soucy. Comparing tree selection as performed by different professional figures. *Forest Science*, 62:213–219, 2016.

H. Stark, A. Nothdurft, and J. Bauhus. Allometries for widely spaced *Populus* ssp. and *Betula* ssp. in nurse crop systems. *Forests*, 4:1003–1031, 2013.

C. L. Staudhammer and V. M. LeMay. Introduction and evaluation of possible indices of stand structural diversity. *Canadian Journal of Forest Research*, 31:1105–1115, 2001.

H. Sterba. Equilibrium curves and growth models to deal with forests in transition to uneven-aged structure – application in two sample stands. *Silva Fennica*, 38:413–423, 2004.

J. H. Sterba and H. Sterba. The semi-logarithmic stem number distribution and the Gini-index – Structural diversity in "balanced" dbh-distributions. *Austrian Journal of Forest Science*, 135:19–31, 2018.

H. Sterba and A. Zingg. Target diameter harvesting – a strategy to convert even-aged forests. *Forest Ecology and Management*, 151:95–105, 2001.

H. Sterba and A. Zingg. Abstandsabhängige und abstandsunabhängige Bestandesstrukturbeschreibung [Distance dependent and distance independent description of stand structure]. *Allgemeine Forst- und Jagdzeitung*, 8/9:169–176, 2006.

G. Stöcker. Analyse und Vergleich von Bestandesstrukturen naturnaher Fichtenwälder mit Lorenz-Funktionen und Gini-Koeffizienten [Analysis and comparison of stand structures of nature near spruce forests using Lorenz functions and Gini coefficients]. *Centralblatt für das gesamte Forstwesen*, 119:12–39, 2002.

V. Stokes, G. Kerr, and T. Connolly. Underplanting is a practical silvicultural method for regenerating and diversifying conifer stands in Britain. *Forestry*, 94:219–231, 2020.

D. Stoyan, A. Pommerening, and A. Wünsche. Rater classification by means of set-theoretic methods applied to forestry data. *Journal of Environmental Statistics*, 8:2, 2018.

K. Streit, J. Wunder, and P. Brang. Slit-shaped gaps are a successful silvicultural technique to promote *Picea abies* regeneration in mountain forests of the Swiss Alps. *Forest Ecology and Management*, 257:1902–1909, 2009.

G. Strobel. Eichen-Biogruppen [Oak bio-groups]. *AFZ-Der Wald*, 8:396–398, 1986.

K. Sturm. Prozessschutz – ein Konzept für naturschutzgerechte Waldwirtschaft [Process conservation – a concept for conservation-oriented forestry]. *Zeitschrift für Ökologie und Naturschutz*, 2:181–192, 1993.

T. P. Sullivan and D. Sullivan. Green-tree retention and recovery of an old-forest specialist, the southern red-backed vole (*Myodes gapperi*), 20 years after harvest. *Wildlife Research*, 44:669–680, 2018.

L. Susmel. Leggi di vairazione dei parametri della foresta normale (*Abies-Picea-Fagus; Picea*) [Laws of variation of the parameters of the normal uneven forest]. Estratto da L'Italia Forestale e Montana, Anno XI Nr. 3, 1956.

L. Susmel. *Normalisazzione delle foreste alpini – basi ecosistemiche – equilibrio modelli colturali – produttività con applicazione alle foreste del Trentino* [*Normalisation of Alpine forests – ecosystem basis – equilibrium management models – productivity – with an application to the forests of Trentino*]. Liviane editrice, Padova, 1980.

R. Susse, C. Allegrini, M. Bruciamacchie, and R. Burrus. *Management of irregular forest*. Azur Multimedia, Saint Maime, 2011.

S. Szymański. Die Begründung von Eichenbeständen in "Nest-Kulturen". Eine wirksame und sparsame Methode des Waldbaus auf wüchsigen Standorten [The establishment of oak stands through planting of oak-nests. An effective and cost reducing silvicultural method on good sites]. *Der Forst- und Holzwirt*, 41:3–7, 1986.

S. Szymański. Ergebnisse zur Begründung von Eichenbeständen durch die Nestermethode [Results of afforestation of oak woodlands through the oak-nest method]. *Beiträge für Forstwirtschaft und Landschaftsökologie*, 28:160–164, 1994.

V. P. Tarasenko. Thinnings in nest-planted oak on light-chestnut soils of the south Ergeni region. *Agrobiologija*, 3:424–429, 1962.

C. M. Taylor. The return of nursing mixtures. *Forestry and British Timber*, 14:18–19, 1985.

Å. Tham. *Yield prediction after heavy thinning of birch in mixed stands of Norway spruce (Picea abies* (L.) KARST.*) and birch (Betula pendula* ROTH *& Betula pubescens* EHR.*)*. PhD thesis, Swedish Univesity of Agricultural Sciences, Umeå, 1988.

Å. Tham. *Crop plans and yield predictions for Norway spruce (Picea abies* (L.) KARST.*) and birch (Betula pendula* ROTH *& Betula pubescens* EHR.*) mixtures*. Studia forestalia suecica 195, Swedish Univesity of Agricultural Sciences, Umeå, 1994.

G. Thelin, U. Rosengren, I. Callesen, and M. Ingerslev. The nutrient status of Norway spruce in pure and in mixed-species stands. *Forest Ecology and Management*, 160:115–125, 2002.

P. A. Thomas and J. R. Packham. *Ecology of woodlands and forests: Description, dynamics and diversity*. Cambridge University Press, Cambridge, 2007.

H. Thomasius. *Waldbau 1. Allgemeine Grundlagen* [*Silviculture 1. General basics*]. Hochschulstudium Forstingenieurwesen, Leisnig, 1990.

H. Thomasius. Prinzipien eines ökologisch orientierten Waldbaus [Principles of ecologically oriented silviculture]. *Forstwissenschaftliches Centralblatt*, 111:141–155, 1992.

H. Thomasius. *Geschichte, Theorie und Praxis des Dauerwaldes* [*History, theory and practice of the Dauerwald concept*]. Landesforstverein Sachsen-Anhalt e. V., Straßfurt, 1996.

R. Thompson, J. Humphrey, R. Harmer, and R. Ferris. *Restoration of native woodland on ancient woodland sites*. Practice guide, Forestry Commission, Edinburgh, 2003.

T. Timmis. Bradford Plan continuous cover forestry: development, history and status quo. *Quarterly Journal of Forestry*, 88:188–198, 1994.

T. Tiwari, J. Lundström, L. Kuglerová, H. Laudon, K. Öhman, and A. M. Ågren. Cost of riparian buffer zones: A comparison of hydrologically adapted site-specific riparian buffers with traditional fixed widths. *Water Resources Research*, 52:1056–1069, 2016.

M. A. Toman and M. S. Ashton. Sustainable forest ecosystems and management: A review article. *Forest Science*, 42:366–377, 1996.

S. Torquato. *Random heterogeneous materials. Microstructure and macroscopic properties.* Springer, New York, 2002.

R. S. Troup. Dauerwald. *Forestry*, 1:78–81, 1927.

R. S. Troup. *Silvicultural systems.* Oxford University Press, Oxford, 1928.

UKWAS Steering Group. Certification standard for the UK woodland assurance scheme. Report, Forestry Commission, Edinburgh, 2000.

R. Valbuena, P. Packalen, L. Mehtätalo, A. García-Abril, and M. Maltamo. Characterizing forest structural types and shelterwood dynamics from Lorenz-based indicators predicted by airborne laser scanning. *Canadian Journal of Forest Research*, 43:1063–1074, 2013.

S. P. P. Vanbeveren and R. Ceulemans. Biodiversity in short-rotation coppice. *Renewable and Sustainable Energy Reviews*, 111:34–43, 2019.

I. Vanha-Majamaa and J. Jalonen. Green tree retention in Fennoscandian forestry. *Scandinavian Journal of Forest Research*, 16 (Res. Supplement 3):79–90, 2001.

K. Vanselow. *Theorie und Praxis der natürlichen Verjüngung im Wirtschaftswald* [*Theory and practice of natural regeneration in managed forests*]. Neumann Verlag, Berlin, 1949.

P. Virgilietti and J. Buongiorno. Modeling forest growth with management data: A matrix approach for the Italian Alps. *Silva Fennica*, 31:27–42, 1997.

V. Vitali, D. I. Forrester, and J. Bauhus. Know your neighbours: Drought response of Norway spruce, silver fir and Douglas fir in mixed forests depends on species identity and diversity of tree neighbourhoods. *Ecosystems*, 21:1215–1229, 2018.

L. Vítková and Á. Ní Dhubháin. Transformation to continuous cover forestry: A review. *Irish Forestry*, 70:119–140, 2013.

L. Vítková, Á. Ní Dhubháin, and A. Pommerening. Agreement in tree marking: What is the uncertainty of human tree selection in selective forest management? *Forest Science*, 62:288–296, 2016.

L. Vítková, D. Saladin, and M. Hanewinkel. Financial viability of a fully simulated transformation from even-aged to uneven-aged stand structure in forests of different ages. *Forestry*, 94:479–491, 2021.

D. Vollmuth. The changing perception of coppice with standards in German forestry literature up to the present day – from a universal solution to a defamed and overcome evil – and back? *Trees, Forests and People*, 10:100338, 2022.

H. Wang, X. Zhang, Y. Hu, and A. Pommerening. Spatial patterns of correlation between conspecific species and size diversity in forest ecosystems. *Ecological Modelling*, 457:109678, 2021.

U. Weihs. Waldpflege. Ein geeignetes Instrument zur nachhaltigen Sicherung der vielfältigen Waldfunktionen [Forest management. A suitable method for sustaining multiple forest functions]. Booklet, Förderverein des Fachbereichs Forstwirtschaft und Umweltmanagemen, Göttingen, 1990.

A. R. Weiskittel, D. W. Hann, J. A. Kershaw, and J. K. Vanclay. *Forest growth and yield modeling.* Wiley Blackwell, Chichester, 2011.

G. Wenk. A yield prediction model for pure and mixed stands. *Forest Ecology and Management*, 69:259–268, 1994.

G. Wenk, V. Antanaitis, and š. šmelko. *Waldertragslehre* [*Forest growth and yield science*]. Deutscher Landwirtschaftsverlag, Berlin, 1990.

P. W. West. Calculation of a growth dominance statistic for forest stands. *Forest Science*, 60:1021–1023, 2014.

P. Whitefield. *Permaculture in a nutshell*. Permanent Publications, East Meon, 2nd edition, 2013.

E. Wiedemann. *Ertragskundliche und waldbauliche Grundlagen der Forstwirtschaft [The production and silviculture basics of forestry]*. Sauerländer, Frankfurt, 2nd edition, 1951.

P. Wikström, L. Edenius, B. Elfving, L. O. Eriksson, T. Lämås, J. Sonesson, K. Öhman, J. Wallerman, C. Waller, and F. Klintebäck. The Heureka forestry decision support system: An overview. *Mathematical and Computational Forestry & Natural-Resource Sciences*, 3:87–94, 2011.

G. J. Wilhelm and H. Rieger. *Naturnahe Waldwirtschaft mit der QD-Strategie [Near-natural forest management based on the QD strategy]*. Eugen Ulmer, Stuttgart, 2nd edition, 2018.

J. L. Willis, S. D. Roberts, and C. A. Harrington. Variable density thinning promotes variable structural responses 14 years after treatment in the Pacific Northwest. *Forest Ecology and Management*, 410:114–125, 2018.

J. L. Willis, C. A. Harrington, L. C. Brodie, and S. D. Roberts. Variable-density thinning promotes differential recruitment and development of shade tolerant conifer species after 17 years. *New Forests*, 52:329–348, 2021.

T. Wilson. Visualising the demographic factors which shape population age structure. *Demographic Research*, 35:867–890, 2016.

E. R. Wilson, H. W. McIver, and D. C. Malcolm. Transformation to irregular structure of an upland conifer forest. *The Forestry Chronicle*, 75:407–412, 1999.

A. Witt, C. Fürst, S. Frank, L. Koschke, and F. Makeschin. Regionalisation of climate change sensitive forest development types for potential afforestation areas. *Journal of Environmental Management*, 127:S48–S55, 2013.

S. Wolfram. *A new kind of science*. Wolfram Media Inc., Champaign, 2002.

W. R. Wykoff. A basal area increment model for individual conifers in the Northern Rocky Mountains. *Forest Science*, 36:1077–1104, 1990.

W. R. Wykoff, N. L. Crookston, and A. R. Stage. User's guide to the Stand Prognosis Model. General technical report int-133, Intermountain Forest and Range Experiment Station, Ogden, 1982.

S. Yachi and M. Loreau. Biodiversity and ecosystem productivity in a fluctuating environment: the insurance hypothesis. *Proceedings of the National Academy of Sciences of the United States of America*, 96:1463–1468, 1999.

R. A. York, H. Noble, L. N. Quinn-Davidson, and J. J. Battles. Pyrosilviculture: combining prescribed fire with gap-based silviculture in mixed-conifer forests of the Sierra Nevada. *Canadian Journal of Forest Research*, 51:781–791, 2021.

M. Yorke. Continuous cover silviculture. An alternative to clear felling. Brochure, Continuous Cover Forestry Group, Bedford, 1998.

L. Zeller and H. Pretzsch. Effect of forest structure on stand productivity in Central European forests depends on developmental stage and tree species diversity. *Forest Ecology and Management*, 434:193–204, 2019.

L. Zeller, J. Liang, and H. Pretzsch. Tree species richness enhances stand productivity while stand structure can have opposite effects, based on forest inventory data from Germany and the United States of America. *Forest Ecosystems*, 5:4, 2018.

N. Zeng. Carbon sequestration via wood burial. *Carbon Balance and Management*, 3:1, 2008.

N. Zeng and H. Hausmann. Wood Vault: remove atmospheric CO_2 with trees, store wood for carbon sequestration for now and as biomass, bioenergy and carbon reserve for the future. *Carbon Balance and Management*, 17:2, 2022.

W. Zucchini and K. v. Gadow. Two indices of agreement among foresters selecting trees for thinning. *Forest and Landscape Research*, 1:199–206, 1995.

Index

Continuous Cover Forestry: Theories, Concepts, and Implementation, First Edition. Arne Pommerening.
© 2024 John Wiley & Sons Ltd. Published 2024 by John Wiley & Sons Ltd.